Mirko Knežević · Ana Topalović · Budimir Fuštić ·
Ljubomir Životić

The Soils of Montenegro

Springer

Mirko Knežević
Department of Soil and Amelioration
Biotechnical Faculty
University of Montenegro
Podgorica, Montenegro

Budimir Fuštić
Podgorica, Montenegro

Ana Topalović
Department of Soil and Amelioration
Biotechnical Faculty
University of Montenegro
Podgorica, Montenegro

Ljubomir Životić
Faculty of Agriculture
University of Belgrade
Belgrade, Serbia

ISSN 2211-1255 ISSN 2211-1263 (electronic)
World Soils Book Series
ISBN 978-3-031-82921-5 ISBN 978-3-031-82919-2 (eBook)
https://doi.org/10.1007/978-3-031-82919-2

This Springer imprint is published by the registered company Springer Nature Switzerland AG
The registered company address is: Gewerbestrasse 11, 6330 Cham, Switzerland

World Soils Book Series

The World Soils Book Series publishes peer-reviewed books on the soils of a particular country. They include sections on soil research history, climate, geology, geomorphology, major soil types, soil maps, soil properties, soil classification, soil fertility, land use and vegetation, soil management, soils and humans, soils and industry, future soil issues. The books summarize what is known about the soils in a particular country in a concise and highly reader-friendly way. The series contains both single and multi-authored books as well as edited volumes. There is additional scope for regional studies within the series, particularly when covering large land masses (for example, The Soils of Texas, The Soils of California), however, these will be assessed on an individual basis.

Preface

Welcome to the enchanting journey through the pages of our book—a compass navigating through space and time, unveiling the Soils of Montenegro, a unique mosaic of natural beauties and historical testimonies. The book is a reflection of profound love for Montenegro that has inspired generations of people, warriors, scientists, artists, and visionaries alike.

From Ancient and Medieval Rulers, the Venetian Republic, the Ottoman Empire, Austria-Hungary, to the present day, a diverse array of individuals has contributed to shaping Montenegro into the multi-ethnic and vibrant country it is today.

In this land, where different worlds converged and clashed, the Montenegrin people have ardently fought for freedom and independence across centuries, leaving indelible marks on the landscape. The lands of Montenegro unfold like a rich, interconnected tapestry, where the stories of nature and people intricately intertwine. From the towering peaks of the mountains to the serene shores of the Adriatic, from the wide plains to the deep gorges, each part of land bears the imprints of time and human destinies that have shaped it.

However, the land is more than mere ground beneath our feet; it is the wellspring of our existence and the bedrock of our civilisation. From the soil, we draw food, materials for our homes, and wealth for our economy. Its innate beauty offers solace for the soul and a sanctuary for relaxation.

An intricate living system, the soil serves as the very foundation of the life itself, nurturing plant growth, supplying oxygen, sustenance, and shelter to animals and humans. With its multifaceted roles encompassing water filtration, carbon storage, and habitat support, soil is integral to ecosystems. Its composition affects vegetation, shaping landscapes. Yet, soil's vulnerability is striking; while its formation takes millennia, human activities like deforestation and urbanisation can destroy it in mere decades or even years. Soil degradation poses a severe threat, especially to vulnerable communities, driving food insecurity and migration. Vital for sustainable agriculture, soil management influences crop productivity and biodiversity. Collaborative endeavours across disciplines, including geology, biology, agronomy, and environmental science are essential to grasp soil's complexities. Recognising soil's significance is crucial in forging sustainable land practices to confront future challenges. Safeguarding this invaluable resource is a collective responsibility, ensuring its preservation for the generations to come.

This book provides a comprehensive overview of soil studies in Montenegro from the early twentieth century to the contemporary era. Chapter 1 introduces the reader to general information about Montenegro and the history of soil research in this country, with a focus on the role of key institutions and the creation of the Soil Map of Montenegro. Chapter 2 analyses Montenegro's geographical and social characteristics affecting soil resources, including the impact of relief, climate, and human activities. Chapter 3 discusses the factors of soil formation, including parent material, climate, organisms, relief, human activities, and time. Chapter 4 elucidates the soil classification system in Montenegro, inherited from former Yugoslavia, and its connection to FAO and WRB classifications. Chapter 5 provides a detailed insight into soil types in Montenegro, including their physical and chemical characteristics, distribution, and usage. Chapter 6 explores Montenegro's soil conservation, land degradation challenges, and sustainable agriculture strategies, underscoring the vital importance of soil and the collective efforts of institutions, legal frameworks, and strategic planning to enhance soil fertility, combat drought, and achieve land degradation neutrality (LDN) through informed decision-making, eco-friendly practices, and integrated national plans.

Each chapter in this book provides important insights into Montenegro's soil resources and the factors influencing them. The book emphasises the importance of understanding the interactions among various components within the ecosystem in the soil formation process and highlights the role of humans in shaping the soil. It also offers valuable information on soil classification, which is essential for proper soil resource management. Consequently, this book serves as an invaluable source of information for anyone involved in soil studies, soil resource management, or environmental conservation in Montenegro.

Beyond its academic utility, this book extends an invitation to explore how people and the land have shaped each other, how they dealt with the challenges over time, and how they are preparing for the future.

In conclusion, we invite you to immerse yourself in the pages of this book, to discover Montenegro—its lands and beauties, to get a deeper understanding of its significance and role in shaping our lives and future.

Podgorica, Montenegro,
2023

Mirko Knežević
Ana Topalović
Budimir Fuštić
Ljubomir Životić

Acknowledgements

The book *The Soils of Montenegro* stands as the culmination of extensive and thorough research that commenced in the 1920s. We extend our heartfelt gratitude to all who, in various capacities, contributed to its inception.

Pedologists, collaborating with scientists and experts across disciplines such as agronomy, geology, ecology, and more, laid the groundwork for our contemporary comprehension of Montenegro's soils through systematic and methodical studies.

Progressing through diverse phases and transformations, from the former Agricultural Institute to the Biotechnical Institute and now the Biotechnical Faculty within the University of Montenegro, numerous scientific research and professional projects in pedology have been implemented. In this context, we acknowledge the invaluable contributions of Grujica Đuretić, Natalija Perović, Mirjana Radulović, Marija Pavićević, Stanka Filipović, Đuretić Mihajlo, Pavle Samardžić, Stanislava Brnović, Mileva Vukčević, Milijana Đurović, Petar Čelebić, Ilija Vukčević, and Dragan Brnović.

Of particular significance to us all is the 1:50,000 pedological maps, a product of dedicated work and collaboration among experts from Montenegro and other regions of the former Yugoslavia. Today, the exchange of knowledge and experiences with "old school" pedologists, former members of JDPZ (Association of Pedologists of Yugoslavia), such as Duško Mukaetov, Borut Vrščaj, Hamid Čustović, Ferdo Bašić, Tihomir Predić, Mihailo Marković, Dragan Rudić, Nevenka Maraš, Ružica Stričević, Gradimir Vasić, Boško Gajić, Milivoj Belić, Vladimir Hadžić, Vesna Mrvić, Vladimir Ćirić, and Jovica Vasin. Collaborating with them has not only enhanced our professional growth but also enriched us on a personal level.

Collaboration with colleagues from ESBN and JRC in Ispra has granted us access to new resources, methods, and techniques, proving pivotal for our research endeavours. Joint workshops and conferences have served as fountains of immense knowledge and experience.

Gratitude extends to institutions in Montenegro, including the relevant Ministries, the Montenegrin Academy of Sciences and Arts, the Geological Survey of Montenegro, and the Institute of Hydrometeorology and Seismology of Montenegro, which have provided unwavering support and resources for soil research over the decades.

A sincere thanks to local communities and residents across Montenegro who warmly welcomed pedologists, granted access to their lands, and generously shared experiences and knowledge, thereby providing a genuine and vivid context for the topics under research.

Appreciation also goes to international institutions, UNCCD, and the Global Mechanism, for their recent substantial support and assistance in adhering to global standards for the protection and management of land.

Special thanks are reserved for all who directly contributed to this book. Colleague Danka Caković provided valuable material and corrected the biodiversity chapter. Thanks to colleagues diligently performing technical and professional tasks at the Centre for Soil and Amelioration: Daliborka Lekić, Mirko Bulatović, and Jelena Vukčević.

Young team members Balša Bajagić and Tatjana Nikolić brought freshness and innovation, working on technical details and improvements. In translating, proofreading, and editing the text, Mia Laušević demonstrated the highest level of professionalism, devoting herself to ensuring each sentence was flawless.

Images often convey more than words, and thanks to the photos provided by colleagues, the pages of this book have come alive. Photographer Goran Roganović, accompanying us in the field and meticulously following our instructions, created professional photographs that enriched our work.

Our collaborators from the publishing house, Dr. Robert Doe, Ms. Corina van der Giessen, along with the series editor, Prof. Alfred Hartemink, contributed their expertise to enhance the quality and professionalism of the book. Besides their expertise, their patience and advice improved every page of this book. We especially owe deep gratitude to Mr. Suresh Rettagunta, Mrs. Amudha Vijayarangan, and Mr. Madanagopal D Deenadayalan for their immense patience with our manuscript delays.

Last but certainly not least, we express gratitude to our families. Their unwavering support has been our anchor. In every sentence, on every page, there is a part of their souls and their love.

Contents

About the Authors

Dr. Mirko Knežević (b. 1971 in Podgorica) completed primary school in Spuž, followed by "Slobodan Škerović" high school in Podgorica. He graduated from the Faculty of Agriculture in Zemun—Belgrade, Department of Land Reclamation in 1995. He commenced his career on 1 February 1998 as a research associate in the Centre for Soil and Amelioration within the Biotechnical Institute in Podgorica. At the Faculty of Agriculture in Zemun—Belgrade, Department of Soil Drainage, he defended his master's thesis in 2000, followed by his doctoral dissertation in 2008. In the period from October 2001 to July 2002 and October 2003 to July 2004, he was on a study stay at the Mediterranean Agronomic Institute of Bari (Italy), Department of Land and Water Resources Management. In the period from 24 May to 29 June 2005, he was trained at the MASHAV International Agricultural Training Center in Shefayim (Israel).

He is Montenegro's representative in the United Nations Convention to Combat Land Desertification (UNCCD) as a scientific consultant and NRC (National Reference Centre) for the field of agriculture to the EIONET reporting network managed by the EEA (European Environment Agency). He is a member of the Council for Revision of Spatial Plans at the Ministry of Sustainable Development and Tourism of the Government of Montenegro.

Since the academic year 2005/2006, he has been engaged as an assistant and/or lecturer at the Biotechnical Institute, now the Biotechnical Faculty, where he teaches the subjects: Pedology, Amelioration, Amelioration and Land management, and GIS in Agriculture. At the Faculty of Philosophy in Nikšić, he was engaged as a lecturer in the subjects Geography of Land and Geographic Information Systems (GIS) at the Study Program for Geography.

Dr. Ana Topalović (b. 1975 in Berane) is Assistant Professor for agrochemistry at the Biotechnical Faculty, University of Montenegro. After completing the undergraduate thesis in chemical technology, she continued her studies at the Faculty of Chemistry, University of Belgrade, where she earned a Master's Degree in 2006, and a Ph.D. in 2012. In 2000, she joined the Centre for Soil and Amelioration in Podgorica, actively participating in the field of soil chemistry, soil fertility and plant nutrition. In the period from September to November 2005, she was trained in Agrochemistry at IAEA—Food Safety Laboratories/Seibersdorf (Austria). In the development of the laboratory for agrochemistry and soil science, her contribution and active involvement concerned renovation, equipping, training of personnel and the introduction of new methods. She has been engaged in education of students on the courses of Chemistry and Agrochemistry at Biotechnical Faculty and on Soil Protection at the Faculty of Metallurgy and Technology. She is an author of one university Practicum for Agrochemistry. For the past 15 years, she has been intensively studying the influence of growing and cultivation conditions on the secondary metabolites of plants. She is a member of the editorial board of national scientific journal *Agriculture and Forestry*.

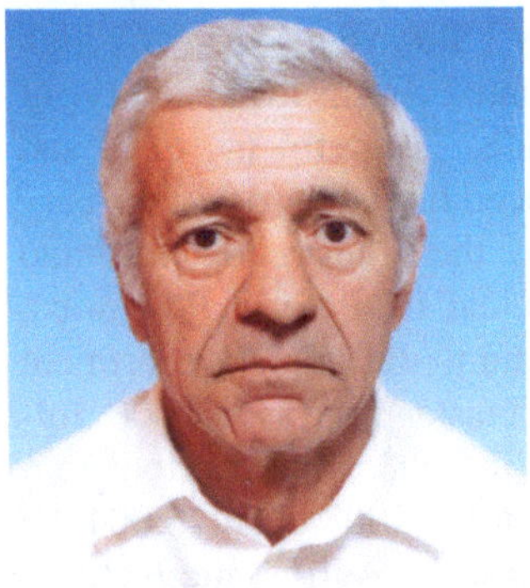

Dr. Budimir Fuštić (b. 1939 in Potrk—Bijelo Polje) graduated from the Faculty of Agriculture in Belgrade in 1964 and later earned his master's in 1974 and Ph.D. in pedology in 1994 from the same institution. After his studies, he worked briefly in Mojkovac in agriculture and forestry. In June 1967, he joined the Institute for Soil Study of the Agricultural Institute in Podgorica. There, he engaged in pedological research and developed a pedological map at a 1:50,000 scale. Fuštić published numerous research papers on soil-related topics, including pedology, soil improvements, and erosion. He also authored important studies, reports, and the monograph Soil of Montenegro and an Atlas of Montenegro's soils. In his pursuit of expertise, he undertook specialised training in Sofia on erosion and soil protection and participated in study visits across Hungary, Czechoslovakia, Austria, and parts of the USA. Throughout his career, he held important positions in various professional associations and governing bodies, including being a delegate for the University Assembly "Veljko Vlahović" and the Assembly of SIZ NP "Lake Skadar".

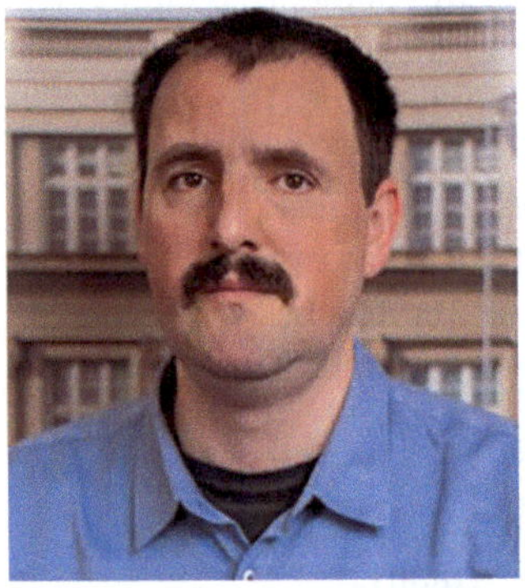

Dr. Ljubomir Životić (b. 1980 in Belgrade, Serbia) is Assistant Professor of Pedology at the Faculty of Agriculture, University of Belgrade. He graduated in 2003 from the Institute for Soil Management, and obtained a Master of Sciences degree in Irrigated Agriculture in 2005, at the Mediterranean Agronomic Institute in Bari, Italy. He has been employed at the Faculty of Agriculture in Belgrade since 2009, where he earned a Ph.D. degree in Soil Science in 2016. So far, he teached Pedology and GIS in Pedology, conducted laboratory exercises and organised field trips. His fields of interest are soil classification and genesis, soil water characteristics, soil and land degradation, irrigation and crop water requirements, climate change in agriculture, application of GIS in environmental sciences and agriculture. He actively contributed to the broader community through membership in working groups in the fields of agriculture and environment. He was the Executive Secretary of the Serbian Soil Science Society from 2019–2023. He was President of the Organising Committee for the 3rd International and 15th National Congress of the Serbian Soil Science Society, held in Sokobanja, Serbia, in 2021. He lives in Belgrade.

List of Figures

List of Tables

1 Introduction

Abstract

This chapter presents general information about Montenegro and a brief retrospective of soil research activities in Montenegro. The first steps in soil investigations date back to the third decade of the twentieth century and primarily focus on soil improvement on agricultural land. The State Experimental Station for Southern Crops was the only institution conducting soil analyses at that time. After the break during Second World War (WWII), soil science activities in Montenegro intensified. The pedological laboratory in Bar was reconstructed, and the research was devoted to areas where amelioration and other agrotechnical measures were required to intensify agricultural production. This chapter further presents the pioneers of soil science in Montenegro, research institutions, and soil-related activities, as well as the steps passed in the creation of the Soil Map of Montenegro. It highlights the role of the Institute of Agriculture from Podgorica, which was a predecessor of the Biotechnical Faculty today—the leading institution in the biotechnical sciences in Montenegro.

1.1 General Information About Montenegro

Thanks to its position, Montenegro represents a link between different influences and cultures. As a country of Southeastern Europe, lying on the eastern coast of the Adriatic Sea, it has communication with the entire world, and the Dinaric karst gives it a unique character, both in terms of nature and humanity (Fig. 1.1).

Montenegro borders Croatia and Bosnia and Herzegovina to the southwest and west, Serbia to the north and the northeast, and Albania to the southeast. With Italy and Croatia, it has a sea border, and with Albania, a border in Lake Skadar and the Bojana River.

With its land area of 13,812 km^2 (the area of the sea aquatorium is about 2540 km^2), Montenegro is among smaller European countries, such as Albania, Slovenia, Macedonia, and Belgium. According to the 2011 Census, Montenegro has a population of 620,029, living in 24 municipalities, 58 urban and 1307 rural settlements, with a population density of 44.9 people per 1 km^2. By area, Andorra, Liechtenstein, Luxembourg, Malta, and San Marino are smaller than Montenegro, and by population, in addition to these, Iceland is smaller. According to the official estimates of the Statistics Office, the population in 2020 was 621,306. Podgorica (Fig. 1.2) is the capital and largest city, while Cetinje is the former royal capital.

Montenegro's territory lies between 18°26′ (Sutorina, Herceg Novi 42°29′; 18°26′) and 20°21′ Jablanica, Rožaje 42°53′; 20°21′) east longitude and between 41°52′ (Ada Island, Ulcinj 41°52′; 19°22′) and 43°32′ (Moćevići, Pljevlja 43°32′; 18°58′) north latitude. The distance between the northernmost and the southernmost points is a 200 km aerial line and between the westernmost and the easternmost is 173 km.

Based on the Spatial Plan of Montenegro, the total forest area is 6225 km^2 or 45%; agricultural land covers about 5145 km^2, or 37%, while settlements, roads, water bodies, bare rocks, and other categories account for about 2442 km^2 or 18% of the territory. The areas up to 200 m above sea level (m a.s.l.) account for 10% of Montenegro's area, 33% are areas from 200 to 1000 m a.s.l, the areas from 1000 to 1500 m a.s.l. occupy 40% of the territory, while areas of high mountains above 1500 m a.s.l. cover about 17% of the territory of the state, of which 1% is at elevations above 2000 m a.s.l.

There are five national parks covering a total area of 100,427 ha (7% of the total area of Montenegro): Biogradska Gora 5650 ha, Durmitor 32,519 ha, Lovćen 6220 ha (Fig. 1.3), Lake Skadar 40,000 ha and Prokletije 16,038 ha. There are three strict nature reserves (420 ha), one is a special nature reserve (150 ha), eight nature parks (79,583.10 ha), 57 monuments of nature (4493.54 ha), and

M. Knežević et al., *The Soils of Montenegro*, World Soils Book Series,
https://doi.org/10.1007/978-3-031-82919-2_1

Fig. 1.1 Geographical position of Montenegro

two are areas of outstanding landscape (196.05 ha). A total of 76 protected areas occupies 185,269.69 ha or 13% of Montenegro's area.

Based on the International Union for Conservation of Nature (IUCN) categorisation, there are three protected areas under category Ia (420 ha), five under category II (100,427 ha), 57 under category III (4493.54 ha), and ten under category V (79,779.15 ha). A total of 75 protected areas occupy 185,119.69 ha or 13% of Montenegro's area.

In 1991, Montenegro was declared an ecological state and became the first country in the world to regulate its ecological commitment by the Constitution. With this proclamation and promotion of the 2009 Sustainable Development Strategy, the concept of preservation of the environment has become the cornerstone of Montenegro's overall development. By setting the equality between development and environmental needs as a basis for sustainable development, Montenegro has joined all those countries that have accepted the UN Rio de Janeiro Declaration (2009) and other important acts adopted by various global and European forums.

Montenegro ratified the United Nations Convention to Combat Desertification (UNCCD) in 2007 and thus assumed the obligation to implement and report on the fulfilment of obligations under the Convention in accordance with its objectives, all in terms of sustainable soil management and prevention of land degradation.

1.2 History of Soil Research in Montenegro

1.2.1 Research in the 1900-1950

Unlike botanical and geological research, carried out by a number of international experts as early as the nineteenth century, pedological research started much later. The first studies date back to the third decade of the twentieth century, when Jovan Kaluđerović (in 1924 and 1925) published papers on the land improvement in Ulcinjsko Field and Lake Skadar drainage, while Professor Đorđe Zloković (1930) published a paper "Pedological research in the Nikšićko Field" in the journal Geological Annals of the

Fig. 1.2 Podgorica—The Capital of Montenegro; A view of bridges over Morača the River (*Photo* G. Roganović)

Fig. 1.3 National Park Lovćen—one of the symbols of Montenegro (*Photo* G. Roganović)

Balkan Peninsula. Zloković then continued his research activities and published several papers by 1940 on the karst field soils during the same name excursion tours (on his own or with students): Grahovsko, Dragalj, Njeguši, and Cetinjsko fields, as well as coastal fields: around Grbalj, Bar and Risan. Dragan Carikov (1930) published a paper "On soil and economic situation in the Lake Skadar area within the borders of our country", and Aleksandar Stebut (1934) published a paper "Climate, hydrological and pedological characteristics of eastern Montenegro".

A State Experimental Station for Southern Crops was founded in 1937 in Bar, which had a chemical-pedological department and a laboratory where chemical and physical soil analyses were performed.

As shown in this overview, Montenegro's pedological research was limited compared to other areas of Yugoslavia before World War II, which had relevant institutions and staff and primarily focused on two agricultural faculties and several agricultural schools. That allowed research on a larger scale and studies on certain soil types in accordance with Dokuchaev's view of the soil formation and genesis presented in the monograph on Russian chernozem. Further to important scientific papers on the soils of Serbia and Croatia, the pedology textbooks were published. The following were of particular importance: Agropedology (1949, 1950, 1953) by above-mentioned eminent pedologist Aleksandar Stebut (Fig. 1.4), Professor at the Faculty of Agriculture in Belgrade, and Pedology (1951) by Mihovil Gračanin, one of the most famous pedologists, plant physiologist and ecologist, Professor at the Faculty of Agriculture of the University of Zagreb.

Stebut introduced several new types into the soil classification, such as: Brown Forest Soil, Smonitza, Humic Gleysoil, etc.

During World War II, activities and research were slowed down or even suspended in many areas, including in the field of pedology. In Montenegro, this break lasted until 1949, when the Federal People's Republic of Yugoslavia (FPRY) Ministry of Agriculture set up and deployed a group of agronomists to Montenegro, tasked with carrying out pedological and amelioration research on some areas of importance for Montenegro's agriculture. The need for a rapid reconstruction of the country required fulfilment of conditions for research and scientific work in various fields, including agriculture, which was then the most important branch of the war-torn economy of the then Yugoslavia.

Upon the arrival of a small group of agronomists to Montenegro, the pedological laboratory in Bar was reconstructed, and graduate engineer Borivoje Pušić was appointed head of the Pedological Department within the Station for Southern Crops. By 1953, under the leadership of Srboljub Vukašinović and Rajko Opačić, the research focused on several areas, mainly on flatter areas, where amelioration and other agrotechnical measures were needed to intensify agricultural production.

Studies were carried out, including making a soil map 1:50,000, a description of natural conditions, formation factors for the present soil types, their genichological, physical, and chemical properties, and appropriate amelioration measures.

On this occasion, the field and laboratory analyses were carried out on soils from the following areas: Ulcinjsko and Štoj fields and Šasko mudflat (5.370 ha); Donja Zeta and Ceklin polders (in Lake Skadar riparian zone—12,000 ha); Sutorina near Herceg Novi (220 ha); Lješkopolje and Lješkopoljski Lug (3000 ha); Crmničko and Orahovsko fields (1500 ha); Mrčevo and Grbalj fields (1200 ha); Kolašin and Mojkovac area (1370 ha); Berane and Bijelo Polje hollows (3700 ha) and Bjelopavlićka Plain (6680 ha).

The characteristics of pedological, agropedological, and hydropedological properties of soil, as well as the evaluation of agricultural values, could have been used for the development of hydro and agro-amelioration projects; nevertheless, this has not been done, just as results obtained have never been published.

In addition to the experts mentioned, the following experts participated in some of the studies: Damjan Pejović, Nada Nastić, and Bosiljka Tupanjanin. Nikola Pavićević (Fig. 1.4) independently carried out extensive research on the soil genesis on the limestones of Montenegro's mountainous region, and following his doctoral dissertation defence (1952), he presented the results obtained in a book titled "Buavica on Montenegrin Karst" (Pavićević 1956). Buavica is the common name of a soil type used at that time to denominate Mountain Black Soils and, later on, Black Soils formed on limestone-dolomite rocks.

Fig. 1.4 Aleksandar Stebut (1877–1952) and Nikola Pavićević (1911–1974)

Later on, N. Pavićević, working as a professor at the Faculty of Agriculture of the University of Belgrade, spent summers studying Montenegro's soils and published several papers on Terra Rossa, Brown Soil on limestone, Brown Forest Soils, Alluvial Soils, and two papers on soil erosion. His work made a considerable contribution to the knowledge of the soil in Montenegro and also in Serbia, where he took part in field and laboratory research, authored several books, and co-authored several monographs on the soils of Serbia.

1.2.2 Research in the 1950-2000

In the next period of pedological research, 1953–1964, only a few pedologists were involved in such work. The research was no longer funded by the Ministry of Agriculture but by local communities, agricultural enterprises, and cooperatives. The character of these studies did not differ significantly from those in the previous period, although their purpose was not only land improvement in order to use it for crops and setting up perennial plantations but also to improve agricultural production on socially owned land.

To that end, the construction of irrigation systems in the Bjelopavlićka Plain, Lješkopolje, Grahovsko, and Crmničko fields began. The construction of the drainage system also began, as well as setting up the trials for soil salinity control, cultivation of new plant varieties, etc.

The Yugoslav Society of Soil Science (YSSS), established in 1953, launched an initiative and made a recommendation for systematic pedogenetic research of soils and making a 1:50,000 soil map, which was accepted by all pedological institutions in Yugoslavia in 1956. In Montenegro, the Department for Soil Research within the Institute for Agricultural Research at that time lacked staff and equipment and was unable to perform such a significant and extensive task. Therefore, the institute involved the Institute for Agroecology Zagreb for the area from the Durmitor and Sinjavina (Sinjajevina) to Nikšić and the Institute for Pedology of the Faculty of Agriculture in Belgrade for the Polimlje area.

The Institute for Agroecology, with Dr. Pavle Kovačević as the person responsible for the task, completed the field and laboratory research in the period from 1958 to 1960 and presented the results in a voluminous study entitled "Pedological and Vegetation Research of Durmitor–Sinjavina and Central part of Montenegro". The study includes documents of pedological profiles, handwritten pedological maps 1:50,000, phytocoenological tables of plant communities and photo recordings of profiles, vegetation, etc.

The Institute for Agricultural Research Titograd, managed by Grujica Đuretić (Fig. 1.5), carried out research in Pljevlja County in the period from 1958 to 1964 and presented the results in the study "Soils of Pljevlja Region" (Đuretić 1964), which, similar to the previously mentioned study, includes a 1:50,000 map, description of soil characteristics and photographs.

Fig. 1.5 Field experiments at Krnovo plateau (*Photo*-archive CZM[1])

[1] Centre for Soil and Amelioration—CZM (Centar za zemljište i melioracije).

These studies were important for the knowledge of the soils of the mountainous and hilly areas, diverse both in terms of types and usage.

As of 1964, the next soil research period began, differing significantly from the previous ones. This particularly concerns pedogenesis research, preparing a soil map 1:50,000, and soil classification. These activities, defined as part of the Yugoslav project, began in all the republics and provinces of then Yugoslavia in the period 1956–1958. However, different methods were initially used, at the discretion of the researchers, so the results and data could not have been compiled into a single entity. In order to overcome these differences, at the initiative of YSSS and the leading scientists in soil science, it was decided to harmonise and unify the approach to the research work. The most important decisions were:

- To perform all tasks using the same methods of work, criteria, and standards adopted by the newly formed Commission, made of representatives of all the pedological institutions of Yugoslavia.
- The Commission monitors and evaluates the work and gives consent and approvals for printing soil maps and other publications, primarily the monographs on soils.
- The nomenclature of systematic soil units and colours is to be marked with uniform symbols adopted and in accordance with the new classification adopted by the YSSS in 1963 at the congress in Ohrid.

Thanks to these decisions and arrangements, the Institute of Agriculture in Podgorica (Fig. 1.6) became the lead institution in the soil research project in Montenegro. From 1964 to 1988, the Institute of Agriculture completed all field and laboratory research and printing of the 1:50,000 soil map for the entire territory of Montenegro, presented on 36 sheets (Fig. 1.7), division by Paris (Paris mapping system based on the first meridian in Paris).

The basic soil map 1:50,000 was and still is an invaluable document for the knowledge of the nature of Montenegro. The map contains data on soil properties, characteristics, and spatial distribution of systematic soil units. The quality of the printed maps is high, considering the period when they were made. Figure 1.8 provides an overview of the sheets of the soil map 1:50,000.

The pedological data describe the mechanical-physical and chemical soil properties. Furthermore, a systematic project, which lasted almost three decades, existed in the past, when several thousand soil profiles were opened throughout Montenegro, and their characteristics were recorded in six volumes (Fuštić 1978).

Summarising the results of these research activities, we can conclude that they provided Montenegro with a complete inventory of the soil on its territory, covering an area of 1,381,200 ha (Fig. 1.9). These and other research activities on specific areas, particularly those suitable for agricultural production, provided sufficient data, presented in

Fig. 1.6 Biotechnical Faculty building—former Institute of Agriculture (*Photo* G. Roganović)

Fig. 1.7 Colleagues from the Scientific Sector together with scientists from the Biotechnical Institute in Podgorica (From left to right: Miladin Vučković, Dejan Milatović, Smiljana Prelević and Budimir Fuštić - *Photo*-archive CZM)

detailed 1:10,000, 1:5000, and 1:2500 scale maps. All of that enabled the publication of a large number of scientific and research papers, including the monograph "Soils of Montenegro" by Budimir Fuštić and Grujica Đuretić (Fuštić and Đuretić 2000).

The obtained data and results of the pedological research have become available to all those concerned with the development, use, protection, and planning of the soils and land of Montenegro. The subsequent partial data digitisation enabled a very rapid use and utilisation of soil maps in various fields and branches of the economy: agriculture, forestry, water management, spatial planning, etc.

The pedological research of Montenegro, together with those in other republics and provinces of the former state, enabled the making of a 1:1,000,000 scale soil map of Yugoslavia, a map of Europe, organised by FAO/UNESCO.

Simultaneously with the research for the soil map 1:50,000 and particularly following the completion of the soil map, a series of research activities (mostly applied and to a lesser extent fundamental) was carried out in Montenegro on certain specific properties and limitations of certain types of soil and lower systematic units.

1.2.3 Overview of Studies

The subsequent section provides an overview of studies undertaken in master and doctoral theses by researchers from the Centre for Soil and Amelioration (Fig. 1.10).

Fuštić (1974) focused on soil erosion in the area of the Tara River basin (Mojkovac Valley), which covers an area of approximately 95 km^2 and is located in a mountainous belt at altitudes between 750 and 2000 m a.s.l. The action of numerous natural and socio-economic factors led to surface and linear erosion. The most prevalent erosion was of a moderate degree, covering 3980 ha, or 42% of the area. According to the amount of eroded material of over 100,000 m^3 per year or about 1300 m^3 per km^2, the area was categorised as one with strong erosion.

Fuštić (1993) investigated the effects of liming and fertilisation on the chemical properties of acid soils (Dystric Cambisols), yields, and quality of hay from natural grasslands (Agrostietum vulgare and Nardetum strictae) in the mountainous area of Montenegro (Kolašin and Krnovo). The study found that liming with $CaCO_3$ in doses of 13.6 t/ha and 15.5 t/ha over a period of 4 years significantly improved the chemical properties of the soils, while fertilisation with N60, P100, and K80 and manure (20 t/ha) had a lesser effect. Liming with fertilisation was more effective for the increase of availability and accumulation of phosphorus, but affected the leaching of potassium. Fertilisation, either by itself or combined with $CaCO_3$ and Na_2MoO_4, changed the botanical composition of grass cover (the rate of share of gramineae and leguminosis) and increased yields and quality of hay. Complex NPK fertilisers were recommended for practical application in order to obtain the highest yields and improve the quality of animal feed.

Perović (1991) focused on the impact of various NPK mineral fertiliser combinations on the Vranac grapevine variety grown in skeletal soils of Ćemovsko Field. The goal was to identify the best combination to optimise both grape yield and quality. Over four experimental years, field trials

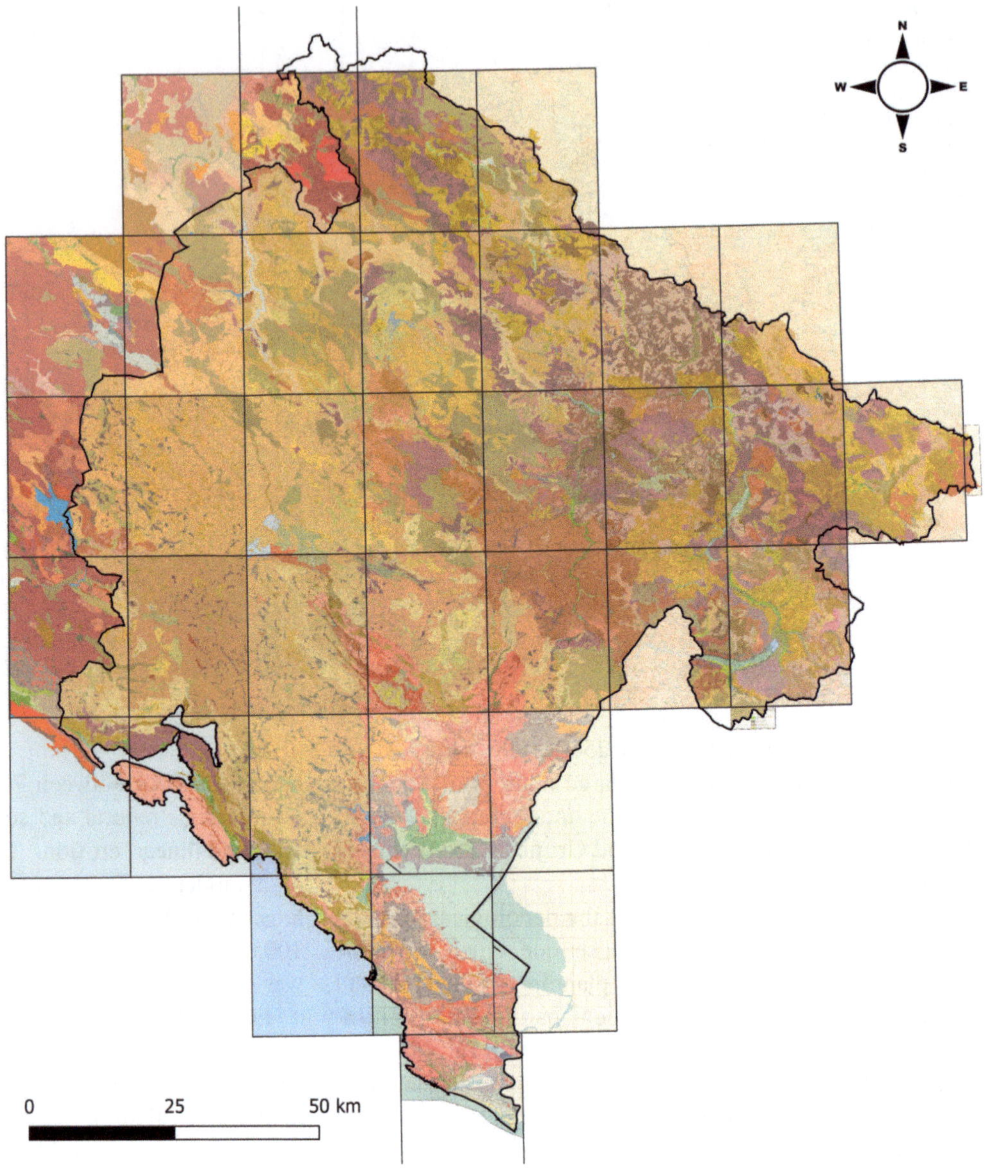

Fig. 1.8 Index Soil map of Montenegro map sheets 1:50,000, by Paris projection

were conducted, examining ten fertilisation variants. The study also looked into foliar diagnostics as a method for assessing grapevine nutrient status, in order to determine the dependence of N, P, K, Ca, and Mg in leaves on the type of fertilisation and the phase of vegetation. Regarding the yield, sugar in grapes and grapevine biomass, K260 variant was the most effective as one with the highest dose of potassium (260 kg/ha) without added phosphorous and nitrogen, followed by K190 variant, then N90K190 and N140P180K190 variant. There was a notable positive correlation between grape yield, quality, and the potassium content in leaves. As the experiment progressed, fertilisation with potassium not only increased its content in the soil and its effect on grape yield, but also amplified the effects of nitrogen and phosphorus. To achieve optimal yields in skeletal carbonate soils, as an important recommendation of this study was that potassium fertilisation should be paired with appropriate magnesium amounts.

Radulović (1993) focuses on the analysis of basic agrochemical characteristics and macro and micro-element content of alluvial, brown shallow, and brown very shallow soils in the Zetska Plain, where tomatoes, peppers, and watermelons were cultivated. The soils were categorised as slightly acidic, neutral, and medium calcareous, except for those under watermelon cultivation, which were slightly calcareous. It was observed that the Alluvial Soil did not provide favourable conditions for vegetable production, unlike the other two soil types, which were found suitable. A key finding was the excessive nitrogen supply for all three crops, indicating a need to reduce the doses of nitrogen fertilisers. The soils were well supplied with exchangeable magnesium, accessible phosphorus, and potassium.

Fig. 1.9 Soil survey during the creation of the Soil Map of Montenegro (1958–1988) (*Photo*-archive CZM)

The accessible iron content varied across the soils, being moderately supplied in brown very shallow soil and very highly supplied in alluvium. Additionally, the soils were found to be moderately to richly supplied with accessible boron, well supplied with zinc, highly supplied with manganese, and richly supplied with copper. Consequently, it was concluded that high productive production of quality vegetables is possible in the conditions of the Zetska Plain.

Radulović (2001) studied the mobility of nickel, chromium, and fluoride in Brown Cambic Soils and Alluvial Soils in Zeta Valley, which is an area with intensive agricultural production and direct influence of aluminium industry (Aluminium Plant Podgorica, abbr. KAP). The study found that the total content of nickel and chromium was high in both types of soil. The mobility of nickel was low, while the mobility of chromium was increased and significantly dependent on clay fraction. The results point to their common geochemical origin. Their accumulation in field plants, as well as in test-cultures (oats, lucerne, and lettuce) from natural resources of the same soils in dishes, is at the normal level. The total content of fluoride is high and significantly higher in Brown Soils. The average content of available fluoride is low, but significantly higher in surface layers of Alluvial Soils. Further, the content of fluoride in the majority of vegetable plants from the field experiments was in the normal range, but found to be increased in lettuce and lucerne, and high in corn leaves grown in Brown Soils at localities close to KAP, because of atmospheric pollution. The effect of testing elements on lucerne and lettuce primarily depended on soil fertility.

Knežević (2000) conducted a study on the link between soil drainage and sustainable development in the Zetsko-Bjelopavlićka Plain by analysing the elements of climate, hydrology, topography, land use, and soil.

A total of 16,679 ha are under the risk due to excess water, with the highest threat found in the areas influenced by Lake Skadar and Malo Blato Lake. The main factors contributing to excess water are changes in water levels and high annual precipitation. The soil types most vulnerable to excess water were identified, namely Alluvial Soils, Eugleys, and Pseudogleys. Pipe drainage systems have proven effective in mitigating excess water, but unfavourable ownership structure and land fragmentation pose a constraint in applying modern agrotechnical measures. The study recommends applying strict protection measures in areas prone to periodical flooding and the utilisation of land in a sustainable manner, as well as the use of modern drainage systems to regulate water flows and mitigate excess water.

Knežević (2008) continued the research with the aim of determining an optimal water regime for modern agricultural production in Bjelopavlićka Plain, on an area of 8120

Fig. 1.10 Employees at the Centre for Soil and Amelioration (2000 year - *Photo*-archive CZM)

ha, covered by six major soil types. The needs for drainage and irrigation under various climate and soil conditions and hydrological factors were investigated.

Climate conditions were defined using daily values of precipitation, temperature, air humidity, and sunshine hours in the period 1961–1990. Hydrological conditions were assessed using monthly water flows of the Zeta River as the main surface water source for irrigation (Fig. 1.11). This study used four scenarios of agricultural production for corn, alfalfa, tomatoes, vines, fruit, and wheat. The Budget model—a soil water and salt balance model (Raes 2002)—was used to process parameters of the water balance during the period of vegetation. The model considers hydrological, soil, and climatic conditions in the calculation of water balance parameters, and estimates relative yield on a daily basis. The results showed the necessity for constructing drainage systems on soils of heavy texture and the significance of applying irrigation across the entire area of Bjelopavlićka Plain. The available water quantity for irrigation in the plain significantly meets the stated needs for appropriate scenarios of agricultural production. The study used GIS to enable spatial data distribution and increase the applicability of this approach at the regional scale.

Spalević (1999) focused on the use of computer graphics methods to study soil erosion in the Berane Valley, on the section from Rijeka Marsenića to Tifran, where the basins of ten tributaries of the Lim River have a surface area between 3 and 136 km^2. Because the measurement, estimation, and monitoring of soil erosion in field plots are difficult and time-consuming, the author developed two computer graphic programs, "Surfaces and Distances" and "Basins". The programs use mathematical models to predict maximum water runoff and soil erosion intensity. These methods are effective and efficient, making it easier to study soil erosion and its impact.

Spalević (2011) studied the impact of land use on runoff and soil erosion in the Polimlje watershed, which covers an area of 2158 km^2 with 57 tributaries. The methodological approach consisted of the collection and processing of data related to climate, geological and pedological features, and physical and geographical parameters of watercourses and river basins. The research involved the development of a new software called "IntErO" that generated data on the prognosis of maximum runoff and the intensity of soil erosion. The study analysed the selected river basins of Andrijevica, Berane, and Bijelo Polje, and proposed

Fig. 1.11 Zeta River in Spuž (*Photo* G. Roganović)

variants of optimum land use based on the criterion of real soil loss with its projections for the year 2030. The results of the study can be used to apply land use practices that minimise soil erosion and runoff in the Polimlje watershed.

Bearing in mind the high prevalence of calcareous soils in Montenegro and the particular importance of defining the fraction of available phosphorus, Topalović (2006) dealt with the chemical speciation of phosphorus by applying an adequate sequence of analytical procedures. The associations features of phosphorus forms with mineral and organic soil constituents were assessed. The obtained results highlighted the role of Al-, Fe-, and Mn-substrate as important factors in the biogeochemical cycle of phosphorus in calcareous soils. Namely, the distribution of total P followed the distribution of phosphorus forms accumulated on Al-, Fe- and Mn-hydroxides, which make up about 3–12% of total P, and which also represents the content of available P in these soils.

The research was continued towards monitoring the influence of foliar fertilisation with P and K on the quality of grape cv. "Cardinal", i.e., its primary and secondary metabolites (Topalović 2012). The reasons for the choice of this topic were: on the one hand, the possibility of low P content in some organs of grapevine grown on calcareous soils, as well as the crucial importance of P and K in the synthesis of carbohydrates and transport of assimilates to storage organs; and on the other hand, great attention on phenolic compounds as secondary metabolites due to their strong human health benefits. An integral part of this study was the investigation of relationships between primary and secondary metabolites in grapes and nutrients in the grapevine leaf, considering growing conditions (climatic and soil parameters). The practical contribution of obtained results is the recommendation for foliar PK fertiliser application due to a positive effect on the accumulation of sugars and anthocyanins. In addition, there are indications that Mg, Cu, Zn, and Mn could positively influence the creation of some flavonols and flavan-3-ols.

In addition to the employees of the Centre for Soil and Amelioration of the Biotechnical Faculty (Institute), other colleagues also dealt with soil issue. Stevan Sinanović, in a dissertation on soil amelioration of Grahovsko and Nikšićko fields, emphasised the importance of soil in these karst fields. In his dissertation, Srđan Šoškić investigated the irrigation of skeletal and skeletal soils in the area of Ćemovsko Field, which were cultivated according to the program implemented by "Agrokombinat—13 jul" from Podgorica.

In recent times, digitising numerical records of soil properties in order to obtain modern digital soil databases has been of considerable interest. Digital soil databases improve the evaluations, interpretations, and decisions in resource management and environmental science.

As for the data on the soils of Montenegro, the work on their preparation for digitisation began earlier, and

Fig. 1.12 Field work with students at a Summer School organised by the Centre of Excellence BIO-ICT (*Photo* A. Topalović)

it included the collection and selection of data manually inserted and stored in notebooks, as well as the entry into Excel files and verified. Digitisation of these data was done within the first Centre of Excellence in Montenegro (Fig. 1.12), with the cooperation of experts in the field of information technology from the Faculty of Electrical Engineering and soil scientists from the Biotechnical Faculty of the University of Montenegro. A total of 7666 profiles were digitised (16,151 horizons and layers), of which 4986 profiles with particle size distribution and 6692 with chemical properties, while 4030 with both mechanical and chemical properties. Exact locations with coordinates exist for 3536 profiles or 6555 horizons.

For the time being, half of the data (about 1800 profiles with about 4000 layers) can be used. Considering that Montenegro's area covers about 13,812 km^2, there is one profile per every 7.7 km^2 on average, which can be considered sufficient.

The data that may be available to the full extent for 1800 profiles are as follows: horizon position and thickness, the content of gravel, coarse sand, fine sand, silt, clay, total sand, total clay, hygroscopic moisture, pH in H_2O, pH in KCl, $CaCO_3$, humus (organic matter), available phosphorus, and potassium. Hydrolytic acidity, sum of base cations, cation exchange capacity, and the base saturation were determined for approximately 47% of profiles (about 850).

References

Carikov D (1930) O zemljištu i ekonomskim prilikama područja Skadarskog jezera u granicama naše države, Glasnik ministarstva poljoprivrede, sv. 31., Beograd

Djuretić G (1964) Zemljišta Pljevaljskog podurčja - studija. Zavod za unapredjivanje poljoprivrede. Titograd

Fuštic B (1993) Uticaj kalcifikacije djubrenja na promjene nekih hemijskih osobina kiselih planinskih zemljišta, prinose i kvalitet sijena prirodnih travnjaka. Univerzitet u Beogradu, Poljoprivredni fakultet Beograd-Zemun, Doktorska disertacija

Fuštić B (1974) Prilog proučavanju erozije zemljišta u dolini Tare (okolina Mojkovca). Magistarski rad, Poljoprivredni institute Titograd

Fuštić B (1978) Prikaz pedoloških istraživanja Crne Gore, Poljoprivreda i šumarstvo XXIV, 2, 53–64, Titograd
Fuštić B, Đuretić G (2000) Zemljišta Crne Gore. Univerzitet Crne Gore, Biotehnički institut, Podgorica
Gračanin M (1951) Pedologija, III deo, Školska knjiga, Zagreb
Knežević M (2000) Odvodnjavanje zemljišta Zetsko-Bjelopavlićke ravnice u uslovima održivog razvoja. Univerzitet u Beogradu, Poljoprivredni fakultet Beograd-Zemun, Magistarski rad
Knežević M (2008) Planiranje i projektovanje sistema za odvodnjavanje i navodnjavanje u Bjelopavlićkoj ravnici. Univerzitet u Beogradu, Poljoprivredni fakultet Beograd-Zemun, Doktorska disertacija
Pavićević N (1956) Buavice na Crnogorskom kršu. Beograd
Perović N (1991) Uticaj različitih količina i kombinacija mineralnih đubriva na vegetativno-produktivni potencijal vinove loze sorte Vranac u uslovima skeletnih zemljišta. Univerzitet u Sarajevu, Poljoprivredni fakultet u Sarajevu, Doktorska disertacija
Radulović M (1993) Snabdjevenost zemljišta i povrtarskih biljaka hranjivim elementima u području Zete- Crna Gora. Univerzitet u Beogradu, Poljoprivredni fakultet Beograd-Zemun, Magistarski rad
Radulović M (2001) Proučavanje mobilnosti nikla, hroma i fluora u nekim tipovima zemljišta Zetske ravnice. Univerzitet u Beogradu, Poljoprivredni fakultet Beograd-Zemun, Doktorska disertacija
Raes D (2002) BUDGET—a soil water and salt balance model. Reference manual. K.U.Leuven, Department Land Management, Leuven, Belgium
Spalević V (1999) Primjena računarsko-grafičkih metoda u proučavanju oticanja i intenziteta erozije zemljišta u Beranskoj kotlini. Univerzitet u Beogradu, Poljoprivredni fakultet Beograd-Zemun, Magistarska teza
Spalević V (2011) Uticaj načina korišćenja zemljišta na oticanje i intenzitet erozije zemljišta u Polimlju. Univerzitet u Beogradu, Poljoprivredni fakultet Beograd-Zemun, Doktorska disertacija
Stebut A (1934) Klimatske, hidrološke i pedološke osobine istočnog dela Crne Gore
Stebut A (1949) Agropedologija. Prvi deo. Naučna knjiga, Beograd
Stebut A (1950) Agropedologija. Drugi deo. Naučna knjiga, Beograd
Stebut A (1953) Agropedologija. Treći deo, Beograd
Topalović A (2006) Hemijska specijacija fosfata u krečnjačkom zemljištu, Magistarski rad, Hemijski fakultet, Beograd
Topalović A (2012) Uticaj folijarnog prihranjivanja na hemijski sastav nekih sekundarnih metabolita grožđa, Doktorska disertacija, Hemijski fakultet, Beograd
UNCCD (1994) United Nations convention to combat desertification in those countries experiencing serious drought and/or desertification, Particularly in Africa (1994), Chapter XXVII. Environment, Paris, 14 October 1994
United Nations Conference on Environment and Development Rio de Janeiro, Brazil. (1993). Agenda 21 : programme of action for sustainable development ; Rio Declaration on Environment and Development ; Statement of Forest Principles: The final text of agreements negotiated by governments at the United Nations Conference on Environment and Development (UNCED), 3–14 June 1992, Rio de Janeiro, Brazil. United Nations Dept. of Public Information.
Zloković Đ (1930) Pedološka ispitivanja na Nikšićkom polju, Geološki anali Balk. poluostrva, knj. X, sv. 1, Beograd

Land Use and Regions

2

Abstract

This chapter explains some of the geographical and social characteristics of Montenegro which markedly affect land resources. Montenegro is located in the southwestern part of the Balkan Peninsula and is characterised by very pronounced forms of hilly-mountainous relief and high altitudinal gradients, and pretty high wilderness of its territory. Only 6% of the country's territory has slopes below 5%, and the country is characterised by a distinct climate, from the Adriatic coast to the high mountains, which influence the division and structure of land use. The chapter discusses also the structure of land use in Montenegro, emphasising the high share of forests, pastures, and natural grasslands in the country and land ownership. The wide areas of the country are characterised by extensive agriculture, skeletal soils, and pronounced erosion processes. The overview of land resources is given by means of three main agroecological regions: (1) Coastal region spreading along the Adriatic coast and covering the area of coastal mountains characterised by the Mediterranean climate—hot and dry summers, but humid winters; (2) central region with Zeta-Bjelopavlići subregion characterised by the modified Mediterranean climate, and Karst subregion with sparse vegetation, scarce in soil and water resources, and (3) northern region, with subregion of high mountains and plateaus, covering the highest and largest part of the country under the mountainous climate, and the continental subregion as hilly-mountain area intersected by the river valleys, with milder climate and more inhabited in this region. The chapter discusses hydrological, biological, meteorological, and geological conditions in three regions, as well as anthropogenic impacts on soils.

2.1 Importance of Soil as a Natural Resource

One of the primary functions of soil is the use for plant production, which provides most of the biomass on the planet. In agriculture and forestry, the soil is a basic resource for production, as it serves to provide food and fibre as the most important source of sustenance for humans and animals.

Given the growing population on the planet (over 7.7 billion) and limited soil resources, the whole world, including Montenegro, is facing the issue of providing food for the population. That is why all countries in the world, in line with the UN Forum and other institutions' recommendations, are taking appropriate measures to preserve the existing soil and maintain and improve soil fertility.

If treated with due care and used rationally, soil can be considered a renewable resource, although the soil-forming processes are very slow and long. According to Montanarella (2007), the slow soil genesis process is reflected in the fact that it takes 500 years to form a 2 cm soil layer in natural conditions. Durn (2003) considers that it takes 2 million years to form a 50 cm terra rossa layer on a compact/hard limestone substratum. We are witnessing the process of irreversible loss of vast areas of land, even high fertility soils, rendered unsuitable for cultivation, caused to a large extent by various degradation processes and/or by use for purposes other than their main function (often more than actually needed).

2.2 Condition and Land Use

The geographical position of Montenegro in the southwestern part of the Balkan Peninsula, the geomorphology of the terrain with very pronounced forms of hilly-mountainous

M. Knežević et al., *The Soils of Montenegro*, World Soils Book Series,
https://doi.org/10.1007/978-3-031-82919-2_2

relief (about 94% with slopes over 5%), and the distinct climate characteristics (from Mediterranean to mountainous) influenced the division and structure of land use.

2.2.1 Agricultural Land Use Structure

Data on the structure of agricultural land use according to MONSTAT (Statistical Office of Montenegro) differ significantly. Table 2.1 provides comparative data according to the old and new methodology (harmonised with EUROSTAT). The old methodology concerned cadastral data, while the new methodology uses data based on censuses and relevant research activities.

According to the old methodology, agricultural land area is 0.83 ha per capita, while under the new methodology, the total utilised agricultural area is 0.41 ha per capita. The data from the old structure (2011) show that the arable land is 9.99 acres per capita (arable fields and gardens, orchards, and vineyards). Data from the new structure (2019) show that arable land is 2.37 acres per capita (utilised kitchen gardens and/or gardens, used arable land, vineyards, orchards—plantations, and extensive nurseries).

In addition to the official MONSTAT data (Table 2.1), Corine Land Cover data (Table 2.2 and Fig. 2.1) are provided as well.

A very high share of meadows and pastures (87%) influences the extensive character of agriculture, as the pronounced orography, primarily of the karst in geological structure, skeletal soils, and the pronounced erosion processes prevent full utilisation of these areas. Irrigation is applied on 12,518 agricultural holdings on an area of 5204.2 ha.

2.2.2 Census data overview

According to the census data published by MONSTAT, the total number of agricultural holdings is 48,870, of which 48,824 are family agricultural holdings.

Table 2.2 Land cover types—CLC 2018

Land cover types (2018)	Area (km^2)	Area (%)
Artificial surfaces	272	2
Agricultural areas	2231	16
Forest and semi natural areas	10,991	79
Wetlands	125	1
Water bodies	264	2
Total	13,883	100

Source Montenegro—land cover—2018

According to the size classes of utilised agricultural area, 15,418 or 32% of the family agricultural holdings are in the range of 0.1–0.5 ha.

There are 23,242 agricultural holdings or 48% of the total number of holdings in the category specialised production type—livestock production. Perennial meadows and pastures exist in 41,715 agricultural holdings, or 88% of the total number of family agricultural holdings, while only 122 holdings, or 0.2%, have nurseries.

Most agricultural holdings that use perennial meadows and pastures are situated in the municipalities of Nikšić, Podgorica, Bijelo Polje, Pljevlja, and Berane. On average, 2.03 persons were employed on agricultural holdings.

Of the total 48,824 holders of family agricultural holdings, 6286 are women or 13%, and 42,538 are men, or 87%.

An average family agricultural holding has 6.0 ha of total available land, 4.3 ha of utilised agricultural land, 0.4 ha of other uncultivated agricultural land, 0.2 ha of uncultivated agricultural land, 0.8 ha of forested land, 0.2 ha of infertile land, and 3.6 livestock units. The total value of the economic size of agricultural holdings in Montenegro expressed in euros is 125,817,765.2 or 2574.54 per agricultural holding, on average.

The municipality with the lowest number of family agricultural holdings is Tivat—169, which accounts for 0.4% of the total number of family agricultural holdings. The municipality with the highest number of family agricultural holdings is Podgorica—7276, which accounts for 15% of the total number of family agricultural holdings.

Table 2.1 Agricultural land use structure (2011 and 2019)

Structure (2011)	Area (ha)		Structure (2019)
Agricultural land	515,740	257,469.6	Total utilised agricultural area
Arable fields and gardens	45,748	2009.8	Utilised kitchen gardens and/or gardens
Orchards	12,007	7204.6	Used arable land
Vineyards	4399	2880.0	Vineyards
Meadows	126,990	1373.3	Orchards—plantations
Pastures	323,953	1214.6	Orchards—extensive
Ponds, fishponds, wetlands	2643	69.8	Nurseries
		242,717.5	Perennial meadows and pastures

Source MONSTAT—Statistical Yearbooks (2012, 2020)

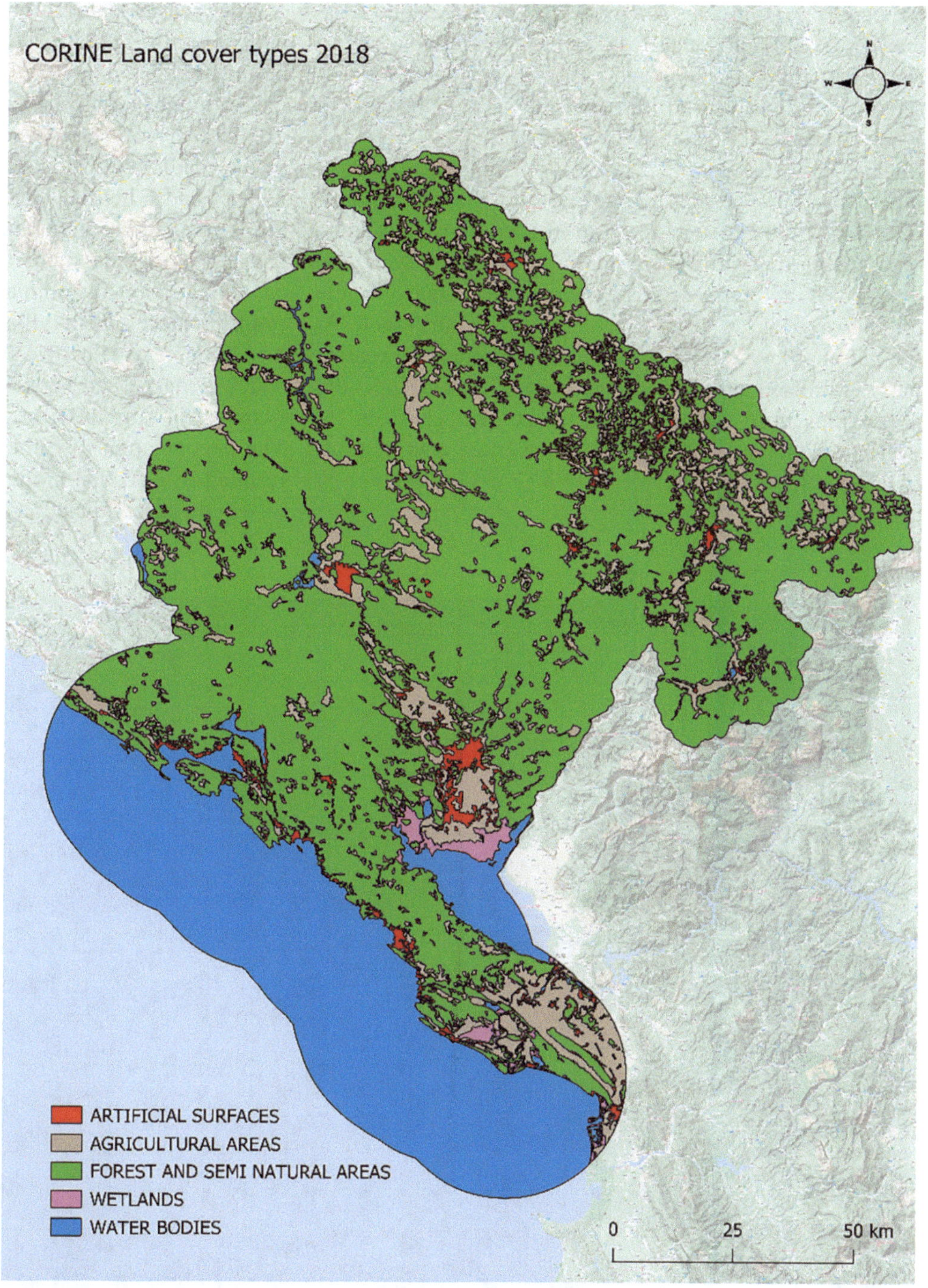

Fig. 2.1 Montenegro—land cover—2018

As mentioned above, there are 48,824 family agricultural holdings in the country . According to the data presented, most of the family agricultural holdings, 15,418 in total, belong to the size class 0.10–0.50 ha, while the lowest number of family agricultural holdings, 312, belongs to the size class 30.00–50.00 ha. Fig. A1 (Appendix A) shows the area (ha) of total available agricultural land by municipality, ranging from 322.7 ha in Tivat to 38,265.3 ha in Nikšić.

Of 48,213 family agricultural holdings utilising their agricultural land, the largest number is situated in Podgorica—7162, while the lowest number of holdings utilising their agricultural land is in Tivat—163. Also, kitchen gardens (gardens) are mainly used by households in the Municipality of Bijelo Polje—5592. Most of the households that use their agricultural land as orchards are in Bar—1276, and most of those that use their agricultural land as perennial pastures and meadows are in Podgorica (5643), and households that use their land as vineyards and arable fields again in the Municipality of Podgorica—1007 and 1824, respectively. Fig. A2

(Appendix A) provides an overview of the data on the number of holdings and areas covered by certain categories, other than meadows and pastures (left out for easier reference—significant differences in area compared to other categories) (Figs. 2.2, 2.3 and 2.4).

The data on family agricultural holdings according to the number of separate parts of utilised agricultural land indicate that most of the holdings have one separate part of utilised agricultural land—26,080. The number of holdings decreases with the increase in the number of separate parts of utilised agricultural land. At the national level, 2.36 is the average number of separate parts of utilised agricultural land used. Fig. A3 (Appendix I) provides an overview of the average number (from 1.3 in Tivat to 4.7 in Plužine) and the average area of plots (from 2648.4 m^2 in Budva to 57,749.7 m^2 in Plužine) by municipality.

The census data show a total of 32,656 households with a total of 112,862 livestock units. The highest number of livestock units is in Bijelo Polje, 16,979.9, while the lowest number of livestock units is in the territory of the Municipality of Tivat, 171.5 (Fig. A4, Appendix A). Most of the holdings in Montenegro have 0–5 livestock units—27,095, and 6 holdings have 100–500 LU. Livestock farming is particularly important as vast areas of meadows and pastures providing more than 90% of roughage (fodder) can be utilised by animal husbandry. Moving livestock (transhumance) from villages to katuni, temporary summer settlements (Laković 2017), i.e., from areas in the river valleys, coastal and karst fields to elevated areas with vast pastures, is a tradition in Montenegro's livestock farming. The depopulation of rural areas and the "ageing" villages resulted in a decline in the number of sheep and goats, but also in the discontinuation of a long tradition.

Fig. 2.2 Olive groves, Džidžarin, Bar (*Photo* B. Bajagić)

Fig. 2.3 Agricultural production in the central part of Montenegro (*Photo* G. Roganović)

2.3 Land Resources and Agroecological Conditions for Agriculture Development

Orography and very diverse climate characteristics resulted in diversified agricultural production. All branches of agriculture, different crops, numerous plant varieties, and animal breeds are present. This reduced not only opportunities for more intensive production but also limited production to very small areas.

Significant land resources for certain aspects of agricultural production are located in several coastal fields and terraces of the hilly hinterland up to about 400 m a.s.l. Such land in the coastal zone covers about 22,000 ha (12,000 ha in plains and 10,000 ha on terraces).

Another important and the largest flat area is the Lake Skadar basin, where the Zetsko-Bjelopavlićka Plain is located. It has an area of about 40,000 ha.

Unlike the coastal and central part of the northeastern part of Montenegro, the soils appear in a larger, almost uninterrupted continuity, thanks to the high share of silicate rocks in geological structure, as well as Quaternary lacustrine sediments around Berane and Pljevlja, and deposits in the valleys of Lim, Tara, Ćehotina and Ibar. The largest flat terrain area is located in Polimlje, from Gusinje and Plavsko Lake to Bijelo Polje, of about 14,500 ha, then around the Tara River 3700 ha, the Ćehotina valley 3700 ha, and the Ibar valley 800 ha.

Further to the areas mentioned above, certain flat areas are located in the karst fields and plateaus at 1100–1400 m a.s.l. The Nikšićko Field (about 6000 ha) is the largest of the karst fields, followed by Dragalj, Grahovsko, Njeguši, Cetinjsko, Gostilje, and Radovče fields.

In addition to soil, limiting factors for agricultural development are relief, climate characteristics, and lack or excess water in some areas. Also, insufficient application of modern technological developments and specialisation of production prevents better utilisation of natural resources.

An overview of land resources and agroecological conditions for the development of agricultural production is given by regions (Fig. 2.5): Coastal, central (with subregions of the Zeta-Bjelopavlići and Karst), and northern (with subregions of high mountains and plateaus, and continental subregion).

2.3.1 Coastal Region

The coastal region consists of two entities, rather different in terms of climate and relief. The first entity is a narrow coastal strip of up to 500 m a.s.l. The second entity consists of a range of coastal mountains, Orjen, Lovćen, Sozina, and Rumija, stretching from northwest to southeast (Fig. 2.6).

Fig. 2.4 Flock of sheep in Žabljak (*Photo* A. Topalović)

2.3.1.1 Narrow strip of Montenegro's coast

The narrow strip of Montenegro's coast, from the border with Croatia to the Albanian border, ends with the Bojana River and Lake Skadar (Fig. 2.7). Along the seashore, with a length of 297 kms, there are several bays and gulfs with larger or smaller areas in the hinterland in the form of valleys and extended fields. Deep Alluvial-Colluvial Soils and sandy soils (Arenosols) are formed here. Their total area is about 12,000 ha, of which about 8000 ha in the area of Ulcinj, 1000 ha in Bar, 250 ha in Buljarica, 500 ha in Mrčevo Field, 1000 ha in Tivatsko Field, 250 ha in Sutorina and 1000 ha elsewhere.

The rest of the coastal strip is made of limestones and dolomites (Fig. 2.8), which sometimes rise up from the shoreline and are intermittently broken by blocks of flysch and other silicate rocks. Limestones and dolomites from the Rumija and Sozina spread to the southern shore of Lake Skadar and Crmnica; flysch elements occur in the limestone roof layer and alluvial-colluvial deposits in the Crmničko and Orahovsko fields (area of 1000 ha).

In the flysch sequences on the coast and in Crmnica, significant areas can be found on human-made terraces, not only in the flysch zone but also in the karst area, using rocks as retention walls and filling them with soil. The terracing as a practice in the territory of Montenegro was taken from Mediterranean countries. Under human influence, the soil in terraces is of anthropogenic origin, with partially present colluviation processes. The anthropogenic soil in terraces in the flysch zone and other silicate substrata belongs to Eutric Brown and Dystric Brown Soil types, while those in the limestone zone are Terra Rossa Soils. In addition to terraces, Terra Rossa occurs in the coastal belt in several valleys in Luštica and Donji Grbalj (from Bigova to Jaz valley) and in smaller flat areas in the back of Dobre Vode, Mala Gorana and Velja Gorana and in sinkholes. In these relief forms, Terra Rossa was formed by the colluviation process (rainwashed from the surrounding elevated grounds) and presents a deep soil suitable for cultivation and various crops. The larger area of the Colluvial Terra Rossa can be found in valleys and terraces on the southern

Fig. 2.5 Regions: Coastal (I); Central with subregions: Karst (II) and Zeta-Bjelopavlići (III); Northern with subregions: High mountains and plateaus (IV), and Continental (V) (*Source* https://gifex.com, modified by Tatjana Nikolić)

Fig. 2.6 Coastal region (*Photo* M. Knežević)

shore of Lake Skadar, from the Albanian border to Virpazar. In the area to the Rijeka Crnojevića on a larger surface, the Colluvial Terra Rossa appears around Ostros to Bobovište and Livar, then from Murići to Krnjice and Seoca, and from Popratnica and Dupilo to Riječani. Higher yields of good quality can be achieved on Terra Rossa, if irrigation is applied.

The water regime in the Alluvial-Colluvial Soils of the coastal and Crmničko fields is not regulated. For that reason, certain hydro and agrotechnical amelioration measures need to be carried out to improve agricultural production. Historically, smaller complexes were partially covered by irrigation and drainage measures: 100 ha of Ulcinjsko Field, 80 ha in Ambul, 120 ha of Sutorina, 220 ha of Mrčevo, and 70 ha of Tivatsko fields. In depressions in the area of Ulcinj, known under the term "knete", in the Buljarica Field, as well as in the lowest part of Tivatsko and Crmničko fields, the hydro and agrotechnical amelioration should be carried out. Drainage is required for certain areas of the Tivatsko and the Crmničko fields, while irrigation is needed in all coastal fields and the Lake Skadar basin. The precondition for this is the provision of water for irrigation, which is available in sufficient quantities only in the area of Ulcinj that the Bojana River runs through.

The climate conditions of the coast are very favourable for the cultivation of various agricultural crops. The climate is characterised by a mean annual temperature of about 16 °C in the coastal belt and Lake Skadar basin. With the increase in altitude, the temperature drops and changes into the mountain climate, though not quite sharp. The precipitation increases from the southernmost Ulcinj towards Herceg Novi (Figs. 2.9, 2.10 and 2.11) and Krivošije (Crkvice is the place with the greatest precipitation in Europe, 4631 mm in the period 1961–1990).

Moderately mild winter, warm summer, as well as autumn and spring, almost without frost and snow, enable the cultivation of olive groves, subtropical fruit (tangerine, orange, kiwifruit, lemon, grapefruit, and pomegranate) and other fruits (apple, pear, peach, fig, raspberry, nuts, and cherry), various field crops and vegetables in open field

Fig. 2.7 Bojana River mouth (*Photo* G. Roganović)

Fig. 2.8 Boka-Kotorska Bay—A view from the Lovćen sideslopes (*Photo* M. Knežević)

and greenhouses. Due to later planting in the open field and frequent irrigation, watermelon and melon cultivation in Ulcinj is very important, along with tomatoes, peppers, and cucumber.

2.3.1.2 Range of coastal mountains

The second entity includes the coastal mountain range, extending from the border with Bosnia and Herzegovina to Albania. From the coast, the Lovćen's (1749 m a.s.l.) very

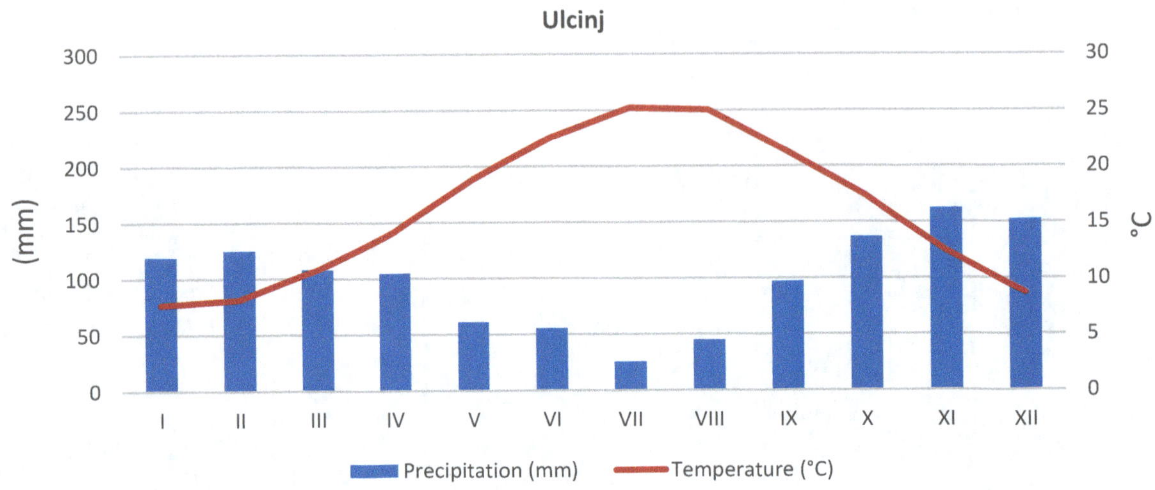

Fig. 2.9 Average monthly precipitation and temperature in Ulcinj (1981–2010)

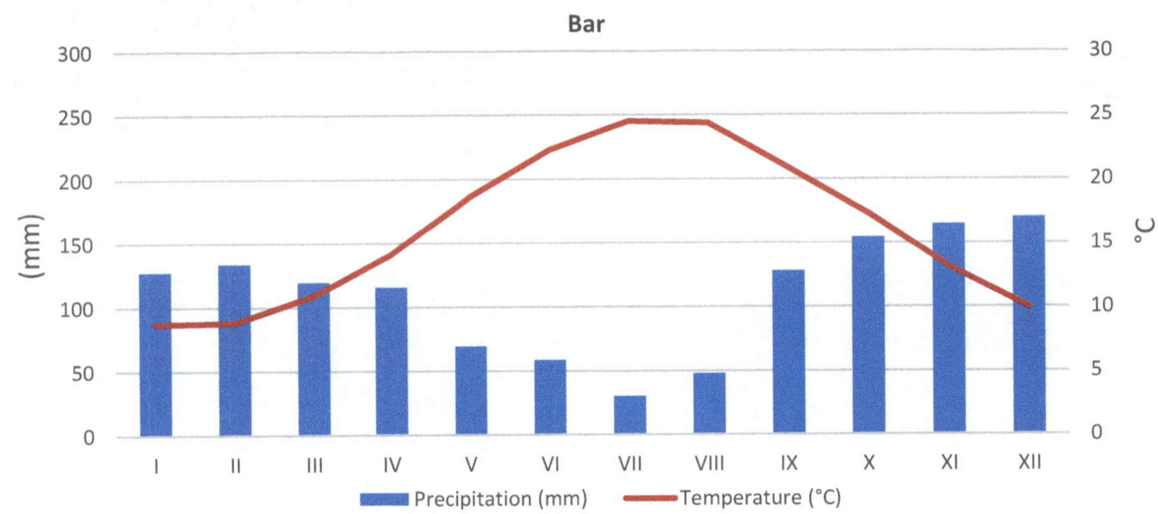

Fig. 2.10 Average monthly precipitation and temperature in Bar (1981–2010)

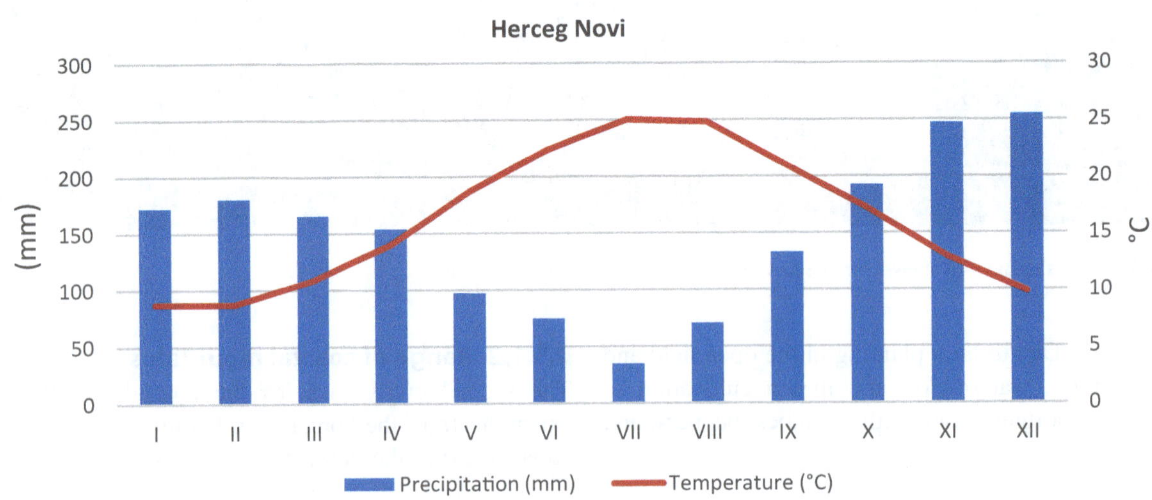

Fig. 2.11 Average monthly precipitation and temperature in Herceg Novi (1981–2010)

Fig. 2.12 Njeguši Karst field (*Photo* M. Knežević)

steep slopes rise, passing into a plateau of the Katunski karst with smaller karst fields—Njeguši (Fig. 2.12) and Cetinjsko fields, and a large number of valleys, sinkholes and flat areas that are almost non-existent on its southern side towards the Coast. The Lovćen glaciation was less pronounced than that of the Orjen (Fuštić and Đuretić 2000), and therefore, glacier traces are fewer but preserved in the direction of Njeguši, Krstac, and Cetinjsko fields (Vasović 1955).

Mountains Sozina, Sutorman, and Rumija with Lisinje are a group of Dinaric mountains extending in the northwest-southeast direction. On the side towards the sea, these mountains lean steeply on the zone of flysch and other sediments, so their slopes and numerous cliffs are almost bare, free from or with very sparse vegetation and a very shallow layer of Black Soils on limestone and dolomites (Calcomelanosol), which is the prevailing soil also on other mountains and passes between them. The situation towards Lake Skadar is completely different, as the mountain range gradually descends over numerous valleys, sinkholes, and flats, especially in the areas of Šestan and Krajina. In the Crmnica area, the Sozina is the only one reaching the Crmničko Field with a steep slope, thus dividing the flysch zone of the northwest from the southeastern part of Crmnica with Anthropogenic Cambisols as soil on the terraces.

From the Rumija and Lisinje, on the slopes towards the sea, the terrain is heavily broken down by numerous faults, with many cliffs in the back and watercourses cutting gorges with steep slopes. Scree and debris occur intermittently, particularly on the southern slopes of Lisinje. Along slightly less steep slopes and some valleys in this area, arable land can be found on terraces, which allowed the following settlements to be formed: Mali Mikulići and Velji Mikulići, Mali Kalimanj, Međureč, Dedići, Gureca, and others. Similar settlements are located in sinkholes and valleys of Somina and Bijela Poljana, where flysch sediments occur intermittently, then in limestone plateau sinkholes below the peaks of Lovćen, as well as in Zalazi. Also, under the Orjen, where in addition to numerous sinkholes, there are also valleys filled with glaciofluvial deposits (Ladenice, Dragalj, Grahovo, Ubli, Kameno, Kruševice, and Vrbanje).

The much more unfavourable climatic conditions at the footslopes of the coastal mountains and their highest parts have limited the choice and variety of crops produced. In the above conditions of this entity, there are almost no orchards or vineyards, so the main crops are potato, carrot, cabbage, and other field and fodder crops.

2.3.2 Central Region (Zeta-Bjelopavlići and Karst Subregions)

2.3.2.1 Zeta-Bjelopavlići subregion

The Zeta-Bjelopavlići subregion includes the namesake plain and rimming footslopes, including the northern coastal zone of Lake Skadar. Due to the proximity of the Adriatic Sea and its openness along the Bojana River valley and Lake Skadar, the subregion's climate is characterised by a modified Adriatic (Mediterranean) climate with features similar to that of the coastal belt. Average annual air temperatures are about 15 °C, but during autumn, winter, and spring, frost occurs (up to − 9.7 °C in Podgorica), while snow is usually short-lived.

Regarding climatic features, the Zetsko-Bjelopavlićka Plain is very similar to the coast. The influence of the Mediterranean climate, which penetrates the subregion along the Bojana valley and Lake Skadar, can be felt around the valleys of rivers Morača, Zeta and Rijeka Crnojevića. The influence is felt in terms of temperatures and precipitation. The Zetsko-Bjelopavlićka Plain is a hollow, so summer temperatures are high, with torrid summers, strong heat, and pronounced aridity. The temperatures in June, July, August, and September are slightly higher than on the coast, while temperatures from December to March are slightly lower, with more frost days than on the coast, where these are very rare. The mean annual temperature in Podgorica is 15.9 °C, and the maximum absolute temperatures exceed 40 °C.

The average annual sum of precipitation is 1637 mm in Podgorica. Despite the abundance, due to uneven distribution, aridity is pronounced during the vegetation period, especially in the three summer months. High temperatures, intensive evaporation, and lack of humidity affect the yields of crops that are not irrigated. The precipitation distribution is unfavourable, as less than 10% of the average annual sum of precipitation falls during the three summer months. That is why the dry period, which in some years lasts from April to October, hampers the development and yields of agricultural crops. The lack of humidity lost due to high evaporative demand caused by high temperatures and strong winds has to be compensated by irrigation.

This subregion covers about 41,000 ha in the plain (Fig. 2.13), which is about 75% of the total plain area in Montenegro. Different soil types are present in this area, with different physical, chemical, and biological characteristics. Brown Soils (Cambisols) prevail, which are mostly very shallow and shallow, like in the Ćemovsko Field, on the banks of the Cijevna, Rujela, Ribnica, Morača, and Sitnica. The medium-deep and deep soils appear along the rim of the Zetska Plain, as in Doljani, Dajbabe, Lješkopolje, Beri, and Farmaci.

In the Zeta area, from the Podhumski Bay in the east to Malo Blato in the west, Cambisols change into Alluvial Soils and Eugley Soils and continue further into the Ceklinsko Field, along the Crnojevića River to the namesake settlement, Rijeka Crnojevića. The coastal areas of Lake Skadar, due to oscillations of its water level, are continuously flooded up to about 5 m a.s.l. elevation, and seasonal floods reach the 9 m a.s.l. elevation level (absolute maximum 10.44 m a.s.l.) at the highest water levels (Fig. 2.14). Seasonal floods prevent timely soil preparation

Fig. 2.13 Zetska Plain—an aerial view (*Photo* G. Roganović)

Fig. 2.14 Panorama of the northeastern part of Lake Skadar (*Photo* M. Knežević)

and its cost-effective use for agricultural production. Floods and high levels of groundwater in the riparian zone of the Lake impose the need to regulate the Bojana River and Lake Skadar simultaneously, for which Albania and Yugoslavia prepared several variants of preliminary designs. They were not implemented due to conflicting opinions from the commissions on both sides regarding the investments. The plan envisioned significantly greater regulation on the territory of Montenegro (12,000 ha) than on the territory of Albania (3500 ha).

Soil flooding in the Lake Skadar riparian zone still occurs, and the area of the Alluvial Soils unaffected by flooding has been reduced to 4500 ha. Cultivation, sowing, and planting of seasonally flooded soils are possible only after the lake level drops and the water recedes, in the period from mid-May onwards. The water withdrawal in some years is slow, taking place gradually, depending on the hydrological conditions in the basin area. The largest part (about 3500 ha) from which water apparently recedes is influenced by high levels of groundwater, and the soil cannot be cultivated but can only be used as meadows and for grazing. The wetlands belt from Podhumski Bay to the Ceklinsko Field transforms into a belt of wetland vegetation growing in permanently shallow water, such as reed, bullrush, water lily, and other species, including willow, alder, and ash. The wetland meadows belt is still present in the Crmničko and Orahovsko fields on the southern shore of Lake Skadar. Thanks to the conservation of the ecosystem, Lake Skadar was listed in the World List of Wetlands of International Importance in 1996 by the Ramsar Convention (Figs. 2.15 and 2.16).

Cambisols on diluvial loams prevail in Lješkopoljski Lug (Fig. 2.17) and Bjelopavlićka Plain (Fuštić and Đuretić 2000). The transitional (B) horizon and clayey surface are impermeable, which is why the perched water table stagnates in the soil layer during the rainy autumn–winter period. For more intensive agricultural production, such soil condition requires drainage and, during the vegetation period, also irrigation. These amelioration measures, along with the enrichment of the arable soil layer with organic matter, can result in a good water–air soil regime and create optimal conditions for the cultivation of field crops and fruits. Following WWII, an irrigation system was built in the Bjelopavlićka Plain for an area of 1800 ha on the right bank of the Zeta River. Furthermore, the main drainage channel was built in Lješkopoljski Lug which collected only excess water from the Mareza water source. Later on, in the social sector complexes in the Bjelopavlićka Plain, a pipe drainage system was installed on an area of 840 ha (Fig. 2.18), while the "Agrokombinat—13 jul" started building an irrigation system, not only for its own land complexes but for the privately owned land as well. The restitution of land from the former social ownership to its original owners (1992) was a major step back in the development of agriculture in Montenegro. Expectations that the private sector will achieve better results than social agricultural enterprises, i.e., individual holdings and complexes, have not been met. The soil, restituted from the well-managed social sector complexes, has been completely neglected, and instead of continuing its intensive use, it has been transformed into a wasteland. The perennial orchards and vineyards were cut, and drainage and irrigation systems became dysfunctional as their regular maintenance was discontinued. Over time, in the absence of agricultural production, the storage and processing capacity, equipment, and agricultural machinery were rendered useless.

Along the rim of the Zetsko-Bjelopavlićka Plain, some areas of importance for agricultural production can be found on terraces, valleys, and smaller flats. East of the Cijevna River (Fig. 2.19) to the Albanian border, in Malesija, on terraces and valleys, Colluvial Terra Rossa can be found, which is a prevailing and the only arable soil also in Kuči (Fig. 2.20) up to about 500 m a.s.l., and then in Bratonožići and the eastern part of Piperi (Gornje Mrke, Petrovići, and Blizna). Along the Morača River, on old river terraces, which are somewhat larger near Bioče, at the confluence of the Mala and Morača rivers, Brown Eutric Shallow Soil formed on glaciofluvial deposits and

Fig. 2.15 Flooded zone of Lake Skadar—territory of the village of Dodoši (*Photo* M. Knežević)

conglomerate can be found. A narrow flysch strip, with some soil on the terraces, is present in the area of Doljani, from Zlatica to Fundina. Flysch extends through Piperi (Zavala, Radeća, Lopate, and Crnci) and Gornji Martinići, then from Slatina to Brijestovo and Vinići, and from Šobajići to Povija in the form of broken strips. A narrow flysch strip also occurs in Pješivački Do (Živaljević et al. 1973), Bjelopavlići and Drenovštica. In all these sites, arable land is mainly found on terraces or smaller flats such as Šobajići, Jovanovići, and Vinići.

Colluvial Terra Rossa as arable soil is widespread in the area of Lješanska Nahija, Dobrska Župa and Progonovići, then in Komani, Zagarač and Zagreda.

Due to favourable climatic conditions, the soils on terraces, valleys, and smaller plains are used to grow various agricultural crops, including the grapevine, which has the highest share, followed by figs, pomegranates, and other fruit species.

2.3.2.2 Karst subregion

Karst subregion is a large area with scarce soil, water, proper communications, and intermittently sparse vegetation. Arable land can be found in karst fields, sinkholes, valleys, and smaller flats. The largest karst field is the Nikšićko Field, covering an area of about 7000 ha. The Nikšićko Field is made mainly of glaciofluvial gravel (Radojičić 1996) and diluvial loam in Glibavac, Mokra Njiva, and Slivlje. Brown Eutric Soil was formed on such parent material, mostly in a shallow layer on the gravel substrate, 20–30 cm (Straševina and Kočani, Kapino, Pac, and Rudo fields). Parts of the Nikšićko Field from the urban area to Vidrovan dispose of better soil quality, so in this area, agricultural production is more diverse than in the parts of the field south of the town of Nikšić. In the southwestern part of the Nikšićko Field, two artificial lakes, Krupac and Slano (Fig. 2.21), and the third, intermittent one, Vrtac, were built, waters of which are mainly used for power generation. The

Fig. 2.16 Floodings in Lake Skadar area—Dodoši village (*Photo* G. Roganović)

Fig. 2.17 Area of Lješkopoljski Lug (*Photo* G. Roganović)

Fig. 2.18 Soil drainage works in the 1980s in the Bjelopavlićka Plain (*Photo* Archive CZM)

lowest southern part of the Nikšićko Field is flooded in autumn and winter when, due to heavy rainfall, the water level of the Zeta, Mrkošnica, and Gračanica rises, and the sink Norin does not have the capacity to receive that water quantity. Then a huge amount of water forms a lake and floods the land of the Suvo Field, Lugovi, Straševine, and Budoške Bare. Flooding of certain areas in the period autumn–winter is pronounced on the banks of the Zeta, upstream of Nikšić. Regulation of the Zeta River bed would prevent flooding in the area of Mokra Njiva, Glibavci, Brezovik, and Miločani.

A smaller karst field near Nikšić, Šipačno, is located immediately by the road Nikšić–Plužine. This field is also covered by glaciofluvial deposits, with a slightly deeper soil layer compared to Kapino and Pac fields.

Nikšićka Župa, an extended valley of the Gračanica River, where the Liverovići reservoir was built (Fig. 2.22), is situated northeast of Nikšić. The valley is filled mainly with glaciofluvial deposits, and for that reason, Brown Eutric Soil prevails, occasionally shallow, as is the case, for example, in the Miolje Polje, and in most cases of medium depth in Gornje Morakovo and Donje Morakovo, Dučice, and Liverovići. Where the valley changes into the mountain area, smaller sequences are made of Deluvial-Colluvial Soils, suitable for cultivation, as well as Brown Eutric Soils.

The Grahovsko and Dragalj fields lie southwest of Nikšić, along the road Nikšić–Risan. Fluvial-glacial deposits cover both fields (Fig. 2.23). The Grahovsko Field is at 690–780 m a.s.l., covering an area of about 6.4 km^2, with Rendzina Soils of various depths (20–150 cm), Brown Eutric Soil (40–70 cm deep) covering an area smaller than rendzina, in the northwestern part of this field. A micro accumulation built at the edge of the Grahovsko Field, at the source of the Grahovska River, which is an intermittent watercourse, provides water not only for the town of Grahovo but also for irrigation. However, the irrigation covered a small area in different periods, although various projects had envisaged irrigation of 300 ha of agricultural land.

The Dragalj Field is lower, at 600–660 m a.s.l., with an area of 6.3 km^2 and a very shallow rendzina layer, slightly deeper and suitable for cultivation at the edges of the field. This field is a typical representative of the so-called dry karst fields because there are no surface waterstreams or water sources.

Several smaller karst fields, Cetinjsko, Njeguši, Trešnjevo, and Ledenice, at the footslopes of the coastal mountains, gravitate towards the karst area.

Fig. 2.19 Characteristic form of the Cijevna River (Cijev–Eng. pipe) (*Photo* J. Lazarević)

Fig. 2.20 Centre of the Kuči area—Medun (*Photo* G. Roganović)

Fig. 2.21 Part of the Nikšićko Field and artificial reservoir—Lake Slano (*Photo* G. Roganović)

The Bajovo and Pejovića fields at the northeastern footslopes of Mountain Golija (1942 m a.s.l.) also belong to the karst area. Rendzina formed on moraine deposits is present on both fields as well as in neighbouring Donja Brezna. As an extension to these three fields, which are located at elevations of 980–1000 m a.s.l., is Gornja Brezna, at the

Fig. 2.22 Artificial reservoir Liverovići (*Photo* S. Radusinović)

footslopes of Mountain Vojnik (1997 m a.s.l.) and above the Komarnica Canyon. In this slightly wavy field, unlike the previous three, the parent material is composed of cherts and limestones, on which Humic Brown Acid Soil was formed (Humic Dystric Cambisols), which, despite its medium depth, is unsuitable for cultivation due to acid reaction and other unfavourable chemical characteristics. That has contributed to its use mainly for grazing and partly as a meadow when fertilised.

In the karst area gravitating towards the Zetsko-Bjelopavlićka Plain, namely the Lake Skadar basin, several smaller karst fields and plains occur. The following plains extend from the Nikšićko Field and Nikšićka Župa from northeast to southeast: Topolovo, Gostilje, Kopilje, Radovče, Trmanja, Momče, and Kržanja. All these fields were formed by moraine deposits over the primordial valleys over which Rendzina Soil was formed as the dominant soil type, with variable depths in different fields but also in the same field. In most cases, these are very shallow and shallow Rendzinas, which prevail in Radovče, Kopilje, and Kržanja. In addition to these karst fields, there are several valleys and plains in a part of this karst plateau: Baršno, Međeđe, Ponikvica, Vukotica, Brajovićka Ponikvica, Stravče, Orahovo, Ubli, Medun, Zatrijebač, and Arza. Ubli, Medun, Zatrijebač, and Arza are situated in the Terra Rossa zone, at an elevation of up to about 600 m a.s.l., so the soil has greater importance for agriculture since climate conditions allow the cultivation of grapevine and figs. Most of the other locations are used as katuni (for grazing).

In the vast karst areas, apart from limestones, dolomites, moraine, and glaciofluvial deposits, other parent materials are insignificant. Only flysch sediments appear in the form of narrow strips, like the one through the Duga Gorge, which, though intermittent, reaches the Nikšićko Field. Another narrow flysch strip extends from Bogmilovići to Pješivački Do and Drenovštica. A smaller flysch enclave enters the area of Montenegro in the vicinity of Krstac, and its larger area extends towards Gacko and Avtovac, from where a large area continues through the Piva, in the Vrbnica River and Pivsko Lake basin.

Fig. 2.23 Soils on Fluvial-glacial deposits in the Grahovsko Field (M. Knežević)

The mountains of Bijela Gora (1554 m a.s.l.), Somina (1601 m a.s.l.), Njegoš (1761 m a.s.l.), Zla Gora (1459 m a.s.l.), Golija (1945 m a.s.l.), Pusti Lisac (1476 m a.s.l.) and Garač (1436 m a.s.l.) rise from the karst plateau. These, as well as the coastal mountains of Orjen, Lovćen, Sozina and Rumija, stretch in the Dinaric direction. The karst plateau with the mountains mentioned above, from the border with Bosnia and Herzegovina to the Skadar hollow, covers an area about 90 km long and 40 km wide. From the limestone plateau, which ends in numerous faults and cliffs along the rim of the Bjelopavlićka Plain, rise the Ostroške Grede (1143 m a.s.l.) (Fig. 2.24), Prekornica (1927 m a.s.l.), Lisac (1563 m a.s.l.), Lebrčnik (1586 m a.s.l.) and Kamenik (1815 m a.s.l.). This plateau and mountains are parallel to the Katunska Nahija all the way to the Morača Valley. From the Morača valley onwards, the limestone plateau is somewhat lower in the area of Bratonožići, Kuči, and Malesija, from where only Siljevik (1147 m a.s.l.) and Žijevo peaks (2184 m a.s.l.) rise. The area of these mountains is poor in soil resources, so it is used as pastures and forests.

The climate of the karst area is characterised by rather high temperatures of 10.1 °C in Cetinje and 11.1 °C in Nikšić. Such high average annual temperatures are conditioned by a relatively small distance of these towns from the sea, which also influences the karst area. This influence reaches Nikšić from the Zetsko-Bjelopavlićka Plain, and Grahovo from Risan and Trebišnjica. Nikšić is further away from the sea, yet with a slightly higher average annual temperature than Cetinje, which is shaded by the Lovćen. Cetinje and Nikšić are at almost the same altitude (620–640 m a.s.l.). During the summer, mean monthly temperatures

Fig. 2.24 Cliffs of Ostrog—Ostroške Grede (*Photo* G. Roganović)

are lower than those in the Zetsko-Bjelopavlićka Plain, but the summers are, nevertheless, warm and dry, which is reflected in the yield of agricultural crops.

The influence of the Mediterranean climate in the karst area is more evident in precipitation than in air temperatures (Fig. 2.25, 2.26 and 2.27).

The climate with mild winter, hot summer, and warm autumn and spring enables the growing of grapevine, fig, kiwifruit, peach, other fruit species, vegetables, and field crops, not only on the plain but also on terraces up to about 500 m a.s.l. Grapevine thrives in Nudo, a small glen gravitating towards the Trebišnjica valley upstream of Trebinje. In other parts of the karst area to the mountain footslope, the climate allows for the cultivation of field crops, vegetables, and continental fruit.

2.3.3 Northern Region (High Mountains and Plateaus and Continental Subregions)

2.3.3.1 Subregion of high mountains and plateaus

The subregion of high mountains and plateaus covers the highest and the largest area in Montenegro. It stretches from the border with Bosnia and Herzegovina to the border with Albania. It is an area of vast limestone plateaus from which several high mountains of the Dinaric chain rise, from the northwest to the southeast. The rivers Tara, Piva, and Morača and numerous tributaries carved their beds in between the mountain peaks. Hollows and gorges are interchanging in the valleys of the main watercourses of the Tara, Piva, and Morača rivers. The sources of these watercourses are on the terrain formed of clastic rocks originating from limestone sediments. Denudation of Paleozoic and Werfen shales, sandstone, and other rocks, as well as of the Durmitor flysch, then silicate and igneous rocks, created extensions or smaller and larger hollows with the most significant agricultural land areas. In limestone formations, valleys of the main rivers and their tributaries were formed by cutting in or ramming, which are of canyon type, narrow, and sometimes almost vertical sides, 800–1300 m high. The Tara Canyon is the second deepest canyon in the world (1300 m a.s.l.), after Colorado.

The first group of high mountains includes Volujak (2337 m a.s.l.), Bioč (2396 m a.s.l.), Maglić (2386 m a.s.l.), Pivska Planina (1930 m a.s.l.) (Fig. 2.28), Treskavac (2093 m a.s.l.) and Vojnik (1997 m a.s.l.) (Fuštić and Đuretić 2000). The Volujak, Maglić, and Bioč are west of Pivsko Lake, forming a spacious mountain area between the Sutjeska, Piva, and Drina rivers that belongs to Bosnia and Herzegovina and Montenegro. Pivska Planina, a limestone plateau between the Piva and Tara, with an average height of 1450 m, is lower. This plateau is characterised by numerous glens, shallow sinkholes, and flats, among which

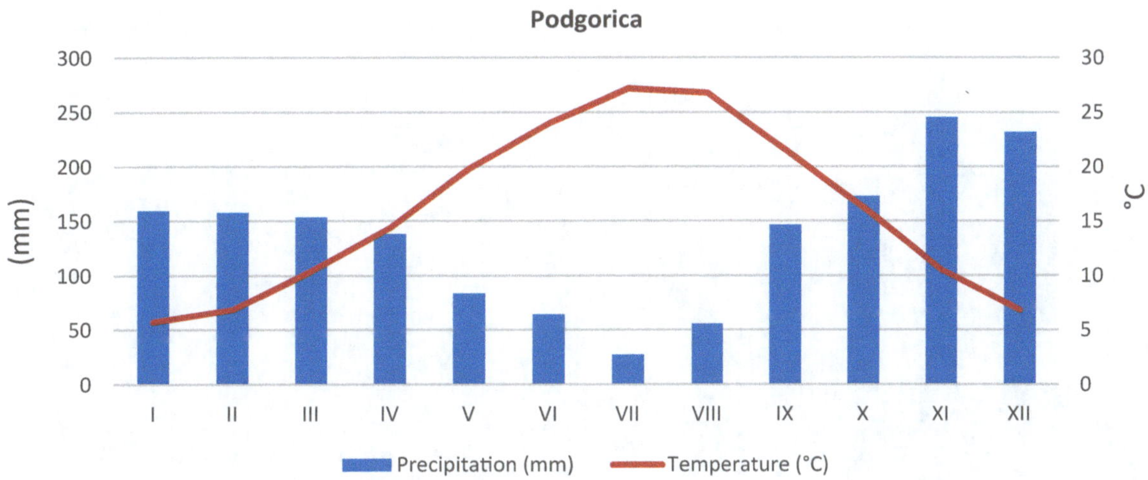

Fig. 2.25 Average monthly precipitation and temperature in Podgorica (1981–2010)

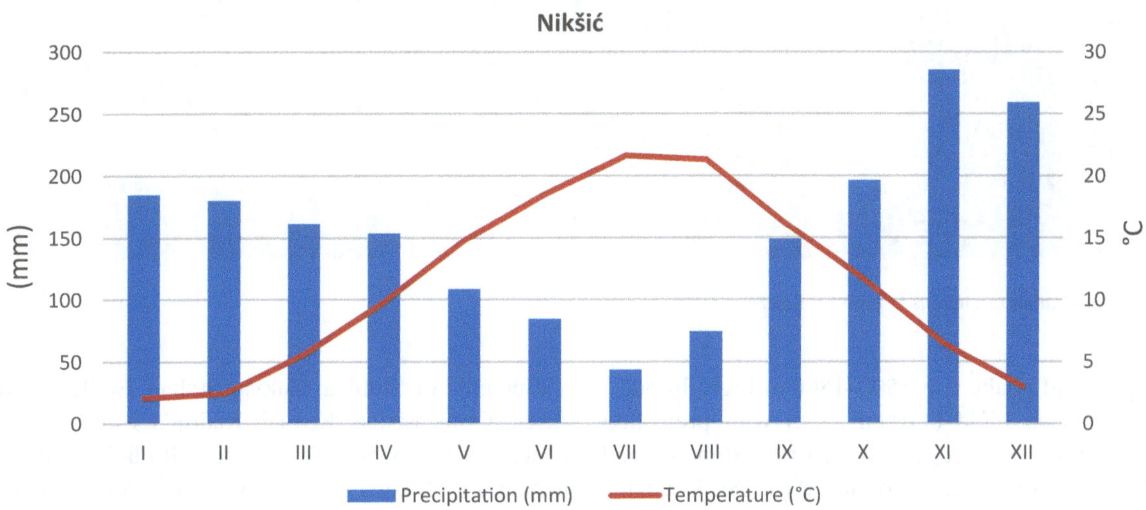

Fig. 2.26 Average monthly precipitation and temperature in Nikšić (1981–2010)

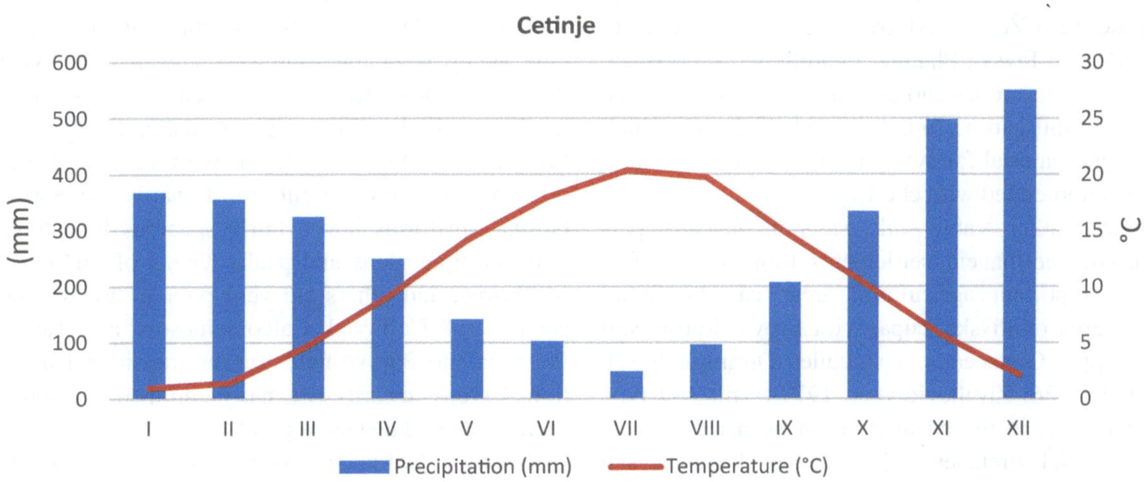

Fig. 2.27 Average monthly precipitation and temperature in Cetinje (1981–2010)

Fig. 2.28 Pivska Planina mountains (*Photo* M. Knežević)

passes, ridges, and peaks rise 1500–1800 m high. In addition to the Crkvice Field (1100 m a.s.l.) with prevailing Brown Eutric Soils on Neogene lacustrine sediments and glens of Unač, Suvodo, Barni Do, and Pirni Do, Rendzina on moraine deposits alternates with Black Soils formed on limestones and dolomites, which is the most widespread soil type on these mountains and foothill plateaus, such as Brljevo and Košara below Bioč, Ulobići and Katunsko Field below Maglić, then Žeično, Nikovići, Kneževići, Nedajno, Trsa, and Pišče on Pivska Planina. In addition to the Black Soils formed on limestones and dolomites, in smaller areas, Brown Dystric Soil can be found formed on cherts, which becomes parent material following the chemical dissolution of limestone interbedded with cherts.

Sporadically, other arable soils occur in plateaus, particularly around permanent settlements formed from former katuni. Important agricultural areas can be found in the lower area of Pivska Župa, with Brown Eutric Soil formed on Upper Cretaceous or so-called Durmitor flysch (Kalezić et al. 1966; Živaljević et al. 1979). This soil type covers larger areas in the basin of Vrbnica, a tributary to Pivsko Lake, which stretches to the border with Bosnia and Herzegovina, i.e., from the plateau Ravno to Plužine and Smriječno. It is the area from Volujak to Golija where the settlements of Stubica, Zukva, Stabna, Sedlari, Miloševići, Jasen, Orah, Lisina, Kovači, Radojičići and Smriječno are concentrated. Around Goransko, Rudinice, and Seljane, narrow flysch belts with Brown Eutric Soils occur sporadically, while the most widespread soil is Calcomelanosol (Black Soils on limestones and dolomites) to the canyon valley of Komarnica and the peaks of Golija. The Golija (1942 m a.s.l.) stretches from Gatačko to Nikšićko fields as a flattened ridge from which slightly larger valleys spread. One valley is in the southwest, towards the Njegoš, that the road Nikšić–Krstac–Gacko passes through. The valley has several smaller glens (Goslić, Gornje, and Donje Srijede) and Duga Gorge, but with very little arable land of importance for agriculture. Towards the northeast, the Golija transforms into a fairly spacious limestone plateau with multiple glens and peaks. Glens of Suvodo, Latično, Miljkovac, and others are very poor in arable land, which consists of Colluvial Calcomelnosols and Rendzinas in Pejovića and Bajovo fields (at the transition from the mentioned glens towards Brezna, mentioned previously in the section on the karst subregion).

Between Golija and Vojnik, there are several valleys, also poor in arable land, which is particularly the case with Papratne Doli and Lipova Ravan, partly for Jasenovo Polje

Fig. 2.29 Komarnica River canyon (*Photo* G. Roganović)

and Praga, on the southern slopes of the Vojnik. The massive appearance of the Mountain Vojnik ends on very steep slopes in the Komarnica Canyon (Figs. 2.29 and 2.30), while the transition towards the flats Mokro and Kruševice is milder. There is an even more gradual transition to the Krnovo plateau, where shallow and deeper soils alternate, as in the Gradačka Poljana, Krnovo Field, and Ćeranića Gora, where Brown Dystric Soil on flysch alternates with Rendzinas on moraine deposits (occurring from Krnovo Field and Ćeranića Gora along the valley of Gornje Vučje and Donje Vučje, then from Ivanje, covering Bukovik, Blaca, Seoca, and Lukovo). Most of Rendzinas are shallow and only in some parts deeper/thicker.

The Krnovo plateau (Fig. 2.31), with an average height of 1450 m, extends to the Lukavica plateau (1500 m a.s.l.), with Dystric Cambisol formed on flysch sediments as the dominant soil type. Southwest of Lukavica is the Konjsko flat, mainly covered with Dystric Cambisols, while Luke Bojovića and Mountain Maganik (2139 m a.s.l.) are south of this plateau. Luke Bojovića is made of moraine deposits with Shallow Rendzinas also used as arable land. In contrast, the area of Lukavica and Konjsko is mainly used for grazing, which is also the case with Calcomelanosols (dominant soil on the Maganik and its slopes, which southward gradually descend into plateau Štitovo, with its cliffs enwrapping Nikšićka Župa). On this plateau, flats of Rekovica, Dugi Do, and Maganičko Field, in the Maganik foothill, and to the south, Suva, Srednja, and Brajovića Ponikvica are important transhumance areas—katuni—for livestock farmers from Nikšićka Župa, Bjelopavlići, and Piperi. In the northeast, the Maganik, over its steep slopes, extends into the Velje Duboko hollow, with Brown Dystric Soils of the flysch sediment belt. This soil covers a larger area of Donja Morača and Gornja Morača. The lowest part of the Velje Duboko hollow is covered by glaciofluvial deposits on which Rendzina Soil, mostly shallow, was formed. A very dynamic relief in the hollow, non-resistant geological substratum and soil, interacting with climate and other factors, resulted in increased erosion activity. This contributed to the significant erosion deposition by the Mrtvica River with its tributaries in the bed and the banks of this watercourse. Below the village of Donje Velje Duboko, the Mrtvica runs through the canyon with the Liješnje flat on its northeast bank, at 1100 m a.s.l., composed of plenty of quality, cultivated soil. The Mrtvo Duboko valley in the lower part of the canyon is poor with quality soils, mainly made of Dystric Cambisol formed on flysch, but it has more favourable climate conditions because it is located at the elevation range between 250 and 500 m a.s.l.

Northeast of Lukavica, Kapetanovo Lake and Velje Duboko are Žurim (2035 m a.s.l.), Lola (2191 m a.s.l.), Stožac (2072 m a.s.l.), Kapa Moračka (2226 m a.s.l.) and Javorje (1675 m a.s.l.), which form the largest group of

Fig. 2.30 Komarnica gorge (*Photo* G. Roganović)

Fig. 2.31 Krnovo plateau (*Photo* G. Roganović)

high mountains at the watershed of the Morača and Piva rivers basins. This large plateau (with a mean altitude of 1750 m a.s.l.) is made of silicified limestones, allowing the formation of numerous glens, valleys and passes between a large number of ridges with 15 peaks over 2000 m a.s.l. The plateaus and ranges of these mountains have very steep slopes, descending into the upper catchments of the Mrtvica and Morača rivers. Numerous intermittent and permanent watercourses have cut their beds along this area. Similarly, the terrain descends steeply towards the Bukovica River tributaries Tušina and Bijela. Milder relief forms usually have a deeper soil layer, which is significant for livestock grazing and summer stays in the mountains.

The Morača formed its valley to the mouth of the Sjevernica tributary, that is, to the entrance into the Platije Canyon, in the area of the Durmitor flysch. Therefore, in the area of Gornja Morača and Donja Morača, despite the pronounced and dynamic relief towards the surrounding mountains, soil depth on the milder slopes enables agricultural production. Thus, in Gornja Morača, at the elevation of 365 m a.s.l. near Mioska up to 1100 m a.s.l. on both sides of the valley, there are significant areas of arable land in the districts of Ljevišta, Bojića, Starča, Redica, Svrka, Zavraca, and Požnje. Donja Morača has milder relief forms, so it has more quality soils, especially on the old river terraces with Brown Eutric Soils and in Međuriječje, at the Morača Monastery and Borje. In Gornja Morača, these terraces were preserved only in Dragovića Field, but their area is negligible. Outside the terraces to the northeast, significant areas with deeper soil of good fertility are found in Vrujci, Đuđevina, and Petrova Ravan, while those in the Sjevernica basin are less suitable for agriculture due to steeper slopes. From the Morača Monastery to the west, arable land is scarce due to the steep terrain and can be found in the area of Osreci, Ocka Gora, and Cerovice, then in Gornji Rovci and Vlahovići. Štavanj, Stupovi, Duške, Slacko, Lijeva Rijeka, and Brskut belong to the Morača basin flysch zone up to the divide with the Tara watershed. In the lowest parts of these areas, Deluvial Soils and some Alluvial Soils can be found apart from Dystric Cambisols.

Mountains Durmitor (2598 m a.s.l.), Sinjavina (2203 m a.s.l.), Bjelasica (2139 m a.s.l.), and Komovi (2487 m a.s.l.) stretch from the Pivska Planina in the Dinaric direction. The Durmitor, with a well-branched structure, and Sinjavina rise from the vast limestone plateau (Bešić 1983) with an average elevation of 1450 m a.s.l. The plateau is mainly built from Triassic and Jurassic limestones, interbedded with cherts in the lowest parts, often covered by moraine deposits. Due to such substratum composition, Calcomelanosols and Rendzina Soils prevail. Glaciation of the Durmitor and Sinjavina was intensive, with numerous trails left behind. In cirques, many smaller and larger lakes were formed, and in valleys and flats, moraines. The watercourses transported glaciofluvial deposits into the valleys of the Komarnica, Grabovica, Bukovica, and Tušina, which form the upper catchments of Piva River, as well as in the Plašnica, Štitarica, and Bistrica in the Tara basin and its valley in Trebaljevo, Sjerogošte, and Mojkovac hollow. Numerous glens and flats can be found in the Durmitor footslopes, from the Black Lake and Nadgora through the Tepačko and Njegovuđe fields and the vast and slightly wavy plateau of the Jezera and further into the valleys of Bukovica, Grabovica, and Komarnica. The Bukovica and its tributaries Tušina and Bijela formed their valleys in the flysch sediments lying on limestone formations. Thus, Dystric Cambisols are present there, occasionally also on sandstones of Javorje, Pašina Voda, and Bijeli Potok, as well as in several places on cherts. Numerous glens, shallow, bowl-like sinkholes, and fields can be found in the Sinjavina (Lipovac 1987a, b), such as Zminica, Zauglina, Pasji Nugao, Đedovo and Smoljen fields, Suvo and Odrag fields, Okruglica, Studenci, Šaranske Gomile, Grkovo, Ružica, Šaransko, Kričačko and Tušinsko fields and others. In all these milder relief forms in the Sinjavina, the Durmitor footslopes, and neighbouring river valleys, where climate conditions are more favourable, there are deeper and good quality soils for agricultural production, limited in terms of the selection of crops in plateaus due to harsher climate.

The Bjelasica and Komovi (Fig. 2.32) are separate entities located in the Tara and Lim watersheds. In terms of relief, the Bjelasica and Komovi, as well as the neighbouring Ključ (1973 m a.s.l.) and Crna Mountain (1981 m a.s.l.), are broken by numerous watercourses (formed in the zone of Paleozoic shales and sandstones, igneous rocks of the Bjelasica and flysch, mainly in the Tara basin from Kolašin to Albania's border and in Plavsko Lake basin), and present a great divide (Cvijić 1921).

In addition to these sediments and rocks, Quaternary alluvial and colluvial deposits are frequently found in river valleys. In the Tara valley (Fig. 2.33), larger areas of alluvial and colluvial soils can be found in expanded parts of the valley from Veruša to Sjerogošte and Mojkovac hollow and also in the valleys of Tara's tributaries. Such soils are also present in the Plav–Gusinje hollow and the Lim Valley and its tributaries, which will be described in more detail in another section. Alluvial and Colluvial Soils and Brown Eutric Soils of old river terraces, most of which are in the Kolašin and Mojkovac Valley, then in Trebaljevo, Sjerogošte, Bistrica, and Dobrilovina, are also the most significant areas for agricultural activities. Unlike these, the Dystric Soils of sloppy terrain are formed on various silicate and igneous parent materials are hardly suitable for field crops due to high slope gradients and erosion processes, so they are mainly used as meadows and pastures and for orchards up to 900 or 1000 m a.s.l. The Tara River has cut its valley into the range of the highest Dinaric

Fig. 2.32 Komovi Mountain (Vasojevićki and Kučki Kom peaks) (*Photo*. G. Roganović)

Fig. 2.33 Tara River valley near Mojkovac (*Photo* G. Roganović)

mountains in Montenegro, which starts in the Prokletije Mountains. It is the longest river in Montenegro, which runs through an 80 km long canyon, along which smaller portions of arable land are found in Dobrilovina, Đurđevića Tara, Tepca, and Čokova Luka. Therefore, the most important arable areas are found in its valley from Veruša to Bistrica, where the Tara Canyon begins.

The southwest slopes of the Ljubišnja Mountain (2238 m a.s.l.) and Obzir (1869 m a.s.l.) belong to Bosnia and Herzegovina, and towards the southeast, they change into a limestone plateau, settled between the Tara Canyon and the Ćehotina valley. From the plateau, with an average elevation of 1450 m a.s.l., several peaks rise, 1550–1850 m high, with numerous sinkholes and glens between them (Meki Do and Pušanjski Do, Krupice, Crvena Lokva), as well as several fields (Bitinsko, Kosaničko, Stupsko, Rudo, and Kameno fields). Dystric Cambisols on cherts prevails in these relief forms, but in the lowest parts of the terrain (950–1350 m a.s.l.), shales, sandstones, and other clastic sediments and igneous rocks appear sporadically. Likewise, they are covered by Dystric Brown Soils, and significant agricultural areas are found in Bobovo, Glibaći, Kosanica, Vaškovo, Premćani, and Barice.

This pronounced mountainous area is broken by the watercourses of the Piva, Tara, and Morača rivers, and their numerous tributaries. The pronounced orography and distance from the sea influence the moderate continental climate in the Morača valley (Gornja Morača and Donja Morača), characterised by moderately cold winter and relatively warm summer. Winters in the Tara and Piva valleys have heavier snowfall, so they are sharper and longer than in the Morača, while the summers are warm. The largest part of the area is at the elevation from 900 to 1450 m a.s.l., characterised by a transient variation between the continental and high mountain climate. It is impossible to draw a clear line between these climate zones since the valleys of the mentioned rivers above 900 m a.s.l. have certain climatic specificities. These are caused by the openness of river valleys and hollows, the proximity and heights of mountains and plateaus rising directly from them, the wind speed and force, etc. When this is taken into consideration, it can be said that the transitional climate is present in the upper catchment of the Tara basin, plateaus around the Piva River, Drobnjaci, and Uskoci, the Durmitor and Ljubišnja footslopes, as well as on the Sinjavina slopes (Šaranci, Krnja Jela, Štitarica).

The climate above 1450 m a.s.l. is subalpine, with very harsh and long winters and short summers. The climate of the high mountains area is characterised by data from Žabljak and Kolašin meteorological stations (Figs. 2.34 and 2.35).

When it comes to precipitation, it is high, increasing towards the peaks of the high mountain region due to the ascending wet air from the south, i.e., from the Adriatic Sea. For that reason, Kolašin gets more rainfall (2108 mm) than Žabljak (1445 mm). Also, other locations closer to the sea, which are south of the high mountain range, get much more rainfall than the northern ones, which are more distant from the sea, and this trend is directed towards the continental part.

The annual precipitation distribution is even, which is particularly important for agricultural production. The minimum precipitation is in July and August, and the maximum is in November and December. Furthermore, fluctuations in precipitation are less pronounced than in temperatures. Low temperatures often occur in spring, limiting the sowing and planting periods for agricultural crops or destroying them during the germination phase. Low temperatures with frosts in the plateaus of this area prevent potato planting before 15 May, and potatoes are grown widely, especially in permanent settlements.

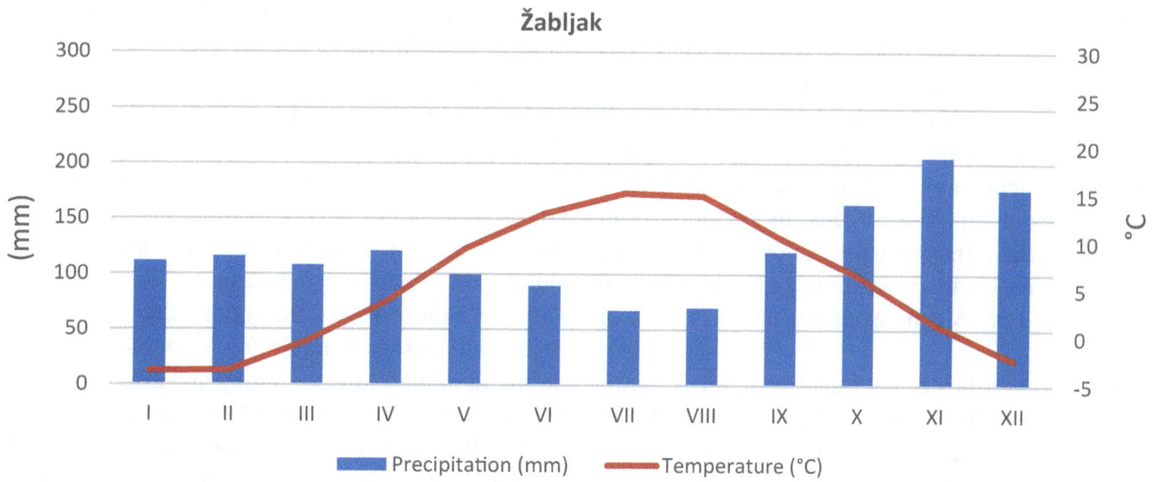

Fig. 2.34 Average monthly precipitation and temperature in Žabljak (1981–2010)

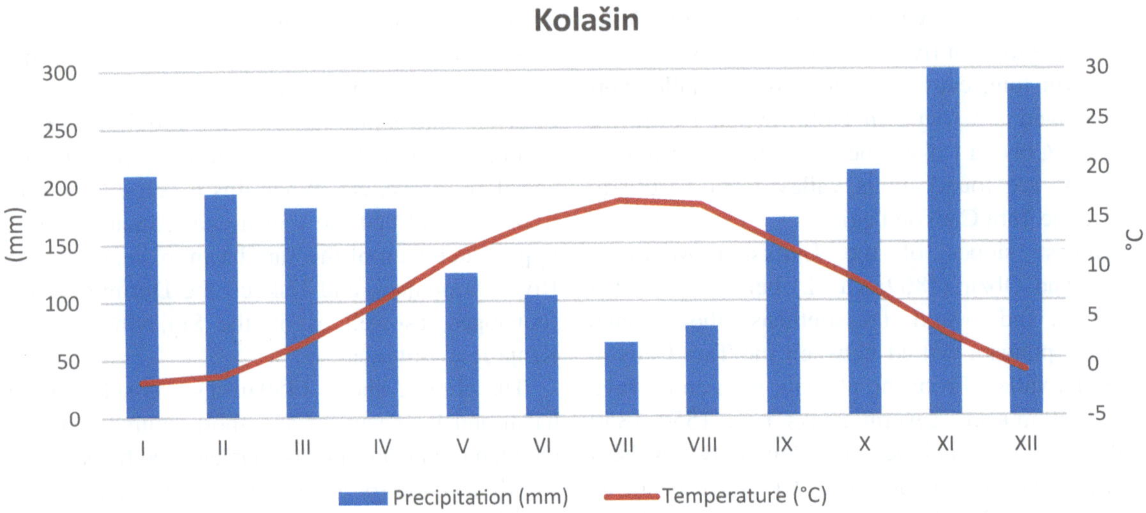

Fig. 2.35 Average monthly precipitation and temperature in Kolašin (1981–2010)

2.3.3.2 Continental subregion

The continental subregion continues on the previous mountain region, although it cannot be separated from it more accurately. It is also a hilly-mountain area intersected by the valleys of the Ćehotina, Lim, and Ibar rivers and their tributaries.

In the Ćehotina valley (Fig. 2.36), the largest extension is the Pljevlja hollow, with a flat area of about 1600 ha and an average altitude of 770 m a.s.l. Upstream of Pljevlja, the valley extends to Vrulja and neighbouring Maoč. The soil in Vrulja is Alluvial, and together with Colluvial Soils, it appears in larger areas downstream of Pljevlja and in the Vezičnica valley. The Maočko Field is formed of the Miocene lacustrine sediments, which occur most frequently in the Pljevlja hollow, Mataruge, Ljutići, and Otilovićko fields. On such sediments are formed Brown Eutric Soils, which can be found in several locations downstream of Pljevlja, mainly in Krć, Sađ, and the Glisnica valley. Eutric Cambisol and Alluvial-Colluvial Soils are of the highest quality for agricultural production. However, in Pljevljsko Field, Borovica, and Šumani, significant areas of very fertile soils were, in fact, utilised by coal mining in open-pit mines. In addition to taking up a certain area and space for mining, power generation, urbanisation, and transport, the soil of the Pljevlja hollow is affected by pollution from various harmful agents. That requires increased efforts and concern to apply adequate prevention and rehabilitation measures in order to protect the environment and health of the population.

Outside the narrow valley of the Ćehotina River, a significant area of arable land in the area of Pljevlja can be found in the Ljubišnja and Ravna Gora footslopes, then along the road Pljevlja–Prijepolje, Pljevlja–Goražde, Strahov Do, and Bukovica. Dystric Brown Soil prevails in these locations, and the soils are formed on sandstones, shales, flysch sediments, other silicate rocks, as well as on diabase-chert formations where limestone with cherts and igneous rocks often occur. Paleozoic and Werfen shale and sandstone occur in a smaller area with Brown Dystric Soils.

Polimlje (Fig. 2.37) and Vraneš are the largest areas of the continental region. The Lim springs from Plavsko Lake and flows at the very beginning through the terminal moraine that the Lake was formed of, then through the old river terraces, which are quite large from Brzojevice to Murino. Eutric Brown Soils are formed on these sediments, appearing on old terraces from Kruševo to Lug and in Andrijevica. There are much fewer old terraces on the right bank of the Lim, but downstream of Novšić, Alluvial Soils occur in a narrow alluvial plain, which is slightly wider in the area of Velika, Ulotina, and Lug. Towards the Lim Valley, from the source to the Sutjeska Gorge, the area of Velika and Gračanica rivers have milder slopes with plenty of Dystric Cambisols used for arable production. In other parts, slopes are moderately steep along the road Murino–Turjak, very steep towards the surrounding mountains of Visitor (2210 m a.s.l.), Zeletin (2126 m a.s.l.), Planinica (1963 m a.s.l.), and Sjekirica (1989 m a.s.l.). Andrijevica hollow is clenched between Balja and Borova Glava, descending steeply to the Lim bed, while Zlorečica and Gnjili brook valleys have milder and moderately steep sideslopes which are the Zeletin, Vujeva Mountain, and Komovi, whose peaks are above 2000 m a.s.l. In addition to Brown Dystric Cambisols, Alluvial-Deluvial Soils can be found in the Zlorečica valley and also sporadically along the Lim valley downstream of Andrijevica.

The Berane basin is the largest expansion in the Upper Polimlje. From the mouth of the Marsenića River to the Tifranska Gorge, the valley is about 10 km long, about

Fig. 2.36 Ćehotina River meanders, Northern Montenegro (*Photo* G. Roganović)

4.5 km wide, and located at an altitude of 650–700 m a.s.l. The lowest part of the basin consists of a rather wide alluvial plain, with gravel-sandy deposits originating from the continuous bifurcation of the unregulated Lim River bed. A small area of loamy-sandy soil can be found in the periphery of the alluvial plain, mainly in the area of Vinicka, Pobijenik, Lug, and Lužac. Above the alluvial plain, the old river terraces are developed from Gradina to Beran Selo on the left bank of the river and from Donja Rženica to Jasikovac and Budimlja on the right bank. The terraces continue on the terrain of Zagorje, Petnjik, Jasikovac, Dapsići, and Budimlje and are made of Miocene lacustrine sediments, which are also found in the neighbouring Polički basin at an elevation of 820–1000 m a.s.l. Brown Eutric Soils distributed on terraces and formed on lacustrine sediments are also of the highest quality for agriculture, which has a long tradition in this area. Along the Lim tributaries—Kaludarska, Dapsićka, and Šekularska River and Bistrica, in their downstream parts, there are very small areas of Alluvial-Deluvial Soils due to narrow valley, and much more Dystric Cambisol on mild and moderate slopes.

Downstream of the Tifranska Gorge, Zaostro is a smaller and narrower valley, and the larger one is where the Lim Valley widens after Bioč, reaching its largest width in Zaton. In this part of the valley, there are more sandy-gravely deposits, which are used as raw material for construction, as in the Berane basin. The Lim, at the time of high water levels, undermines the layers of fertile soils on its banks in Brzava, Femića Krš, and Zaton, which imposes the need to regulate the Lim River bed in sections where the soils are at the highest risk. The old river terraces are preserved from Poda to Srđevac and are far larger than those below Štitari and Brzava on the left bank of the Lim. Brown Eutric Soils on terraces, Alluvial, and Deluvial Soils at the foot and toeslopes of the surrounding hills are very fertile and suitable for the cultivation of agricultural crops. In addition to these, also Brown Dystric Soils are present on mild and moderately steep slopes from the Zaton hollow to the Korita plateau and are suitable for the cultivation and growing of fruit and field crops.

From Ribarevine, the Lim valley gradually widens between rather rounded hills, and a larger widening is downstream of Bijelo Polje town. There is practically no alluvial plain in this part of the valley, as the Lim has cut its bed into older river terraces all the way to the mouth of Bistrica, the right bank tributary of the Lim River. These

Fig. 2.37 A typical house in the Polimlje contryside (*Photo* M. Topalović)

terraces have a low content of coarse gravel-sandy deposits. Therefore, earthy material and soil gravel, which originate from silicate sediments, cover a large area from Bjelasica to Korita, south of the Pešter plateau, and then from Bjelasica, Buren, and Stožer to the border with Serbia. Eutric Brown Soils on terraces, as well as some larger areas of Alluvial-Deluvial Soils at the footslopes of the surrounding hilly-mountainous terrain and along the valleys of numerous Lim tributaries, are among the highest quality agricultural soils in the Polimlje region. In the valleys of the Lješnica, Bistrica, and particularly Ljuboviđa, in addition to the Eutric Cambisols, there are also significant areas of Alluvial-Deluvial Soils, somewhat less pure Alluvium, while the slopes are covered by District Cambisols, with significant portions of agricultural areas. The Bijelo Polje basin is located at an elevation range between 550 and 580 m a.s.l., while the Zatonska hollow is about 50 m higher, so they belong to a warmer continental climate area.

Overall, the Polimlje area has extraordinary soil resources of around 15,000 ha, favourable climate conditions up to 900 m a.s.l., which, along with the abundance of water, streams, and rivers, provides opportunities for the development of contemporary agricultural production (fruit growing, field crops and livestock farming, which can, to a large extent, also rely on the surrounding mountain area).

The Ibar Valley has a rather small area with good soils for cultivation and somewhat larger areas in Županica and Đurđevića Luke, in the village of Bać, next to the border with Serbia. The moderately mild slopes are above the Ibar Canyon-like gorge (150–300 m) between Rožaje and Bać. On both sides of the Ibar Valley, the terrain gradually rises, sometimes across milder and sometimes across moderately steep sideslopes of Murgaš (1868 m a.s.l.), Hajla (2278 m a.s.l.) and Žlijeb (2382 m a.s.l.), which are in the border area with Kosovo, i.e., to Krstača (1755 m a.s.l.) and Gradina (1762 m a.s.l.) at the border towards the Pešter plateau.

The continental subregion is characterised by a sharp winter period and warm and dry summer. Autumn is warmer than spring, while mountain climate prevails in the mountains, most of which are in the upper catchments of the Lim, Ibar, and Ćehotina. The climate data for Berane and Bijelo Polje are relevant for the Polimlje and those for Pljevlja for the Ćehotina valley (Figs. 2.38, 2.39 and 2.40).

The data show that the winter temperatures are negative at all locations. During the summer, i.e., from June to August, the average temperatures range between 16.6 and 19.6 °C. However, the absolute maximum temperature in some basins reaches 39–40 °C, while the absolute minimum temperature is even more pronounced, ranging from − 27.6 °C in Bijelo Polje to − 28.3 °C in Berane and − 29.4 °C in Pljevlja. The town of Pljevlja is specific, with almost even values of the average monthly and annual temperatures, because it is most open to the influence of continental climate and is located in a basin. Due to air pollution, the fog lasts longer in the autumn–winter period than in the Lim, Tara, and Ibar valleys. The elevated areas and mountains are characterised by higher temperature fluctuations, as the air is cooled by cold currents penetrating from the north, often accompanied by snowfall.

The annual sum of precipitation is not the same in all areas of the continental region. As already noted, they are decreasing from the south to the north. However, the precipitation schedule is even, although the precipitation is low in the three summer months in all areas.

When considering significant land resources, the abundance of forests and water, and climate conditions, it can be concluded that the continental subregion of Montenegro has outstanding conditions that enable faster development of agriculture. This industry has a long tradition. The population migration from the north towards the coast and the central part over the last 30–40 years, caused by various factors, has prevented the development of this area.

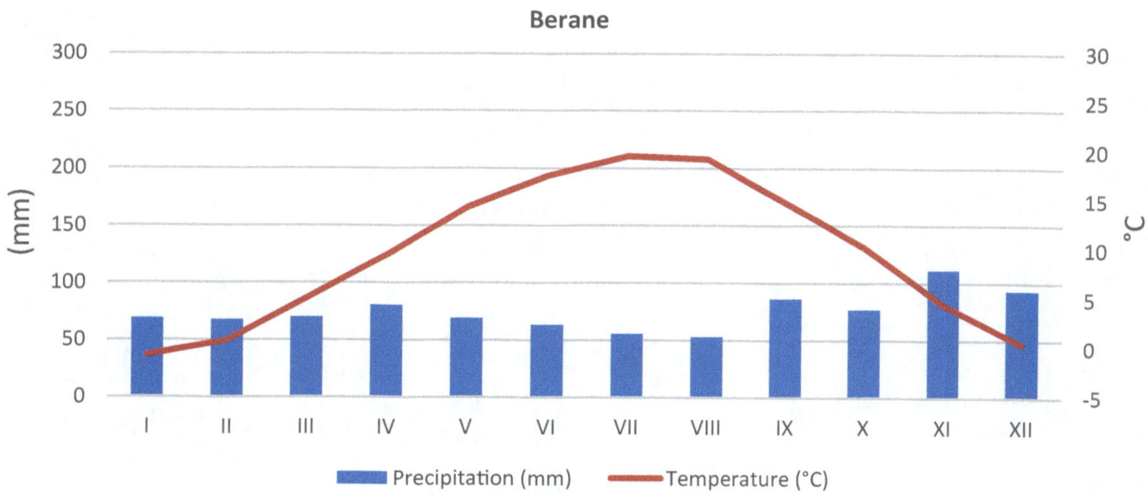

Fig. 2.38 Average monthly precipitation and temperature in Berane (1981–2010)

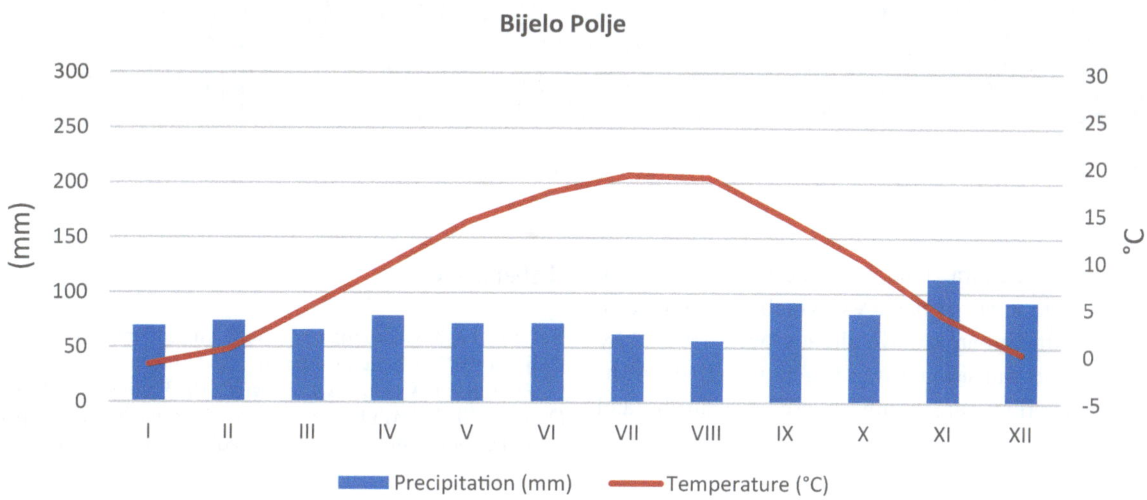

Fig. 2.39 Average monthly precipitation and temperature in Bijelo Polje (1981–2010)

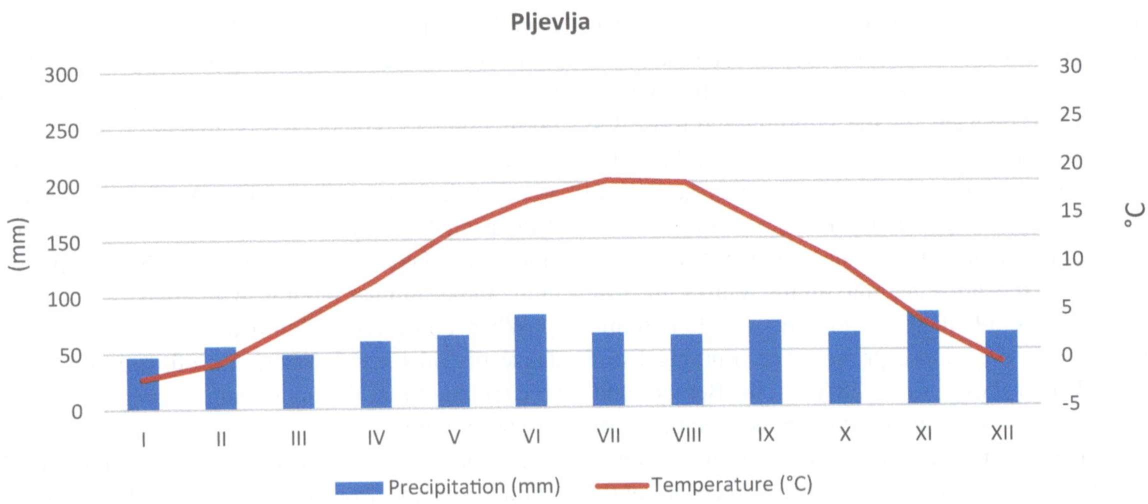

Fig. 2.40 Average monthly precipitation and temperature in Pljevlja (1981–2010)

Fig. 2.41 Abandoned land in the North of Montenegro (*Photo* G. Roganović)

Most of the agricultural land is abandoned (Fig. 2.41), especially the large areas of pastures, which are converting to forests, as significant areas have already been covered by shrubs, bushes, and transitional tree-vegetation. Even natural, unmowed meadows are being covered by shrubs and converting into forests.

References

Bešić Z (1983) Geologija Crne Gore, knj. 3, Titograd

Cvijić J (1921) Ledeno doba u Prokletijama i okolnim planinama. Glasnik SKA , knj. XCI, Beograd, 1913, knjiga XCIII, Beograd

Durn G (2003) Terra Rossa in the Mediterranean Region: parent materials, composition and origin. Geologia Croatica 56:83–100

Fuštić B, Đuretić G (2000) Zemljišta Crne Gore. Univerzitet Crne Gore, Biotehnički instutut, Podgorica
Kalezić M, Mirković M, Škuletić D (1966) Osnovna geološka karta 1:100.000. Tumač za list Šavnik, Zavod za geol. istraživanja Crne Gore, Titograd
Laković I (2017) Katuni Kučke planine, Monografija, Univerzitet Crne Gore, Podgorica. ISBN 978-86-7664-143-7
Lipovac N (1987) Sinjajevina, Stručna knjiga, Beograd
Lipovac N (1987) Sinjavina, stručna knjiga, Beograd
MONSTAT (2012) Statistical yearbooks 2012, statistical office of Montenegro (Montenegro statistical office), Podgorica
MONSTAT (2020) Statistical yearbooks 2020, statistical office of Montenegro (Montenegro statistical office), Podgorica
Montanarella L (2007) Trends in land degradation in Europe. In: Sivakumar MVK, Ndiang'ui N (eds) Climate and land degradation. Environmental science and engineering. Springer, Berlin. https://doi.org/10.1007/978-3-540-72438-4_5
Radojičić B (1996) Geografija Crne Gore: prirodna osnova, Unireks, Nikšić
Vasović M (1955) Lovćen i njegova podgorina, Naučno društvo NR Crne Gore – odjeljenje za prirodno matematičke nauke, Cetinje
Živaljević M, Pajović M, Djokic V, Škuletić D (1973) Osnovna geološka karta 1:100000. Tumač za list Titograd. Zavod za geološka isra̋živanja, Titograd
Živaljević M, Stijović V, Mirković M (1979) Osnovna geološka karta 1:100000. Tumač za list Ivangrad. Savezni geološki zavod, Beograd

3 Soil-Forming Factors

Abstract

The pedosphere forms at the intersection of the lithosphere and atmosphere, incorporating elements from the hydrosphere and biosphere. This chapter examines soil formation factors in Montenegro, including parent material, climate, organisms, relief, humans, groundwater, floodwaters, and time. Soil evolves continuously through pedogenic processes like organic and mineral transformation, migration, and human activities. The chapter details the spatial distribution of limestones and dolomites, igneous and metamorphic siliceous rocks, flysch sediments, alluvial, colluvial, and glaciofluvial deposits, and Neogene lacustrine sediments, and their influence on soil formation and characteristics in Montenegro. It also covers the Adriatic, modified Adriatic, and Mountainous climates, and Montenegro's hydrological conditions. Relief significantly impacts soil formation, with 87% of Montenegro having slopes steeper than 15%, affecting soil water regime and genesis. Soil and vegetation have a symbiotic relationship, with extensive forests, pastures, and grasslands contributing to soil formation. The impact of groundwater and floodwaters, especially around Lake Skadar and river valleys, and human influence are also discussed.

The soil as a separate sphere (Pedosphere), which is characterised by fertility, was formed under the influence of the geosphere, biosphere, hydrosphere, and atmosphere. This chapter provides an overview of the pedogenic factors which significantly impact soil formation. As said, Montenegro's territory consists of the Adriatic coast, with the pronounced influence of the Mediterranean climate, the central region with the modification of this climate type, and continues to the area of the high mountain region, which gradually transitions into the continental hinterland. Due to such geographical position and the sharply pronounced climate, the geological substrate and relief had and still have a strong impact on the formation or the genesis, evolution, and properties of the soil cover. The impact of hydrological and biological factors and the increasing anthropogenic impact, along with the time (age of land surface) as an unavoidable factor, further influence the soil characteristics.

Despite the basic pedogenic factors mentioned, due to the relatively small area of Montenegro (13,812 km^2), the occurrence of zonal soil types can be discussed only conditionally. In soil formation, there are no clear patterns, either in the occurrence or spread of zonal soil types, in difference to the vast areas of continental and northern Europe, Asia, and America. Therefore, on the territory of Montenegro, only Terra Rossa, formed up to about 600 m a.s.l. in the Southeast region, and Caclomelanosols in the higher areas can, conditionally, be placed into zonal soil formations. All other soil types, subtypes, and varieties show the imprint of genesis, characteristics, and properties, which, in addition to the above-mentioned, are also influenced by other factors—natural vegetation, landscape, and the living world.

3.1 Geological Composition of the Terrain and Parent Rocks

The area of Montenegro is made of rocks of different ages and petrographic compositions. Generally, it is obvious that carbonate rocks, i.e., limestones and dolomites are present in the southwest and silicate rocks in the northeastern part of Montenegro. Hard carbonate rocks are mostly of Triassic, Jurassic, and Cretaceous ages. In the silicate rocks, Paleozoic sandstones and shales prevail, but there are also plenty of Mesozoic or shaly sediments of Werfen and diabase–chert formations accompanied by conglomerates, breccia, quartzite, marls, and limestones.

The flysch sediments, present in greater continuity in the northern area of Montenegro, are represented by sandstones, shales, conglomerates, breccia, claystones, and other members of the Upper Cretaceous series. In geological literature, this

M. Knežević et al., *The Soils of Montenegro*, World Soils Book Series,
https://doi.org/10.1007/978-3-031-82919-2_3

Upper Cretaceous flysch is also known as the Durmitor flysch. Unlike the Durmitor flysch, the Eocene flysch is sporadically occurring in the central part of Montenegro, and much more on the coast, made of marls, claystones, various limestones, while sandstones and shales are less present (Fig. 3.2).

Igneous rocks have reached the surface mainly through volcanic activity during the Triassic and Jurassic periods. Volcanic eruptions had affected most of the areas of northern Montenegro, namely Mountain Ljubišnja and the area of Pljevlja, Bjelasica, and the region between Berane and Rožaje.

The most important igneous rocks are andesites, dacites, keratophyres, melaphyres, while diabases, spilites, and various tuffs occur more rarely.

Quaternary sediments, the youngest geological formations, are the most important substratum for soil formation, as those soils are most suitable for cultivation. The alluvial and colluvial deposits are found in the valleys of watercourses and coastal fields and can also be found in the riparian zones of lakes Skadar, Plavsko, and Šasko. After these, the most frequently present are glaciofluvial deposits, formed by glaciation, which was quite intensive in the mountains of Montenegro. The deposits were preserved in the form of moraines on slopes, glens, flats, and karst fields but were also deposited along the banks of watercourses (old river terraces). A negligible area is covered by Neogene lake sediments, i.e., clays, marls, and marly limestones that form coal roofs. The largest area of these sediments is found in the areas of Pljevlja and Berane. In addition to these substrata, diluvial clays deposited in the waters of larger and smaller reservoirs in the area of today's Bjelopavlićka Plain, Lješkopolje, and parts of the Nikšićko Field also serve as parent materials.

The term "parent rock" means more or less consolidated material containing mineral soil-formation components. The differences in the parent rock in terms of lithological and petrographical composition, combined with the effects and other factors, have contributed to the creation of a diverse pedological cover on Montenegro's small area. The composition and characteristics of the parent rock certainly played the most important role in the forming and evolution of a large number of different soil types, subtypes, and varieties. This is quite understandable since soils inherit characteristics of the parent rock. The physical and chemical properties of soil are closely related to it.

The chemical and mineralogical composition of the parent rock determines the soil properties. The mineralogical composition also influences the weathering rate and the quality of the weathered products. When weathering of the rocks results in sandy material, it will form sandy soil in the same and similar conditions, and clayey material will form a clayey soil, heavy and poorly permeable. The acid rocks, poor in alkaline cations, will yield the material that only acidic reaction soil, poor in alkaline cations, can be formed on.

Mineral content is not the same in different rocks. Although originating from different geological periods, limestones and dolomites are characterised by high levels of $CaCO_3$, and $MgCO_3$ and low content of the insoluble residue that the soil is formed of. Cretaceous limestones (Fig. 3.1) stand out in terms of purity, as they are free from any silicate ingredients but contain 98–99% of $CaCO_3$. Chemical dissolution (under the carbonic acid action) produces calcium hydrogen carbonate, which is leached, and the insoluble residue accounts for only 1–2%, so the soil genesis on hard limestone is a very long process. Limestones and dolomites of the Triassic and Jurassic ages are often silicified or contain interbeds or incrustations of silicate components, most often cherts.

The silicate rocks are of a more diverse mineral composition, but mostly quartz (SiO_2), while the share of various coloured and other minerals is variable (some rocks have more and some less). The situation is similar for igneous rocks appearing on the territory of Montenegro (Fig. 3.2).

In general, it can be concluded that soil genesis in compact rocks is slow and gradual compared to the unconsolidated ones that disintegrate faster and are physically better fragmented. The speed of these processes and the number of products formed determine the soil depth. This is why the soils are shallow on consolidated rocks and deeper on unconsolidated rocks that produce plenty of broken material of the soil's parent material.

The natural supply of nutrients accessible to plants in the soil directly correlates with the content of biogenic elements in the parent material. That is extremely important for the natural vegetation, as a better supply of biogenic elements will be reflected in better growth and lusher vegetation, and vice versa.

3.1.1 Coastal Region

In Montenegro's coast, alluvial and colluvial deposits are present in all coastal fields from Bojana to Sutorina, including the Mrkovsko Field under the Mountain Rumija. Also, these deposits are found in smaller bays along the sea coast as well as in the narrow valleys of some waterways (the tributary of the Bojana, Rikavac, Željeznica, Kutska River, and Sutorina).

Near the Bojana River (Fig. 3.3) from Sukobin to the estuary, there are very deep sand-silty alluvial deposits, which have spread in the area of Gornji Štoj and Donji Štoj, where they have been partly shaped also by sea waves. The alluvial deposits in the Bojana delta formed two separate morphologically distinct entities—the island of Ada and Štojska Greda. The island of Ada has an area of about 550 ha and Štojska Greda about 2300 ha. Along the shoreline, sandy beaches stretch over both entities, with Velika Plaža beach in Ulcinj 12.5 km long and the width of the

Fig. 3.1 Non-typical house on limestones (*Photo* G. Roganović)

sandy strip susceptible to aeolian erosion of 250–500 m. From Velika Plaža to the salt pans of the saltworks "Bajo Sekulić", which was built on the location of the former Lake Zoganje, and to the Bojana, the thickness of alluvial deposits is 25 m (Mirković et al. 1968). Coarse-grained and fine-grained sands are present in some layers, while in the surface layer, sand-silty fraction prevails, which is very important for the cultivation and agricultural crops.

The alluvial colluvial deposits of Ulcinj, Anamalsko, and Mrkovsko fields in the area of Ulcinj, then the Barsko Field, Buljarica, Mrčevo, and Tivatsko fields, Kuti and Sutorina, as well as smaller glens (Fig. 3.4) were formed by small, short, and torrential watercourses. Due to pronounced historical denudation and more intensive erosion recently, numerous torrential streams and rivers have brought huge amounts of material to flat parts of the terrain. Dispersion and flooding resulted in the thorough mixing of alluvial and colluvial deposits, rendering their separation difficult, if not impossible. The colluvium often contains more coarse gravel sandy material, while the alluvium is layered, with prevailing loamy-clayey composition.

In the lowest parts of the coastal fields, the groundwater level is high, which is why some areas are swamps and saline (Salina in the Tivatsko Field, Buljarica, and knetes near the Bojana). The swamps occur sporadically on Lake Šasko shores.

The hilly terrain of the coastal region, terraced to a large extent, is formed mainly of carbonate rocks, then of flysch sediments of the Eocene period, and other Triassic silicate rocks. The flysch belt stretches from Debeli Brijeg to Buljarica, then from Sutomore to Albania. The flysch sections are intermittently broken with limestone that sometimes descends to the shoreline, as is the case from Morinj to Risan, from Perast to Kotor, from Buljarica to Sutomore, and from there to Šušanj near Bar and to Ulcinj. In the flysch zone of Sutorina and Grbalj, claystones, marls, and sandstones prevail, and in the southwest part of the zone, also foraminiferal limestone. The flysch sequence from Morinj, towards the Orjen and Morinj contains intercalations of breccia and conglomerates, which is also the case in the sequence around Risan, Orahovac, and Kotor. As a member of the flysch sequence, breccia is widely distributed along the edges of the so-called Budva–Bar trench. Eocene flysch from Sutomore to the Albanian border, namely from Sukobin and Kravari to Ulcinjsko Field, contains marls, sandy and clayey limestones, sandstones, and claystones. Such a lithological composition

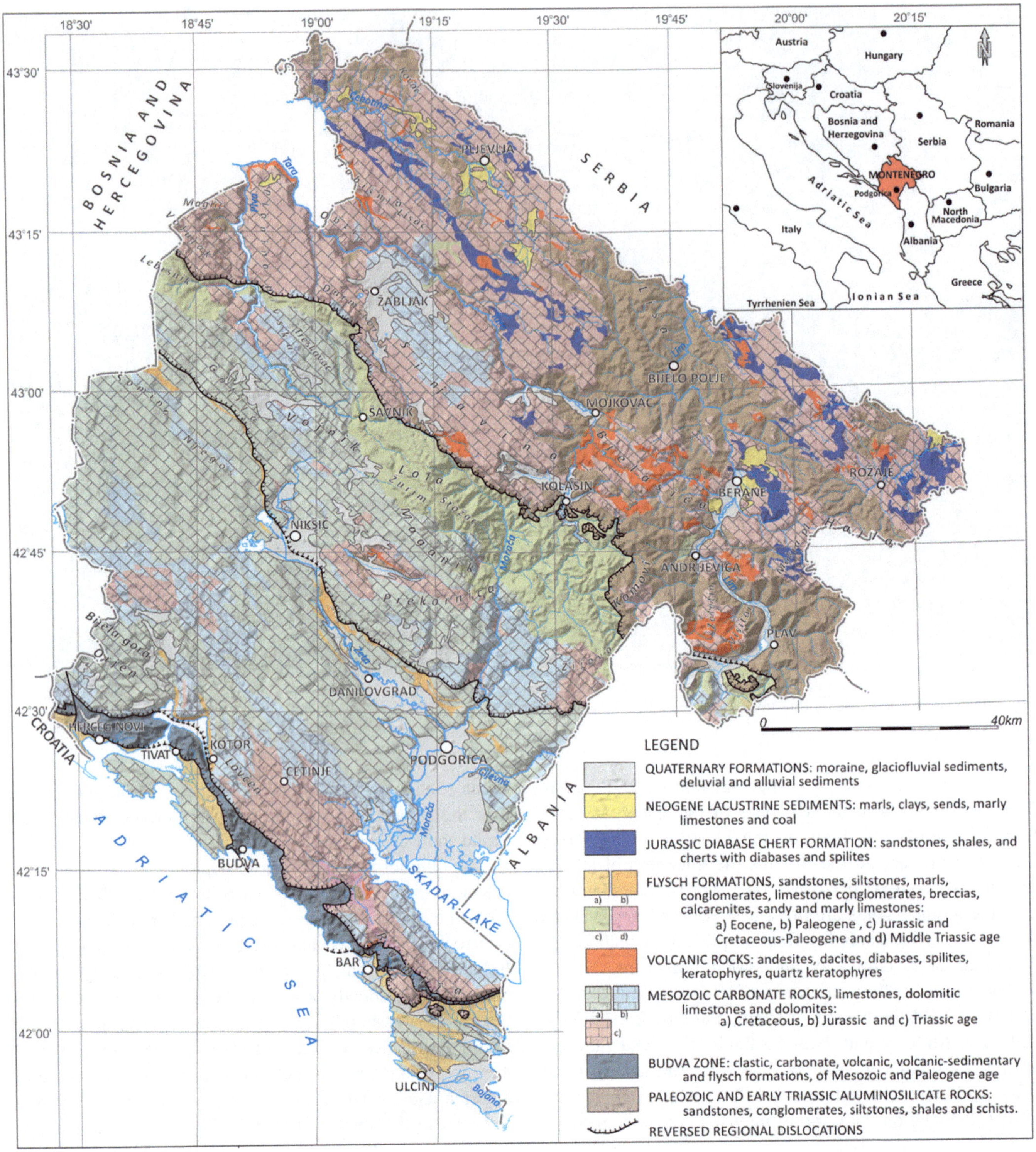

Fig. 3.2 Lithological map of Montenegro (Radusinović and Pajović 2024)

of sandstones with a higher share of carbonate component and clay compared to quartz, 40–50% (Mirković et al. 1968), is largely reflected in the higher share of the clay fraction in the soil, including the colluvial in the coastal fields.

The Middle Triassic sediments are of similar lithological composition, as they are represented by conglomerates and sandstones, shales and sandy marls, and limestones (Antonijević et al. 1969).

Fig. 3.3 A view to alluvial plain of River Bojana at the mouth to Adriatic Sea; an area of alluvial deposits in Gornji Štoj and Ada Bojana (*Photo* G. Roganović)

Fig. 3.4 Colluvial deposits in Bigovo Bay and the surrounding limestones (*Photo* G. Roganović)

The Cretaceous limestones descend from Mt. Orjen to the seashore (Morinj, Risan, Orahovac). The Orjen (1895 m a.s.l.), a quite distinct mountain massif, rises up from the limestone plateau with prominent features of the barest karst, as the karstification process, according to Cvijić (1924, 1926b), is fully manifested there. The karstification process generated various karst relief forms such as valleys, sinkholes, pits, sinks, caves, towers, and smaller karst fields, but the most pronounced form are sinkholes (Bešić 1969). Glaciation has also contributed to the shaping of the Orjen relief, and the consequences are the presence of glacial deposits in valleys and sinkholes in the vicinity of Kameno, Ubli, towards Vrbanja, Grahovsko and Dragalj fields, and Ledenice. The Orjen spurs descend steeply to the coastline between Morinj and Risan and even steeper from Perast to Kotor (Fig. 3.5), with very shallow soil and sparse vegetation. They form the Luštica Peninsula and Donji Grbalj to Jaz beaches, and then the terrain from Barsko Field to Ulcinj. The Cretaceous compact limestones are quite pure, without silicate material.

The Jurassic limestones along the coastal region can be found in several locations, from Kruševice below Mt. Orjen, across Dobroštica; they form the terrain to Verige and Morinj, extending to Vrmac, and from there, they appear intermittently through Pobori and Paštrovići to Mt. Vrsuta and Rumija. The Jurassic limestones are not pure but contain cherts, which become soil substratum after the chemical dissolution of limestone. The soil formed on such a substratum is widespread from Zelenika and Kuti to Verige (Devesilje hill), on Vrmac and Maine.

The Triassic limestones cover most of the Lovćen sideslopes, from the village of Mirac and Čavor to Brajići and further to Buljarica and Šušanj. From there, above the Barsko Field and flysch belt, the Triassic limestones form the Lisinj, Rasovac, Rumija, and Taraboš ridge on the border with Albania. The Triassic limestones are also very silicified, interbedded by cherts, and less by other components.

Igneous rocks appear sporadically and in small areas and together with the Triassic sediments, from Mažići to Bečići, around Mišići, Tuđemili, Sustaše, and Lisinje footslopes. Andesites and dacites prevail, while diabase, porphyry, and keratophyre are rarer.

3.1.2 Central Region

The Zetsko-Bjelopavlićka Plain as a subregion is characterised by a large area of alluvial and colluvial deposits and diluvial sediments. The alluvium fills the Morača valley, downstream of Botun and the riparian zone of Lake Skadar, as well as parts of the Crmničko and Orahovsko fields. Parts of these fields, the flooded Crnojevića River bed, the Ceklin and Donja Zeta polders are either permanently

Fig. 3.5 Part of limestone massive on Montenegrin coast—Boka-Kotorska Bay (*Photo* G. Roganović)

or seasonally flooded, and a large share of them is swamp (10,000 ha, from the elevation level 4.5–6.5 m a.s.l.), very important as a bird reserve, the largest in Southeast Europe. The alluvial deposits in this area are, for the most part, of a sandy-loamy composition, and in the Morača bed, from Botun to Bistrica, it is a sandy and gravely, almost sterile deposit used as construction material.

The rims of the Crmničko and Orahovsko fields are more of a colluvial deposit, mixed with alluvial, difficult to distinguish.

Upstream of Botun, the Morača River cut its bed into the terrace sediments (Fig. 3.6), represented by gravely sandy material, often cemented into a conglomerate. The terrace sediments of the same composition stretch along the Cijevna River, spreading in a fan-like shape into the Ćemovsko Field, all the way to the alluvium in the riparian zone of Lake Skadar. Along the Morača, the terrace sediments occupy larger areas around Podgorica (Dajbabe, Donji Kokoti, Donja Gorica and Gornja Gorica, Lješkopolje, Rogami, Masline) and around Bioče. The terrace sediments are of glaciofluvial origin, deposited in the plain around the Morača and Cijevna. They are very important for soil formation in the flat terrain around Podgorica.

The largest surface of the Bjelopavlićka Plain is formed of diluvial carbonate loams that the Zeta and Sušica rivers cut their beds into. The thickness of this parent substratum is the greatest in the central part of the plain (about 80 m) and significantly lower in peripheral areas (Živaljević et al. 1973). In the valleys of the Zeta tributaries, the streams Rimanić in Martinići and Donje Selo, Suvodo in Viško Field, and Smrdan in Zagorak and Frutak brought and deposited in the then-lake sandy and gravely alluvium which, over time and in some places, has been cemented into a conglomerate. The loams were deposited in the neighbouring Lješkopoljski Lug, Strganica and Drezga, but their thickness is much lesser, as brackish lakes had existed in these locations before.

Flysch sediments occur along the rim of the Lake Skadar basin, in the limestone zone, most often in the form of smaller islands. One rather narrow flysch strip covers the Crmnica area from Podgore to Limljani. Lithologically, it is similar to those in the Paštrovići and other coastal locations. In the area of Bjelopavlići, the narrow flysch strips extend from Pješivački Do to Stubica, and from Frutak to Drenovština along the road Danilovgrad–Nikšić. The larger flysch sequences can be found from Povija to Ostrog, from there to Piperi (Fig. 3.7), and from Zlatica to Fundina. Among flysch enclaves, limestone formations of the Jurassic and Triassic ages can be found.

The southern riparian area of Lake Skadar is made of Jurassic limestone, from Virpazar to Albania (Šestani and Krajina). In the elevated area of Rumija, Sutorman, and Sozina, these are replaced by Triassic limestone. Both have partial chert incrustations, which is why Brown Dystric Soil

Fig. 3.6 Morača River banks formed of conglomerate rocks (*Photo* G. Roganović)

Fig. 3.7 Piperi—the area of Flysch Sediments (*Photo* G. Roganović)

(Livari and Ostros) was formed on them instead of Terra Rossa and Calcomelanosols. In Jurassic sediments around Virpazar, Rijeka Crnojevića, and Cetinje, dolomites prevail, which, apart from chemical dissolution, are subject to physical wear and fragmentation, generating more detritus than harder and more compact limestone (Fig. 3.8).

Limestone and dolomites in the area of Kuči (Fig. 3.9) and Malesija (to the Albanian border) and then around Bioče, Rovci, and the Piperi mountains of Kamenik, Lisac, and Prekornica, and further towards Ozrinići and Župa Nikšićka are from Jurassic–Cretaceous age.

The Triassic limestone covers the northern and northeastern slopes of the Piperi mountains including the Platije Canyon to Lutovo in Bratonožići. Intermittently, Jurassic–Triassic limestones and dolomites (Gostilje, Radovče, and Broćnik) are found along with the Triassic ones.

The karst subregion is characterised by a very large share of limestones and dolomites, which form not only the territory of Montenegro but also of Herzegovina and Dalmatia. It is an area of extremely bare karst or "holokarst" as called by Cvijić (1926a). Mesozoic limestones and dolomites (from the Triassic, Jurassic, and Cretaceous periods) are present in this region. Cretaceous limestone, with nearly 98–99% $CaCO_3$, extends from Herzegovina to Montenegro along a broad belt from Gatačko Field and Ravno. They form mountains Golija, Njegoš, Vojnik, and Garač, then a part of Banjani, Čevo, Bijela Gora and Orjen, from which they descend to the Adriatic Sea.

Jurassic limestone in the karst subregion covers the area from Bilećko Lake and Trebinje to Vilusi and Pusti Lisac, the area of Banjani, Crkvice, and Koprivice on the north side, and over Nudo, Grahovac, and Trešnjevo towards Cetinje. A narrow sequence of these limestones extends from Krstac to Goslić and Donje Srijede. A Jurassic limestone sequence also extends from Seljane and Miljkovac to Duba and Brezne and, again, over Javorak to Jasenovo Polje and Praga.

The Triassic limestones spread from Rijeka Crnojevića and Cetinje towards Virpazar, then towards Lovćen, Njeguši, and Milunovo Počivalo in Ćeklići. The Triassic limestone with dolomites sporadically occurs around Lipova Ravan and Jasenovo Polje, then around Vidrovan and Brezovik. In a larger sequence, the Triassic limestones extend from Ozrinići and Nikšićko Field towards Dragovoljići and along Nikšićka Župa around Suva, Srednja, and Brajovića Ponikvice and Trmanje, including the Platije Canyon to Lukovo.

In the vast karst subregion, moraine deposits often occur as soil substratum, mainly on the slopes of mountains Orjen, Bijela Gora, Lovćen, Golija, Vojnik, and Prekornica. The moraine deposits from Bijela Gora were transported to Nudo glen and the northwest part

Fig. 3.8 Jurassic limestones spreading from Virpazar to Albanian border (*Photo* G. Roganović)

Fig. 3.9 Limestones and dolomites in the area of Kuči (*Photo* G. Roganović)

of Grahovsko Field. From the Orjen, these deposits reached the Grahovsko and Dragalj fields and Ledenice, although plenty remains still on the slopes from Crkvice to the mentioned fields. From the Golija, several moraine deposits descended to Pejovića and Bajovo fields, Zaborje, and Donja Brezna. Numerous moraines from the Vojnik moved to Krnovo plateau to Lukovo and from the Prekornica they occur from Topolovo and Kovačka Dolina to Radovče. The moraine deposits are substratum for Rendzina Soils, which are mostly shallow and very shallow, and occur as thicker only in the Grahovsko Field. The share of flysch in this subregion is small. A very narrow strip of the Eocene flysch stretches from the Nikšićko Field (Kusida through the Duga Gorge). Larger areas of the flysch sediments stretch from Krstac towards the Gatačko Field (Čarađa) and from Ravno, on the border with BiH, into the Piva basin to Plužine. Marly and sandy limestones with marls, claystone, and sandstone have the largest share in this flysch area.

3.1.3 Northern Region

In the subregion of high mountains and plateaus, Triassic limestones prevail. They are mixed with dolomites and dolomitic limestone. Although they belong to hard carbonate rocks, they are subject to more intensive chemical weathering, unlike compact Cretaceous and Jurassic limestones. Since they sometimes contain more and sometimes less silicate, mainly cherts, karst phenomena are not as pronounced as in the pure Cretaceous and Jurassic limestones. That is why Cvijić (1926a) called them transitional karst type, so-called merokarst.

The Triassic limestones form mountains Vlasulja, Bioč, and Maglić, then lower terrains in the Vrbnica and Pivsko Lake basin (Žagrica, Goransko, Dubljevići, and Borkovići), and mountains Pivska Planina, Durmitor, Sinjavina, Bjelasica, and Komovi (Fig. 3.10).

Numerous traces and products of glaciation in the form of cirques, glacial lakes, and moraines have been preserved in the footslopes of these mountains. The moraine deposits consist of materials of various sizes, i.e., large semi-rounded blocks and pieces of carbonate rocks, and to a lesser extent, igneous and silicate rocks, in the mass of finely fragmented gravel and sandy silty particles. Although unsorted, the moraine deposits are a very important soil formation substratum in the high mountain region. The moraine deposits cover large areas under Maglić, Pivska Planina, Treskavac and particularly below the Durmitor (Kovačka Dolina, Žabljak, Motički Gaj, Jezera, Tepačko, Ćukovo, and Njegovuđe fields) (Fig. 3.11). The moraine deposits can also be found in the glens and karst fields of Sinjavina (Zauglina, Suvo, Smoljen, Šaransko, and Odrag fields, Okruglica, etc.).

Fig. 3.10 Pastures on limestones on Sinjavina Mountain (*Photo* Lj. Životić)

Fig. 3.11 Glacial sediments—moraine deposits near Žabljak (*Photo* B. Jovanović)

The glaciofluvial deposits were maintained in smaller and larger glens that the streams from the melting glaciers had been flowing through. The watercourses also transported deposited fluvial or terrace deposits, preserved mainly in the lower parts of the river valleys. These deposits are better processed, made mostly of rounded or semi-rounded coarse and finer gravel, and sand with soil-like sediment. There are plenty of such deposits in the valley of Komarnica, Grabovica, Bukovica, and Tušina, then in the Plašnica valley and in the Tara River upstream (Veruša).

Upper Cretaceous or Paleogenic flysch sediment can be found in the high mountain region, also known as the Durmitor flysch (Bešić 1948), stretching from the border with BiH to Albania. From the direction of Gacko and Volujak, flysch extends from Piva, covering the lowest and central parts of the basin of Vrbnica, now Pivsko Lake tributary. From Plužine, flysch covers the Pirni Do valley, stretching from there in narrow strips in the southern slopes of the Durmitor, from Todorov Do to Stožina and Ranisava, and then expands significantly through the valley of Bukovica and Tušina to Semolj, eastward, to Grabovica and Šavnik in the south. From the Vojnik and Šavnik, the flysch sediments spread along the Bijela River, and from Mokro to Krnovsko fields, Ćeranića Gora and Lukavica to Kapetanovo Lake. From there and Semolj, flysch covers the Morača basin to the Sinjavina and Platije Canyon, then to the Tara basin to the line Crkvine–Brajići–Vranještica–Kraljske Bare–Komovi in the northeast and to Platije, Jablan, Brskut, Lake Rikavac in the south. From there, flysch occurs intermittently from the Komovi to Mt. Vujeva and ends in the Dolja rivulet basin in Gusinje. The northeastern flysch boundary overlaps with the Durmitor tectonic dislocation, so the Triassic formations are thrust over the flysch sediments, i.e., Sinjavina limestones to Donje Vučje and Bistrica, and Paleozoic sandstones and shales from there to the Komovi. Along the southern line, flysch lies, for the most part, on the Cretaceous, Jurassic, and Triassic limestones.

Igneous rock outcrops occur in several locations in the high mountain region. Most of these are smaller outcrops covering small areas in the Ljubišnja and Bjelasica areas. In the Piva area, andesites and keratophyres occur in the Mratinje River basin, then in Ulobići, Kruševo, and Jasen, as well as in the Tara Valley (Baširev Do and Đemska Luka). Diabases are present around Seljanovo and the Piva Monastery. Around Durmitor, minor igneous rock sequences are present around Lake Zminje, Bosača, and Nadgora. Smaller igneous rock sequences are found in the Bukovica basin (Timar, Krnja Jela) and consist of andesites and dacites, also forming the Semolj Mountain.

The larger masses of mostly andesites are found in the Ljubišnja area, around Šuplja Stijena, from where they spread on the Goli Vjeternik, Kolijevka, and Bobovo. On the Bjelasica, they spread from the Tara River (Gradac), along the Jezerštica and Biogradska River valleys, and then in the Mušovića River basin. Also, igneous rocks form a narrow strip from Brskovo along the Bjelojevića River, covering a larger area from Lake Šiško to Šestarevac, around Lake Pešića, Kurikuće, and Lubnica, as well as in the Jelovica basin. Igneous rocks of the Bjelasica are represented by keratophyres and quartz keratophyres, although they are often accompanied by tuffs and pyroclastic rock types (breccia, cherts, marls and limestones) generated as volcanism products, where eruptions get in contact with sedimentary rocks.

The largest areas of the oldest sediments, i.e., the Paleozoic sandstones and shales with accompanying members of the sequence, can be found in the Tara and Lim basins. From their watershed below the Komovi, along the Tara valley, they cover a larger area to Ključ and Kolašin, then around Trebaljevo and Sjerogošte, and then from Štitarica to Gojakovići, Jelin Vrh, and Buren. In the Lim River watershed, they cover a wider zone from the Komovi to the basins of Zlorečica and Kraštica that flow into the Lim near Andrijevica and from there extend further to the Triassic masses of the Bjelasica.

On the limestone plateau of Sinjavina, in several locations, limestone dissolution formed shales and sandstones that form the soil substratum (Ružica, Mutna Lokva, Studenica, Grkovo, and Savina Voda). Even larger areas of shaly sediments occur from Kosanica to Stožer, in the area of Krupice and Barice (Premćani, Đurđevića Tara, Selac, Vaškovo, Barice and Stožer) and from Kosanica to Pusti Lisac (Kosanica and Bitinsko fields, Glibaći and Lever Tara). Under the Durmitor, the sandstones were found only around Pašina Voda and Virak.

In addition to the Paleozoic shales and sandstones, these sediments are present but of the Triassic age, Werfen section. They usually occur in the collision or contact zone with limestones and igneous rock.

In the valleys of the Lim, Tara, Morača, and Piva, certain areas are covered by the youngest Quaternary sediments. Alluvial and alluvial-colluvial deposits prevail, composed mainly of coarse gravel or sand. Such composition and the continuous inflow of new deposits have an adverse impact on the genesis, which is in the initial soil formation stage. The genesis begins after the deposit stabilisation and the growth of the vegetative cover (Fig. 3.12).

Along steep sideslopes and especially around the faults under the vertical cliffs, scree, brash, and debris occur. The good permeability of these materials makes them dry, especially during the vegetative period. That is why scree and debris are free from vegetation and soil, indicating that the soil formation process is very slow and long. The debris material is partly fragmented, so it is a slightly more favourable substratum for the soil than scree, brash, and hard rocks, such as limestones and igneous rocks.

A broad belt of Paleozoic shaly sediments spread in the central part of the continental subregion, i.e., from the northern slopes of the Bjelasica to the border with Serbia. Sandstones and various shales (phyllites, argillaceous

Fig. 3.12 Alluvial and alluvial-colluvial deposits in the Tara River valley (*Photo* G. Roganović)

schists, etc.) dominate. Towards northwest, the Paleozoic sediments extend from the limestone plateaus Buren–Stožer–Kovren–Kamena Gora, from where they enter the territory of Serbia, towards Brodarevo and Prijepolje. From the eastern slopes of the Bjelasica, intermittently broken by other types of rock, the Paleozoic sediments spread along the valley of Lješnica, the Lim's right bank tributary in the direction of Turjak, Rožaje, and Peć (Antonijević et al. 1968). The second branch covers an even larger area, from Andrijevica to Murino and Čakor, from which it extends to Metohija and also to Plav and Albania. The northeastern border of the Paleozoic sediments is the limestone plateau of Korita, the area gravitating towards the Pešter plateau.

Accompanying members of the Paleozoic sandstone and shales, aleurolites, limestone, dolomitic limestone, breccia, and conglomerates appear locally in smaller areas. Quartz sands and conglomerates with breccia cover larger areas in the area of mountains Starac, Pusti Vrh, Bogićevica, Maja Korolac, and Vukov Krš. These rocks are more compact, appear in a larger mass, and numerous rocky peaks, escarpments, and boulders occur on the terrain surface. The high compactness makes disintegration and fragmentation of quartz conglomerates and quartzites difficult, which is why larger boulders and rocks prevail, while soil occurs in the initial development stage, highly skeletal.

The northwestern part of the continental region is the area of the Municipality of Pljevlja. Paleozoic sediments appear in several places in the Ćehotina basin but also in smaller areas. Phyllitic shales and quartz sandstones are present around Odžak, Zenica, Zabrđe, and Kruševo. Shales and quartz sandstones are present around Vrulja, Slatka, and Vodno, and also from Komina to Gornja Brvenica, as well as in the Poblaćenica River and the border zone with Serbia (Mirković et al. 1972, 1974; Živaljević 1987).

In addition to the Paleozoic sediments in the Pljevlja area, larger areas are covered by shaly sediments from the younger geological periods, i.e., Triassic and Jurassic. The Triassic ones contain more sandstone and less shales and aleurolites. They are mostly made of mica and quartz and are accompanied by sandy and marly limestone, claystones, breccia, etc. Larger areas of these sediments can be found in the area of Bukovica, from Boljanići to Metaljka and Čajniče in BiH, around Vrulje and the Kozička River, Podborova, Gradina, Brvenica, and other locations.

Sandstones (Fig. 3.13) and shales are accompanied by diabase-chert formations, while claystones, breccia, and conglomerates occur intermittently. A larger section of these rocks covers the northern slopes of Ljubišnja from Kosanica to the border with BiH, including the villages of Kakmuža, Meljak, Šula, Vitina, Petina, Gradac. Also, north of Pljevlja, diabase-chert formations are found in the area of Gotovuša, Jugovo, Crljenica, Adrovići, and Mihailovići.

A larger area in the Pljevlja region is made of the Triassic limestones. They also formed Ljubišnja, Pusti Lisac, and Kraljeva Gora and the limestone plateau on the right Tara bank to the Mojkovac basin. Also, these rocks form most of the terrain northwest and northeast of Pljevlja. In the Ćehotina valley, the share of Triassic limestone is high, particularly downstream of Maoč, from where they spread broadly to Kamena Gora, Jabuka, and Boljanići. They occur in larger areas downstream of Gradac, where the Ćehotina flows through the canyon, and can also be found in Bukovica, Kovač, and Gradina, i.e., to the border with BiH and Serbia. The Triassic limestones sometimes contain more and sometimes less cherts and incrustations of other silicate ingredients, due to which silicate materials in a limestone terrain often form soil substratum, enabling Brown Dystric Soil formation.

Miocene lacustrine sediments with coal have a significant share in the geological substratum in the Pljevlja area. Marls, clays, marly limestone, and others form most of the soil substrata. These sediments are sometimes covered with a thinner or thicker layer of alluvial or colluvial deposits, as is the case in the Maočko Field at the confluence of Vezičnica and Ćehotina and in Glisnica. Except in the Maočko Field, the Neogene lacustrine sediments cover a larger area in Mataruge and Ljutići, in the Otilovići and Pljevljsko fields, Šumane, Borovica, and Glisnica. The Neogene appears in smaller areas around Bušnja, Bakrenjače, Gotovuša, Krć, Vojtina, Rađevići, and Tvrdakovići, but in these locations, yellowish, porous limestones accompanied by sandy clays, marls, and marly limestone prevail (Fig. 3.14).

Larger masses and broader areas of igneous rocks occur in the area of Ljubišnja, from Šuplja Stijena to Goli Vjeternik, Kolijevka, and Bobovo. Andesites have the largest share, while keratophyres, and other rock types are rarer (Mirković et al. 1974; Živaljević et al., 1987). Rhyolites occur in the Kozička River and andesites in Obarde. The Triassic andesites and dacites occur in several locations around Mironići, Kovač Planina, Bandere, Gradina, and Kamena Gora. Pyroclastic rocks occur together with igneous rocks in the mass or as accompanying rocks, usually breccia, tuffs, and tuffites composed of fragments of igneous and other rocks and cherts.

Diabases and spilites have the highest share in Jurassic diabase-chert formations. Larger masses of purely igneous rocks in the Pljevlja area are located in the area from Kosanica and Drljno Brdo to Stančani, Varina, and Bujačka rivers. The chemical composition of igneous rocks is diverse but is characterised by high contents of SiO_2 and Al_2O_3, while the shares of plagioclase, coloured, and other minerals are much smaller. Quartz and aluminosilicate minerals of igneous rocks also prevail in soils formed on these rocks. Their higher content brings increased acidity to the soil and also other unfavourable chemical properties.

Fig. 3.13 Triassic (werfen) red sandstones, Kovačevići near Pljevlja (*Photo* M. Đaković)

A very small share of alluvial and colluvial deposits is found in the valleys of the Ćehotina and some of its tributaries. Their largest area is found downstream of Pljevlja, Židovići, Gornja, and Donja Brvenica, Glisnica, and Gradac. A smaller area of the alluvium can be found around Vrulja in the Maočnica valley and slightly larger in the Vezičnica valley (from Odžak to the confluence with the Ćehotina).

A diverse geological structure characterises the northeastern part of the continental region. Limestones, igneous, and silicate rocks alternate with each other. The limestone of the Triassic age, often broken by silicate rocks and especially with sediments of the diabase-chert formation, forms a significant part of the terrain of the Korita plateau and the border zone towards Serbia from Mojstir and Đalovića Gorge to Jablanica in the Ibar valley. Towards the south, limestones stretch from Turjak and Rožaje, separated from a larger limestone and dolomite sequence with a Paleozoic sandstone and shale belt, then continue further towards the Berane valley and the Andrijevica, then on Cmiljevac, Hajla, and Žljeb.

The Paleozoic sediments in this area are often in a collision or are covered with the Triassic sandstones, shales, and limestones, then diabase-chert formations with the presence of limestone, igneous, and shaly sediments (Fig. 3.15). More significant sequences of Paleozoic sediments are from Srđevac in the Lim valley along the Lješnica basin, on Turjak, and further in the direction of Rožaje and Peć. The second branch occurs in Andrijevica, Murino, Čakor, and further towards Metohija, including the Lim and Ibar watershed. The diabase-chert formations sediments cover a larger area in the Ibar Valley, particularly downstream of Rožaje, covering the areas of Grahovo, Balotići, Njeguši, Besnik, and Jablanica. Towards Pešter, the diabase-chert formation occurs in the areas of Jablanovo, Mojstir, and Bistrica, from which they stretch further towards Negobratina, Ličina, Kruščica Planina, and Moravac. They appear, with breaks, south of these sites, around Godočelj, Lagator, Kalica, and the Vlahovi pass from where they continue in the basin of the Ibar and its tributaries—Grahovska and Paučinska rivers. A larger sequence of diabase-chert

Fig. 3.14 Marls and marly limestones above coal deposits—Potrlica near Pljevlja (*Photo* S. Radusinović)

formations covers the basin of Dapsićka and Kaludarska rivers, particularly the areas of Rovci and Kaludra, as well as the slopes of Divljak, Sjekirica, and Mokra mountains. The zones of diabase-chert formations made of sandstones, shales, and cherts are also accompanied by igneous rocks. Thus, andesites prevail around Sipanje, Godijevo, Kruščica Planina, and Savin Bor. On the Sjekirica mountain, larger areas are covered by keratophyres with andesites, while quartz keratophyres prevail in the Cmiljevica (Zagorska River), Donja Lovnica, Grižica, and Crni Vrh.

Diabases and spilites together cover very small areas in Gornja Lovnica, Ćosovići, and Njeguši, and spilites only occur in a larger mass and on a somewhat larger area in the areas of Jablanovo, Mojstir, and Mokri Lug, from where they stretch onto the Pešter.

The Quaternary sediments in the Ibar valley are represented by moraine, alluvial, and diluvial deposits. Moraines were preserved only near Rožaje (Ibarac and Bandžovo Brdo) and Bijela Crkva. The alluvium is scarce in the Županica and Bukovica valleys. Mixed colluvial-alluvial deposits occur more in the narrow Ibar valley (to Rožaje and near Bać) and its tributaries, Grahovska and Paučinska rivers, and other small watercourses.

3.2 Climate and Hydrological Conditions as Pedogenic Factors

Climate conditions make a sharp shift from the Mediterranean to the moderate continental and subalpine type. The coastal zone has a Mediterranean or Adriatic climate. The influence of this climate spreads along the Bojana valley and Lake Skadar, and in terms of temperatures and precipitation, it is also felt along the watercourses of the Morača, Zeta, Cijevna, and Crnojevića rivers. In the northern and northeastern parts of Montenegro, moderate continental climate prevails to about 1000 m a.s.l. The pronounced hilly-mountain relief influences climate

Fig. 3.15 Diabase-chert formation, Dragaši, near Pljevlja (*Photo* S. Radusinović)

modifications, so several climate zones can be distinguished, making the transition from the Mediterranean towards the continental and high mountain climate-type.

In the rather narrow coastal zone and Lake Skadar basin, the Adriatic climate gradually shifts into a modified variant, characterised by Mediterranean and mountain climate elements. A range of the coastal mountains of Rumija, Sozina, Lovćen, and Orjen rise between the coast and the Lake Skadar basin. These mountains act as a screen to the direct penetration of the Mediterranean climate in the karst plateau area, stretching along the rim of the Lake Skadar basin and the Nikšićko Field. However, its influence can be felt along the valley of the Trebišnjica and from the Boka Kotorska Bay, which is why the karst plateau area and the coastal mountains have the characteristics of a modified Mediterranean and mountain climate. Its features are warm and dry summers, humid autumn–winter periods, and moderately cold winters.

The central and northern part of Montenegro is a distinct hilly mountainous area, broken by the valleys of Piva, Morača, Tara, Lim, Ibar, and Ćehotina. A pronounced oro-relief beyond river valleys is a strong climate modifier. For that reason, the continental climate, characterised by cold winter and warm and dry summer, changes into a modified pre-mountain variant to about 1400 m a.s.l. Above this height, on slopes, flats, and plateaus between the surrounding mountains, mountain climate elements prevail with a subalpine climate in the highest mountains.

The territory of Montenegro is influenced by air currents from the Mediterranean, bringing wet weather during autumn and winter. The southern, mostly warmer winds, known as scirocco, are saturated with water vapour, bringing an abundance of precipitation falling during the upslope movement from the Adriatic Sea towards the hinterland. The precipitation is the highest in the southwestern part of Montenegro, particularly the area around the Orjen Mountain (3000–5000 mm). Crkvice (1000 m a.s.l.) village below the Orjen is the site with the European maximum rainfall, an average of 4631 mm in the period 1961–1990. The central parts and the area of the Zetsko-Bjelopavlićka

Plain get 1500–2500 mm, while the northeastern and the northernmost parts get the least, 700–1000 mm (Burić et al. 2012).

Air currents from the northwest last for most of the year and are characterised by the penetration of cooler air masses brought by a cold, very strong, and dry wind—bora. In the northern regions, the wind is accompanied by precipitation, whereas in the south, the rainfall is rare.

When considering the impact of the climate on soil formation and characteristics, it should be noted that temperatures, precipitation, and wind are particularly important, followed by relative air humidity, evaporation, and water surpluses or water deficits. Based on rainfall and temperatures as the most important climate elements for soil genesis, annual and (by Lang) monthly rain factor (Gračanin, 1950) is calculated, as well as climate characterisation (arid, humid, etc.).

Many climate parameters, not only the overall sum of precipitation but also the intensity, precipitation distribution, etc., determine certain processes such as the weathering of hard rocks and their dissolution, the eluvial-illuvial processes, acidification, or alkalinisation, migration of eroded particles and their transformation by water and aeolian erosion. The air temperature regime and wind influence the water balance, evaporation and transpiration, and the water deficit or surplus.

Surface and ground waters, as well as their connections, are determined by morphological and hydrogeological conditions of the terrain. Even 95% of water sources are formed within Montenegro itself.

The watercourses belong to the Adriatic and Danube basins. Most of the coastal and central regions belong to the Adriatic and the Northern region to the Danube basin. For this reason, the description of climate and hydrological characteristics is presented in two sections.

Montenegro's Adriatic basin area is about 6560 km^2, or 47% of the country's territory. The Morača (Figs. 3.16 and 3.17), with its most significant tributary Zeta, flows towards Lake Skadar along with the Sitnica, Ribnica, Cijevna, Orahovštica, and Crnojevića rivers. The water

Fig. 3.16 Morača River; close to the minimum water discharge (*Photo* M. Knežević)

Fig. 3.17 Morača River; water discharge near maximum (*Photo* M. Knežević)

from Lake Skadar flow into the Adriatic Sea through the Bojana River.

The Black Sea basin area is about 7260 km^2 or 53% of Montenegro's territory; the Ibar River flows towards the West Morava River, while the Tara, Piva, Lim, and Ćehotina rivers are directed into the Drina River. The typical discharges and water levels of rivers and Lake Skadar are also shown in Tables B.1 and B.2 (Appendix II).

Montenegro's territory is rich in water, even according to world standards. However, a significant part of Montenegro's area belongs to the continental karst area, without permanent water flows, with numerous sinks that the water drains into and flows further under the ground towards watercourses or the sea (Fig. 3.20). Most surface watercourses in Montenegro are torrential.

A great contribution to the study of the quality of fresh natural waters of Montenegro was made by Prof. Dr. Stanka Filipović, who spent part of her working life at the Centre for Soil and Amelioration of the Biotechnical Institute of the University of Montenegro. She participated in almost all limnological researches on the Lake Skadar, starting from the 1970's (Karaman and Beeton, 1981).

There are 31 lakes in Montenegro, of which 29 are mountain lakes, and the two largest ones are in the plains in the southern part of the territory. Lake Skadar is also the largest lake in the Balkans. Three-fifths of the Lake Skadar area belongs to Montenegro. At the highest water level of about 9.85 m a.s.l., the Lake's area is about 525 km^2 (Fig. 3.18). Lake Šasko is the second-largest lake, situated between Lake Skadar, the Bojana River, and the Adriatic Sea. Both lakes represent significant natural reserves of migratory birds and numerous fish species, which, along with other diverse lake and riparian flora and fauna, illustrate the values of this area as well as the influence on the hydrogenic (hydromorphic)[1] soil formation.

The Black, Plavsko, and Biogradsko lakes are also natural reserves, as typical examples of glacial lakes (Fig. 3.19).

[1] In further text the term hydrogenic will be used.

Fig. 3.18 Lake Skadar and Vranjina island (*Photo* G. Roganović)

Fig. 3.19 Biogradsko Lake (*Photo* G. Roganović)

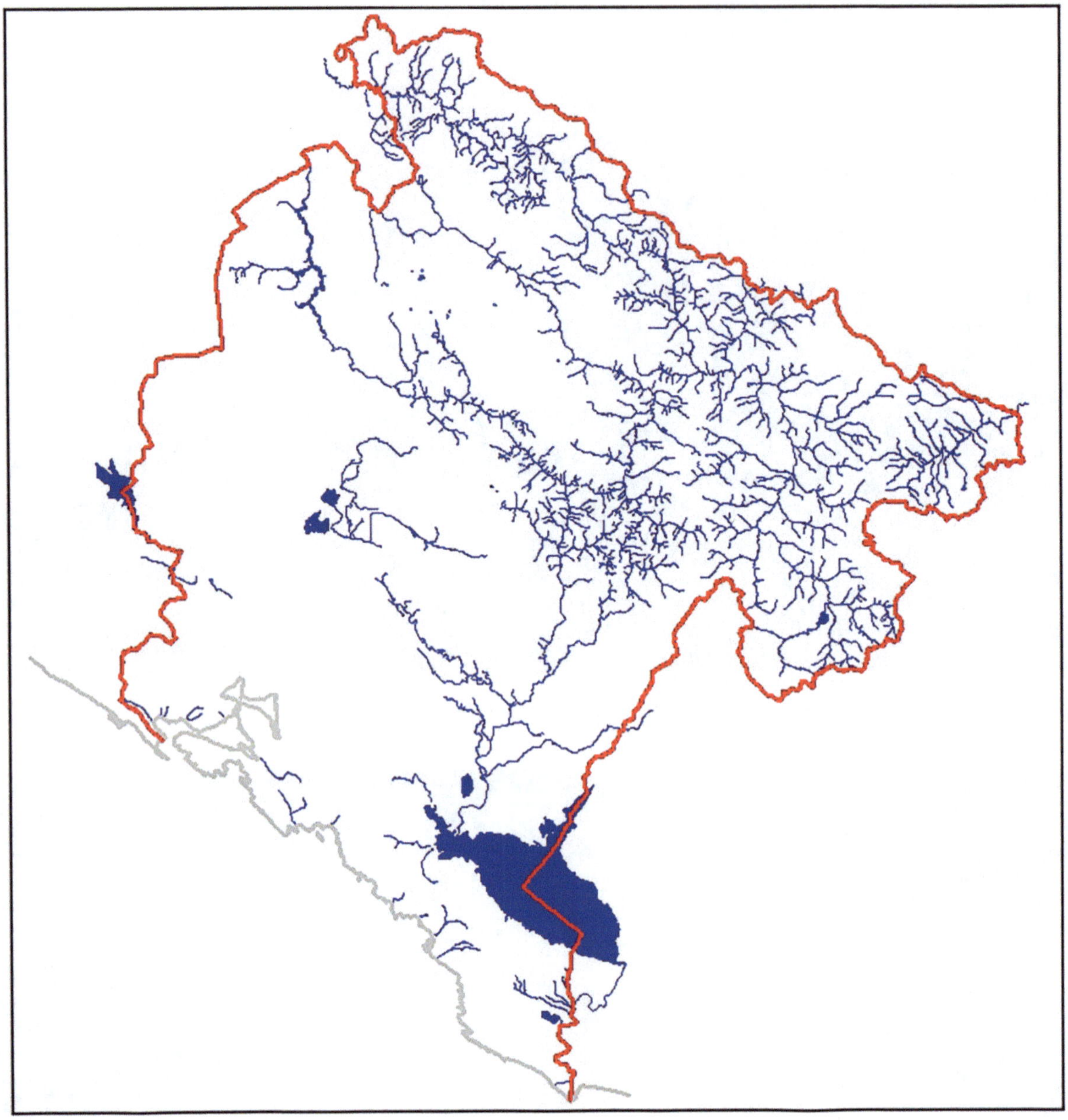

Fig. 3.20 Hydrological map of Montenegro (*Data processing*: Mirko Knežević)

The accumulations (artificial lakes) were constructed on: Piva, Ćehotina, Zeta (Nikšićko Field), and Grahovska River (Grahovo), and a part of the accumulation on the Trebišnjica (BiH territory) is located in Montenegro.

The Adriatic Sea is about 200 km wide between Montenegro and Italy and forms a part of the South Adriatic Pit, where the highest depths of the Adriatic Sea are measured—about 1400 m. About 80% of Montenegro's coast is rocky, with deep waters usually immediately by the coast, while the rest is shallow with a sandy-gravel bottom. The largest beach is Velika Plaža (Long Beach) in Ulcinj. The Adriatic Sea is a relatively warm sea. The dominant sea current flow is parallel to the coast, towards the northwest. The salinity of the South Adriatic waters (38.6‰) is slightly lower than the average for the Mediterranean Sea (39‰).

3.2.1 Coastal Region and Central Region

The Coastal Regions and Lake Skadar basin have some characteristics of the Mediterranean climate. Average annual temperatures are slightly declining in the direction from Herceg Novi to Ulcinj. Also, the precipitation is decreasing (Table 3.1 and Appendix III). That is undoubtedly the result of the position of Herceg Novi and Budva, sheltered by the mountain range Orjen–Lovćen, with

Table 3.1 Climate indicators for the coastal region (1981–2010)

ULCINJ	Jan	Feb	Mar	Apr	May	June	July	Aug	Sep	Oct	Nov	Dec	Annual
Precipitation (mm)	119.0	125.1	107.8	104.3	60.7	55.2	25.3	44.6	96.9	136.2	162.0	151.9	1189.1
Air temperature (°C)	7.6	8.1	10.8	14.2	18.8	22.6	25.2	25.0	21.3	17.3	12.4	8.7	16.0
Mean minimum temperature (°C)	4.7	5.0	7.6	10.6	15.0	18.6	21.1	21.1	17.6	14.0	9.3	5.9	12.5
Mean maximum temperature (°C)	11.5	12.2	15.0	18.4	23.0	26.7	29.6	29.7	26.2	22.0	16.7	12.3	20.3
Relative humidity (%)	66	65	67	71	70	68	63	66	68	70	70	69	68
BAR	Jan	Feb	Mar	Apr	May	June	July	Aug	Sep	Oct	Nov	Dec	Annual
Precipitation (mm)	127.2	133.5	119.5	115.4	69.2	58.4	29.9	47.6	127.7	154.0	164.3	169.6	1316.4
Air temperature (°C)	8.7	8.8	10.9	14.1	18.6	22.2	24.5	24.3	20.9	17.3	13.1	9.9	16.1
Mean minimum temperature (°C)	5.0	5.0	7.1	9.8	13.8	17.4	19.6	19.7	16.6	13.4	9.4	6.4	11.9
Mean maximum temperature (°C)	13.0	13.1	15.3	18.5	23.0	26.5	29.1	29.1	26.0	22.3	17.7	13.9	20.6
Relative humidity (%)	65	64	67	71	71	69	66	68	68	70	69	67	68
HERCEG NOVI	Jan	Feb	Mar	Apr	May	June	July	Aug	Sep	Oct	Nov	Dec	Annual
Precipitation (mm)	171.9	179.7	165.4	153.4	96.9	74.1	34.3	69.9	132.6	191.7	246.5	254.2	1770.6
Air temperature (°C)	8.7	8.7	10.9	14.0	18.6	22.4	25.0	24.8	20.9	17.2	12.9	9.8	16.2
Mean minimum temperature (°C)	5,56	5.4	7.4	10.2	14.4	17.9	20.4	20.5	17.1	13.8	9.7	6.8	12.4
Mean maximum temperature (°C)	13.2	13.1	15.1	18.2	22.9	26.5	29.7	29.8	26.0	22.1	17.5	14.0	20.7
Relative humidity (%)	73	70	71	73	73	70	66	67	72	76	75	74	72

Data source RHMZ

upslope south winds cooling faster, bringing more rainfall. Contrary to these places, the area of Bar–Ulcinj is wider and lower, sheltered by Rumija and Sutorman, which are about 250 – 350 m lower than the Orjen and Lovćen, which influences the precipitation, while the northeastern wind from the Prokletije influences the temperatures.

In the Lake Skadar basin, average annual temperatures are slightly lower, but summer is torrid, with many sunny days. Here, frosts and snow are more common than on the Coast. Although the climate in Podgorica is humid from the annual viewpoint, the three summer months are highly arid, May is semiarid, and other months are either humid or perhumid. That suggests a greater torridity of summer and the second half of spring. April and October have a humid climate, and the period from November to April is perhumid. Due to the influence of the surrounding hills, the precipitation in the Bjelopavlićka Plain is even more pronounced. Although only about 20 kms from Podgorica, the year-level precipitation difference is 500 mm on average. The said climate conditions in the Bjelopavlićka Plain, on the substratum of diluvial loams, cause the shift of dry and wet phases, leading to soil pseudogleysation and lessivage. For that reason, Illimerised Soils prevail in the plain and Pseudogley Soils in its central part. Illimerised Soils are present in Lješkopoljski Lug, in the immediate vicinity of Podgorica, as well as in Strganica and Drezga, where the soil was formed on the diluvial loams.

Under such climate conditions, two contradictions are pronounced in the Coastal region and Lake Skadar basin—a noticeable surplus during the period autumn–winter (Fig. 3.21), causing erosion for a longer period (in some years from April to September), and a deficit that needs to be offset by irrigation during the vegetation period.

Water plays an essential role in the soil formation and genesis processes, as it influences the physical disintegration or wear of the parent rock, resulting in rough regolith forms transported down the slope by erosion and further fragmented. Then follows the chemical weathering and formation of clay minerals of importance for plant nutrition and the transport of nutrients into the soil solution and washing away. In the soil evolution process, weak, often minor acidification occurs at the first stage, which is more intensive, followed by eluvial-illuvial migration in the next stage, and in the final stage, clay minerals are destroyed in the podzolisation process. Extreme weather events can significantly accelerate these processes (Fig. 3.22).

The stagnation of water in a horizon, layer, or entire soil profile causes the development of redox processes, the transformation of organic substances in anaerobic conditions, which impair the original properties of the soil. Water-saturated horizons prevent normal root penetration and limit the physiologically active soil layer. The origin, duration, and intensity of soil moistening determine the needs and ways for the implementation of hydrotechnical soil drainage measures.

The redox processes and the need for drainage are most notable in Alluvial Soils on the banks of the Bojana, Šasko, and Skadar lakes. These processes are also present

Fig. 3.21 Boka-Kotorska Bay (winter period)—a view from the Institute for Marine Biology (*Photo* M. Knežević)

in Alluvial-Colluvial soils, especially in the area of Ulcinj, Buljarica, and Tivatsko fields, then in Crmničko and Orahovo fields (Fig. 3.23). Along the Bojana, Gornji Štoj and Donji Štoj, redox processes are accompanied by salinisation with seawater, which during high tide reaches Reč along the Bojana or is filtered through highly permeable sand deposits (on the banks of the Bojana, Gornji Štoj and Donji Štoj and Velika Plaža). In depressions called knete, salt water persists on the surface throughout the year. The lowest parts of Buljarica and Tivatsko fields (Saltpans), as well as the hinterland of the beach Jaz, are exposed to salinisation due to infiltration of seawater into soil layers.

In the karst area (Table 3.2 and Appendix III), the parameters show that the climate is also warm during the summer with a moderately cold winter period. Average annual temperatures are quite high (10.1 °C in Cetinje and 11.1 °C in Nikšić). Seven months have perhumid climate, and May and September (except in Cetinje) are humid. June is semihumid in Cetinje, and semiarid in Nikšić. Aridity is present in these towns in July, while August is semiarid. Average annual rainfall is high but unevenly distributed (with the least rainfall in three summer months).

The water surplus occurs due to heavy rainfall during the autumn and winter, and it affects only the soils in the Nikšićko Field, in those soils that are formed on diluvial clays and exposed to flooding. The high water levels of the Zeta occasionally flood the soils of Miločani, Brezovik, Glibavac, and Mokra Njiva, and Slivlje is flooded longest due to the inability of the sink Norin to accept all the water that reaches it.

The water surplus poses a problem for Cetinje as an urban settlement because the sink in the lowest part of the Cetinjsko Field cannot accept all the surplus water due to the lack of regular maintenance.

The soil moisture deficit is caused by rapid runoff and infiltration through a shallow layer of soil and a very permeable substratum of glacio-fluvial deposits of karst fields and carbonate rocks. Due to the absence of surface watercourses (with the exception of the Nikšićko Field), the moisture deficit cannot be compensated, which is undoubtedly reflected in the crop yields. The moisture deficit also affects the natural vegetation where scrub and natural (karst) pastures prevail, followed by forests that usually occur at northern exposures with slightly longer moisture retention in shallow soil.

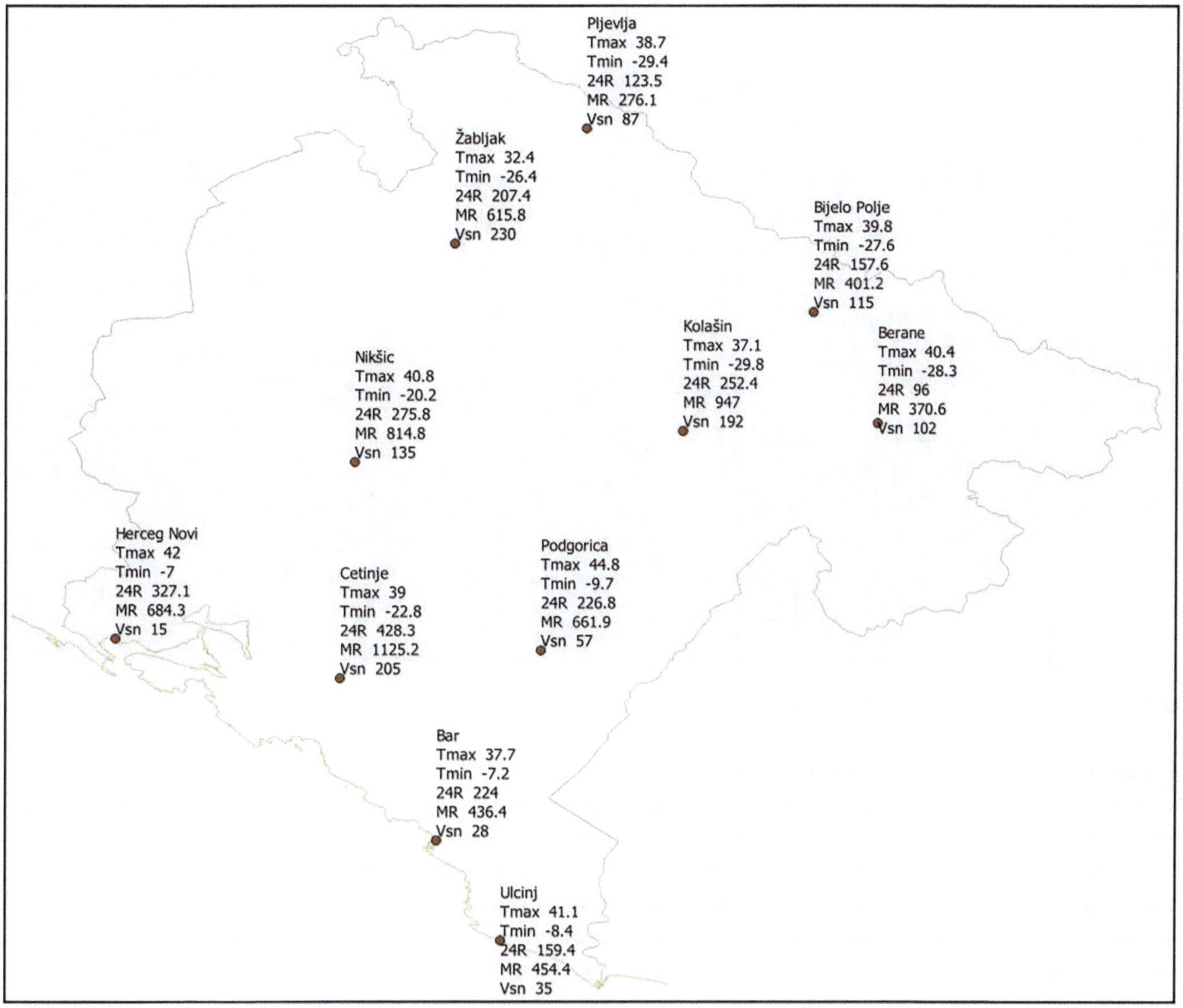

Fig. 3.22 Extreme temperatures and precipitation values for the meteorological statins in Montenegro for measurement period 1948–2020 (Tmax—maximum daily temperature (°C), Tmin—minimum daily temperature (°C), 24R—maximum 24 h rainfall (mm), MR—maximum monthly rainfall (mm), Vsn—maximum snow cover height (cm))

The surplus of water in the karst is due to rapid runoff, while infiltration into the ground layers enables soil moistening, resulting in the formation of the Terrestrial (Autogenic) soil order. The excess moisture (exceeding the field capacity) in soils of the karst area is rare, as their depth is low, except in sinkholes and some karst fields, where it can reach 1 m or above. There is no true groundwater in the Nikšićko Field besides flooding, so there are no hydrogenic soils. The excess perched water table in Glibavac, Mokra Njiva, and Slivlje causes soil saturation but for a shorter time. That is why pseudogleysation processes did not result in the formation of Pseudogley Soils.

3.2.2 Northern Region

The high-mountain subregion (Table 3.3. and Appendix III) is rather heterogeneous in terms of climate, as it covers a large area of mountain massifs from the BiH border to Albania, as well as their slopes towards the valleys of the watercourses of the Piva, Tara, and Morača, including the plateaus. Mountain massifs form a barrier to the ingression of the warmer Mediterranean climate influence, which is particularly notable for temperatures and much less in terms of rainfall. The climate parameters for Kolašin (944 m a.s.l.) and Žabljak (1450 m a.s.l.) show that the area, as a

Fig. 3.23 Crmničko Field under flooding (*Photo* M. Knežević)

Table 3.2 Climate indicators for the Central region (1981–2010)

PODGORICA	Jan	Feb	Mar	Apr	May	June	July	Aug	Sep	Oct	Nov	Dec	Annual
Precipitation (mm)	159.7	157.8	153.9	138.3	83.3	64.2	27.3	55.8	146.7	172.9	245.5	231.9	1637.2
Air temperature (°C)	5.8	6.9	10.4	14.4	19.8	24.0	27.2	26.8	21.5	16.3	10.6	6.8	15.9
Mean minimum temperature (°C)	1.9	2.9	5.9	9.4	14.2	18.0	21.1	21.1	16.6	12.0	6.7	3.4	11.1
Mean maximum temperature (°C)	10.8	12.0	15.7	19.9	25.5	29.7	33.3	33.2	27.9	22.4	16.0	11.5	21.5
Relative humidity (%)	70	65	64	64	60	56	48	50	60	70	73	73	63
CETINJE	Jan	Feb	Mar	Apr	May	June	July	Aug	Sep	Oct	Nov	Dec	Annual
Precipitation (mm)	367.2	355.4	324.6	249.8	141.7	103.0	49.3	96.3	207.7	334.0	498.9	550.4	3278.2
Air temperature (°C)	1.0	1.4	4.8	9.2	14.2	18.0	20.4	19.7	14.9	10.4	5.7	2.2	10.1
Mean minimum temperature (°C)	−3.7	−3.3	−0.3	3.6	7.4	10.6	12.2	12.0	8.5	5.0	1.0	−2.2	4.2
Mean maximum temperature (°C)	7.5	8.0	11.2	15.1	20.9	25.3	28.7	28.7	23.4	18.3	12.3	8.1	17.3
Relative humidity (%)	85	83	79	78	76	74	68	71	80	85	86	86	79
NIKŠIĆ	Jan	Feb	Mar	Apr	May	June	July	Aug	Sep	Oct	Nov	Dec	Annual
Precipitation (mm)	184.8	180.2	161.4	153.7	108.7	84.5	43.7	74.5	149.0	196.5	285.8	259.2	1882.0
Air temperature (°C)	2.1	2.4	5.6	9.8	14.8	18.6	21.7	21.3	16.2	11.7	6.5	3.0	11.1
Mean minimum temperature (°C)	−1.6	−1.5	1.4	5.1	9.2	12.4	14.8	14.7	10.8	7.1	2.7	−0.4	6.2
Mean maximum temperature (°C)	6.5	7.2	10.6	14.7	20.3	24.3	28.1	28.4	22.8	17.7	11.6	7.3	16.6
Relative humidity (%)	71	69	67	68	67	65	57	58	68	75	75	74	68

Data source RHMZ

whole, has a perhumid climate. Average annual temperatures are low (7.5 and 5.4 °C), as the short summer period is cool and moderately warm, and the winter is harsh and long, practically extending to a part of spring and autumn. Temperatures decrease with altitude, while the precipitation is fairly high and evenly distributed. Such climate

Table 3.3 Climate indicators of the subregion of high mountains and plateaus (1981–2010)

KOLAŠIN	Jan	Feb	Mar	Apr	May	June	July	Aug	Sep	Oct	Nov	Dec	Annual
Precipitation (mm)	209.7	193.9	181.8	180.7	124.8	105.3	64.0	77.8	171.3	212.8	300.5	285.8	2108.3
Air temperature (°C)	−1.3	−0.7	2.5	6.7	11.6	14.8	16.8	16.3	12.4	8.5	3.4	−0.3	7.5
Mean minimum temperature (°C)	−5.4	−5.1	−2.0	1.4	5.5	8.3	9.6	9.5	6.8	3.6	−0.7	−3.9	2.3
Mean maximum temperature (°C)	3.5	4.5	8.3	12.9	18.4	21.9	24.7	25.1	20.1	15.4	8.9	3.8	14.0
Relative humidity (%)	82	79	77	75	75	76	74	75	80	81	83	85	79
ŽABLJAK	Jan	Feb	Mar	Apr	May	June	July	Aug	Sep	Oct	Nov	Dec	Annual
Precipitation (mm)	111.5	115.9	107.9	120.5	99.5	89.3	67.0	69.3	120.1	163.0	205.0	176.2	1445.2
Air temperature (°C)	−3.5	−3.5	−0.4	3.8	9.4	13.1	15.2	14.9	10.5	6.5	1.3	−2.3	5.4
Mean minimum temperature (°C)	−7.7	−7.8	−4.6	−0.6	4.0	7.2	9.0	9.1	5.6	2.2	−2.5	−6.0	0.7
Mean maximum temperature (°C)	2.0	1.9	4.8	8.7	14.5	18.0	20.8	21.1	16.5	12.4	6.7	2.6	10.8
Relative humidity (%)	81	80	79	76	73	71	69	70	76	79	81	83	76

Data source RHMZ

conditions also influence the soil genesis and evolution, as they cause acidification, the accumulation of organic matter in the form of raw humus, and slower mineralisation of organic matter. The rainwater and external water surplus cause the occurrence of Hydrogenic soils, especially in plateaus (Mokro near Šavnik and Jezera near Žabljak). The precipitation, in terms of total sum and intensity, including the several-day duration, has an effect on erosion, resulting in the loss of nutritive elements and the formation of Colluvial and Alluvial Soils at the toe and footslopes and in the valleys of smaller and larger watercourses.

The climate limits the choice of plant species grown, and moisture deficit reduces yields. Mountain pastures (*suvati*) are mostly distributed in very shallow soils and dry habitats, exposed to the wind, which is why high-quality grazing grass yield is minimal (Fig. 3.24). Mountain meadows in flat and lower positions are more favourable in this respect, not only for cattle grazing but also in terms of the quantity of hay, which is the most important feed in this area.

In the high sum of precipitation, the share of snow is significant, shortening the vegetation in the spring and autumn due to frost (which can occur even in the summer).

Fig. 3.24 Dry mountainous pastures in the area of Žabljak (*Photo* G. Roganović)

The perhumid and humid climate are most favourable for the forest vegetation, which covers larger areas in the pre-mountain zone up to about 1700–1900 m a.s.l. Soil acidification, due to the accumulation of the decomposing forest litter, affects the dissolution of the clay minerals and the rinsing of certain components, causing podzolisation.

Some of the forest complexes of the high mountain region are well preserved, which is very significant from the perspective of protecting soil against erosion. Some of these were declared reserves or national parks (Durmitor, Biogradska Gora, and Prokletije).

The continental subregion is characterised by heterogeneous values of certain climatic elements (Table 3.4. and Appendix III). Pljevlja has an average annual temperature of 8.7 ^{0}C, Bijelo Polje 9.4 ^{0}C, and Berane 9.6 ^{0}C. Pljevlja has negative average temperatures in December, January, and February, while Berane and Bijelo Polje have negative average temperatures only in January. May and three summer months in Pljevlja have a semiarid climate, while in Berane, July and August are arid, and May and June are semiarid. Likewise, in Bijelo Polje, July and August are arid, and June is semiarid. September is semihumid in Pljevlja and Berane, as well as May and September in Bijelo Polje. In other months, the climate is both humid and perhumid.

Notably, precipitation is declining in the direction of the northwesternmost and northeasternmost parts of Montenegro. In the Lim valley, precipitation is relatively even—with 892.6 mm in Berane and 923.8 mm in Bijelo Polje. The decrease in precipitation is pronounced in the direction of Kolašin (2108.3 mm)–Mojkovac–Bijelo Polje. It is similar in the direction Žabljak (1445.2 mm)—Pljevlja (778.7 mm).

The semiarid climate of the summer period shows that dry periods can also occur in the continental region due to lack of moisture, which imposes the need to irrigate plants in the middle of the vegetation period.

Rainfall poses a great risk of erosion for soils with sparse or destroyed vegetation cover and on very steep slopes (Figs. 3.25 and 3.26). This applies in particular to the high-intensity precipitation, in the form of brief showers causing more intensive surface erosion (wearing away soil particles) and deep in (furrows, ravines, and gullies). More intensive erosion is also enhanced by human activity, which is manifested in irregular cultivation, absence of anti-erosion (soil conservation) measures, unplanned felling, grazing, etc.

The water surplus causes occasional but short-term floods and torrential erosion. The longer-term flooding and high groundwater levels are pronounced on the banks of Plavsko Lake and its tributary Ljuča. This is why Swamp-Gley Soil (Eugley) is formed in certain areas or occasional ponding of Alluvial and Alluvial-Colluvial deposits occur.

The floods caused by Lim, Tara, and Ćehotina in some parts of the basins cause the risk of collapse and undermining of the river banks and removal of the fertile soil layers. River beds are widening due to bifurcation, increasing the area with sterile gravel and sandy deposits. The

Table 3.4 Climate indicators of the Continental subregion (1981–2010)

BERANE	Jan	Feb	Mar	Apr	May	June	July	Aug	Sep	Oct	Nov	Dec	Annual
Precipitation (mm)	68.4	67.3	69.6	80.2	68.6	63.3	55.2	52.7	86.2	76.6	111.3	93.1	892.6
Air temperature (°C)	−0.7	0.8	5.1	9.5	14.4	17.6	19.6	19.2	14.8	10.3	4.5	0.4	9.6
Mean minimum temperature (°C)	−4.7	−4.1	−0.4	3.2	7.3	10.3	11.9	11.7	8.4	4.7	0.1	−3.1	3.8
Mean maximum temperature (°C)	4.6	7.0	11.7	16.0	21.6	25.0	27.7	27.9	22.9	18.0	10.4	4.9	16.5
Relative humidity (%)	81	76	71	69	69	69	68	69	75	78	81	83	74
BIJELO POLJE	Jan	Feb	Mar	Apr	May	June	July	Aug	Sep	Oct	Nov	Dec	Annual
Precipitation (mm)	69.3	74.0	65.4	78.7	71.5	71.5	61.6	55.9	91.1	80.7	112.8	91.4	923.8
Air temperature (°C)	−1.0	0.6	5.1	9.5	14.2	17.2	19.1	18.9	14.6	10.0	4.1	0.2	9.4
Mean minimum temperature (°C)	−4.5	−3.8	−0.2	3.5	7.7	10.8	12.4	12.3	9.2	5.5	0.5	−2.9	4.2
Mean maximum temperature (°C)	3.7	6.6	11.6	16.3	21.5	24.7	27.3	27.6	22.6	17.4	9.5	4.2	16.1
Relative humidity (%)	81	78	74	74	74	75	73	73	78	80	83	84	77
PLJEVLJA	Jan	Feb	Mar	Apr	May	June	July	Aug	Sep	Oct	Nov	Dec	Annual
Precipitation (mm)	46.5	56.1	48.5	60.2	64.8	82.7	66.1	63.9	76.1	65.5	83.2	65.1	778.7
Air temperature (°C)	−1.8	−0.2	4.0	8.3	13.4	16.6	18.5	18.3	13.9	9.7	3.8	−0.4	8.7
Mean minimum temperature (°C)	−5.7	−5.0	−1.3	2.5	6.8	10.1	11.6	11.5	8.2	4.5	−0.3	−3.8	3.3
Mean maximum temperature (°C)	3.1	5.9	10.5	14.8	20.5	23.7	26.0	26.5	21.7	17.1	9.5	3.9	15.3

Data source RHMZ

Fig. 3.25 Soil water erosion processes in the Kosanica area in 1984 (*Photo* B. Fuštić)

Fig. 3.26 Soil water erosion processes in the Kosanica area in 2023 (*Photo* G. Roganović)

significant areas of such deposits are found in the vicinity of Andrijevica, Berane, Zaton, and the confluence of the Bistrica and Lim, in the Tara valley around Mateševo, Kolašin, and Mojkovac, and much less in the Ibar and Ćehotina valley.

3.3 Relief as a Soil-Formation Factor

Unlike the geological substratum and climate that have a direct impact on soil formation and genesis, the relief has an indirect influence on soil formation. The relief balances other factors, to some extent, primarily climate conditions. The relief is defined by the appearance of each surface characterised by horizontal and vertical dimensions. High inclination contributes to the rapid transport of matter over the surface, which takes place by surface runoff, but also under the influence of gravity. Inclination contributes to faster water flow, depending on the degree and length of the slope but also on the condition of the vegetation cover. Once the transportation power of the moving water drops, deposit settling and disposal take place at the foot and toeslopes or in the flat terrain.

3.3.1 Role of relief in water flow and soil erosion

The relief has an effect on the distribution of plant species or plant communities. Due to the climate and relief conditions, the zonal character of numerous plant communities is notable in Montenegro. Thermophilic species prevail in southern exposures. Specific plant species occur in gorges and canyons, which is also the case with karst, under the cliffs where scree and debris occur.

Flat relief forms receive the entire rainfall quantity but also suspended solids and soil material washed away from the slopes. That creates multi-layered sediments or deposits where genesis starts only after deposit stabilisation. Soil cultivation on slopes causes erosion (Fig. 3.27). The loss of soil, i.e., its mineral and organic particles, prevents the development of any vegetation for a longer period. Thus, the end stage of erosion is keeping the soil in the initial stage of development and evolution. Understandably, the disposal of erosion material at the base of the slope covers the existing soil, forming a new substratum (regolith) on which the genesis of new soil—Colluvial or Alluvial—begins.

Relief is an important factor for the hydrothermal soil regime, as its aspect and inclination influence solar energy distribution. This way, southern, southwestern, and southeastern exposures are warming up and drying up faster, followed by western, eastern, and northern the least. The warming is most intensive when the sun's rays fall onto the surface at the right angle.

Thus, the relief has a corrective effect on climate and solar energy and, therefore, also on the pedoclimate. With its configuration, or specifically, with inclination, slope length, and aspect, the relief creates conditions for the development of a local microclimate, which can have a stronger effect on soil formation than macroclimate. In micro depressions, i.e., in the lowest flat relief forms, in the river valleys, on the banks of the lake, and on the coast, the relief is a significant factor of excess water in the soil. Consequently, redox processes prevail in such locations, which are reflected in the emergence of various types of hydrogenous and peaty soils. In the coastal zone (in some coastal fields adjacent to the shoreline), the redox processes are also accompanied by salinisation by the seawater, which directs the genesis towards the saline (Halogenic) soil formation.

The relief forms with high inclinations result in the movement and displacement of smaller or larger amounts of soil from the place of formation. The pronounced impact of the relief is illustrated best by the fact that the slope gradient of 1,202,588 ha (about 87%) of the area of Montenegro is greater than 15% (Fig. 3.28).

3.3.2 Macro relief and its influence on soil formation

Macro relief in Montenegro is very pronounced, formed by the action of endogenous forces, i.e., tectonic movements. The oro-relief influences the shift of bioclimatic zones, which are particularly notable in the form of vertical vegetative cover zones, and together with the inclination and exposure, has a significant influence on soil formation (Fig. 3.29). The zonal character of the soil in Montenegro, given its small territory, can only be discussed provisionally. Under the influence of the climate, relief and substratum, Terra Rossa Soil was formed up to about 600 m a.s.l in the southern area of Montenegro and Calcomelanosols in its mountainous region.

The mesorelief and microrelief forms influence rainfall distribution, hydrothermal regime modifications, and effects the flood and groundwater actions.

The soils formed on a flat relief are normally moistened, i.e., the soils are moistened individually, which is not the case in a wavy relief. Convex forms and steeper slopes cannot collect or absorb all the rainfall; thus, some of the water will runoff, potentially causing erosion. Such terrain will be dry, i.e., it will have a drier pedoclimate compared to concave forms, where soil will be moistened more and have a more humid pedoclimate.

There are several distinct entities in Montenegro's relief from the Adriatic Sea towards the continent. There are three orographic and climatic entities, as follows: the southern part with a narrow coastal belt, Lake Skadar basin, and Nikšićko Field; the high mountain area with some flat terrain in the river valleys and plateaus; the third entity is the continental area, with more flat terrain in river valleys and plateaus above which the medium and high mountains rise.

3.3.3 Soil formation in different relief zones

In the southern area, on the flat relief of coastal fields, glens, and valleys of watercourses, alluvial deposits are common with young soils without genetically developed

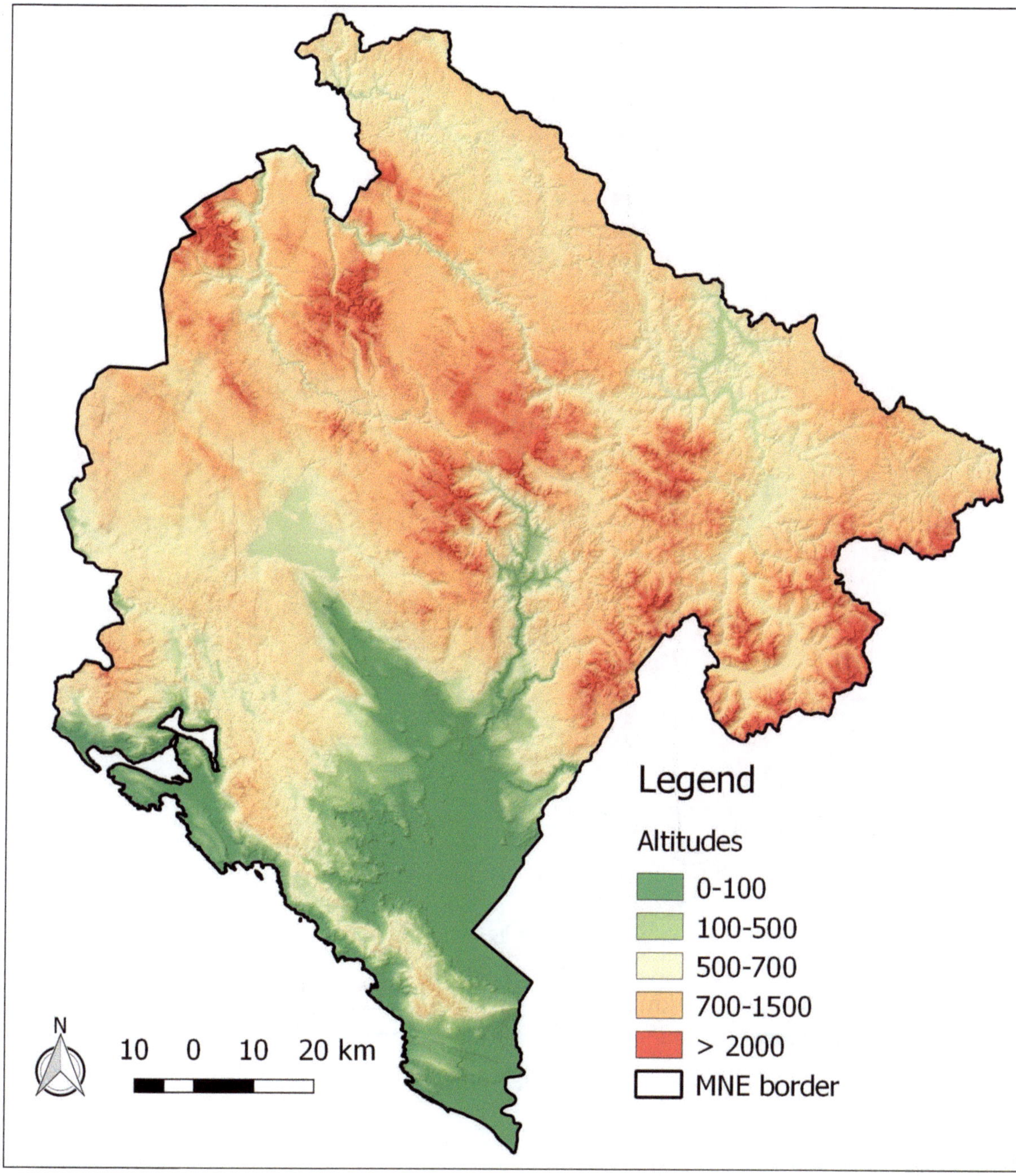

Fig. 3.27 Digital elevation model of Montenegro reclassified into altitude classes (*Source* MARD 2017)

layers. The situation is similar in the area of Lake Skadar basin and in Nikšićko Field, where, apart from alluvial and colluvial deposits, diluvial loams occur in parts of the Bjelopavlićka Plain, Lješkopoljski Lug, and the Nikšićko Field. Under almost identical conditions, such substratum drives the soil genesis towards the development of specific processes and the formation of soils that vary in terms of profile depth, physical, chemical, and biological characteristics. Flat relief enables even soil moistening, but clayey substratum and clayey (B) horizon prevent in-depth percolation of water, which then stagnates in the soil layer, causing pseudogleysation and poor physical, chemical, and other soil properties.

In the southern part of Montenegro, numerous terraces of the flysch belt of the Coast and Skadar basin occur as forms of flat relief created by human activity by raising stone walls (drywall) and colluviation processes (Fig. 3.30). In the karst area, there are also numerous sinkholes, glens, and smaller flat areas where Colluvial Terra Rossa's are distributed up to about 500–600 m a.s.l. and Calcomelanosols

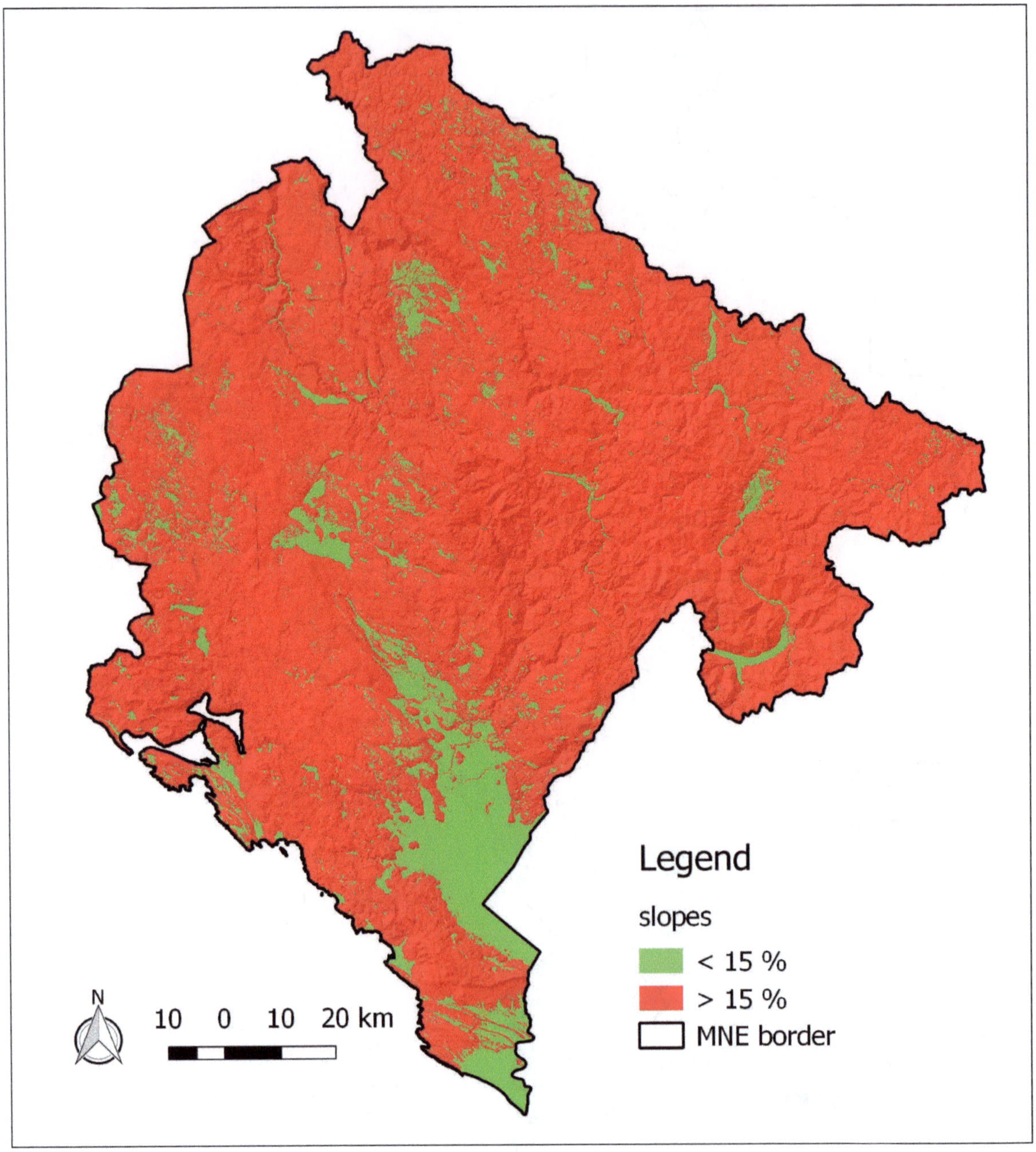

Fig. 3.28 Map of slopes greater than 15% (*Source* MARD 2017)

in higher elevations, as well as with Rendzinas, formed on glaciofluvial and moraine deposits of most of the karst fields (Cetinjsko, Njeguši, Dragalj, Grahovsko, Nikšićko, Gostilje, Kopilje, Radovče, Trmanje, and Kržanja).

The prevailing relief of the southwestern area of Montenegro influences the presence of some plant species adapted to a dry habitat. The habitat aridity is caused mainly by the limestone rocks and relief, and according to Pavićević (1956), it has a crucial impact on soil formation in the karst terrain. Soils in karst relief depressions are, most frequently, formed of materials washed away from the surrounding, elevated areas and deposited in depressions. Thus, the soils remain young and undeveloped in ever-present erosion and colluviation processes.

Beyond the river valleys, the central and high-mountain area has small areas with flat and slightly inclined relief forms. Most of them occur in plateaus of the Durmitor, Bioče, and Ljubišnja, then on Sinjavina and the Pivska Planina where shallow sinkholes, glens, smaller plains, and flat fields occur in larger numbers (Nedajno, Trsa, Pišče, Jezera, Okruglica, Suvo, Odgar, Šaransko Field, etc.). Similar relief form sequences can be found on the plateau

Fig. 3.29 Javorje Mountain (*Photo* Lj. Životić)

on Tara's right bank, known as *do* (Burenski Do, Meki Do, and Pušanjski Do) and fields (Rudo, Kameno, Stupsko, Kosaničko, Bitinsko, Glibaći).

The high mountain range and their slopes towards the valleys of numerous watercourses have numerous extremely steep sideslopes, which facilitate intensive erosion and dislocation of large amounts of soil material.

In the continental part, the hilly mountain relief is the dominant form, but there are also milder inclinations on slopes of up to 1200 m a.s.l. Thus, the soil is less susceptible to erosion, which is manifested mainly by surface sheet erosion and rarely by gully erosion forms. Milder relief forms prevail in the region of the Neogene lacustrine sediments around Berane (Zagrad, Petnjik, Budimlja, Dapsići, and Polica) and Pljevlja (Maoče, Mataruge, Pljevlja Hollow, Glisnica and Rađevići). In addition to these, mild slopes are present on the terrain of shales and sandstones of Paleozoic age and diabase-chert formations. These sediments, usually the most widespread in the continental region, cover mainly the area towards the surrounding mountains that are predominantly made of limestones. Smaller flat area with deeper soils, formed on a mixed substratum of carbonate-silicate origin, can be found on limestone plateaus. The limestone hills, ridges, peaks of the mountains, and cliffs along their slopes occur in the bare karst of the Southwestern and High mountain regions of Montenegro. In mountains of the continental region, intensive dissolution of limestone rocks is not as present as in the high mountain area, so there are no large areas of bare karst and rocks. That is why there are no major faults and scree from the mountain peaks along their slopes, which are typical in the areas of high mountains and canyons (Fig. 3.31).

The valleys of rivers Ibar, Ćehotina, and Lim, in particular, are much more broadened than those in the high mountain region. Such appearance and their relief contributed to the presence of larger areas with alluvial and colluvial deposits, i.e., with more fertile soils. The largest areas with such soils can be found in Polimlje, the valleys of Lim tributaries (Fig. 3.32), Lješnica, Ljubovida, and Bistrica, and also in the Ćehotina valley (Vrulja, Maoče,

Fig. 3.30 Terraced landscape at Lastva Grbaljska (*Photo* G. Roganović)

Židovići, Gornja Brvenica, and Donja Brvenica), its tributary Vezičnica and others. In the valley of the Ibar River and its tributaries, such soils cover a negligible area (Županica, Paučina).

In the Old Royal Capital Cetinje (within the Biljarda complex), a relief map of Montenegro at a scale of 1:10,000 was made in 1916 after the Austro-Hungarian army occupied the city (Fig. 3.33). The relief, which includes parts of Bosnia and Herzegovina, Croatia, Albania, and Serbia, is made of concrete on a stone base, based on precise Austrian topographic maps and land measurements. Designated as a cultural monument in 1948, this unique relief map, besides its unquestionable military-strategic purpose during wartime, has gained significance over time for its aesthetic and historical value. Today, it stands as an indispensable tourist attraction.

3.4 Biological Factors

The soil genesis is initiated by the colonisation of the first organisms (lithobionts). Organisms (silicate bacteria, lichens, and plants) contribute to the formation of a thin earthy layer created by the weathering of rocks and the decomposition of organisms.

3.4.1 Impact of flora and fauna on soil formation

The impact of flora and fauna on soil, which is reflected in the mixing of different soil material, and the impact on mineral weathering, includes the transformation of organic matter and the enrichment of soil by their content. The creation of different organo-mineral components and compounds is of paramount importance. Various soil fauna feed on organic and mineral matter, and their excrements help maintain a stable soil structure (e.g., earthworms and other worms). Moreover, bioturbation processes, i.e., the mixing of organic and mineral soil matter, take place, as well as migration of matter over the thickness of soil profile. Root systems stabilise the plants in the soil, and the soil itself, and influence the decomposition of rocks, which leads to the transfer of macro- and micro-elements from deeper soil layers to the surface. The dead plant parts (roots, leaves, branches, needles, bark, etc.) increase the organic matter content in the soil and affect other soil characteristics (pH, CEC, soil density etc.).

The vegetation substantially modifies the impact of climate on the soil. Areas with well-preserved vegetation provide good soil coverage, longer moisture retention, and lower soil temperature. Conversely, thinned vegetation

Fig. 3.31 Komovi, Kurlaj, katun of Vujadinovići (*Photo* I. Laković)

Fig. 3.32 Plavsko Lake, source of the Lim River (*Photo* M. Knežević)

Fig. 3.33 Relief map of Montenegro (*Photo* G. Roganović)

causes faster warming, fast water runoff, and increased erosion, leading to the transfer of significant amounts of soil material from its place of origin to lower areas.

The origin and distribution of certain plant communities, associations, and species in Montenegro is influenced by its geographical position and a multitude of other factors, including soil, and vice versa. As a result, botanists and other researchers have observed a specific vegetation zonation, with typical plant species. Zonation can be traced from the coast to the highest mountain peaks on one side and from river valleys to surrounding mountains on the other. This alternation of vegetation zones is primarily influenced by climatic conditions, although it is also impacted by the relief, including altitude, inclination, and exposure, and types of parent materials (Blečić 1982; Lazarević 2012; Vuksanović 2016).

Given these factors, Montenegro's vegetation profile is highly complex. In addition to zonal vegetation types influenced by general climatic conditions, various types of azonal, intrazonal, and extrazonal vegetation are present, reflecting the specific ecological factors of a given habitat (Milanović 2021). The section below provides an overview of the most important vegetation types, i.e., those with a wider distribution or that are significant from the aspect of the Habitats Directive.

3.4.2 Vegetation of the Mediterranean region and its influence on soil

In the Mediterranean area, forest vegetation is degraded, and well-preserved stands are relatively rare.

The climatogenic (zonal) vegetation of the first belt, ranging from the sea coast to an altitude of around 300 m (or 350 m a.s.l.), in the Mediterranean area, is typically represented by evergreen hard-leaved forests. This type of vegetation is now rare in the Mediterranean, having been largely replaced by degradation stages—by maquis as the first degradation stage, or subsequent ones in the form of garrigue, or rocky pastures. This is also the case in Montenegro, where small stands of preserved hard-leaved forests can only be found in the foothills of the southern slopes of Mountain Rumija—between Bar and Ulcinj (Petrović, 2011).

In most of Montenegro's coastal region, maquis is represented by a plant community with holm oak (*Quercus ilex*) (Fig. 3.34), manna ash (*Fraxinus ornus*) and myrtle (*Myrtus communis*) as edificators. Maquis is dominated by shrubs, while trees are less present. An interesting maquis community, very rare on the Adriatic coast of the Balkan Peninsula, is found in the Ulcinj area. Its edificators are kermes oak (*Quercus coccifera*) and ash (*Fraxinus ornus*). Despite the

Fig. 3.34 Quercus ilex in the Mediterranean part of Montenegro (*Photo* D. Caković)

limited distribution of this community in the Ulcinj area (the best stands have developed on Bijela Gora and Pinješ), these are among the best-developed kermes oak stands in the Adriatic coast. The vegetation is developed in the low maquis form, with a pronounced shrub layer.

An interesting azonal forest vegetation type on the Montenegrin coast are chestnut forests, developed in the Boka Kotorska Bay (Stoliv, Lepetane, the area above Tivat, Kostanjica between Morinj and Kamenari, and Savina). These forests belong to the *Lauro-Castanetum sativae* community, with stands at a low altitude, ranging from around 10 m a.s.l up to 200 m (Fig. 3.35).

Erica arborea plays an important ecological and phytocoenological role in the thinned communities of sweet chestnut and laurel. The chestnut forest stands in Donji Stoliv are particularly dense and well-preserved. (Janković 1966; Karaman 1997; Odalović et al. 2013).

The northern aspects are dominated by thermophilic deciduous forests and scrub, descending to the very shoreline, while the southern aspects are home to evergreen hard-leaved vegetation. Thermophilic deciduous forests are represented by numerous plant communities, with oriental hornbeam (*Carpinus orientalis*) being the edificator or an important constituent species. The association comprising butcher's broom and oriental hornbeam (*Rusco-Carpinetum orientalis*) as edificators is widely distributed, with degradation stages. A typical form of this association is characterised by well-expressed stratification, with oriental hornbeam (*Carpinus orientalis*), downy oak (*Quercus pubescens*), and manna ash (*Fraxinus ornus*) as the dominant species in the tree layer. Oriental hornbeam (*Carpinus orientalis*), mock privet (*Phyllirea media*), myrtle (*Myrtus communis*), hawthorn (*Crataegus monogyna*), traveler's joy (*Clematis vitalba*), and maples (*Acer monspessulanum, Acer campestre*) prevail in the shrub layer. The herb layer is comprised of species such as rough bindweed (*Smilax aspera*), ivy (*Hedera helix*), comfrey (*Symphytum tuberosum*), and *Vincetoxicum huteri*, etc. In the Mediterranean zone, the association is represented by a subassociation with myrtle (*Rusco – Carpinetum orientalis subas.*

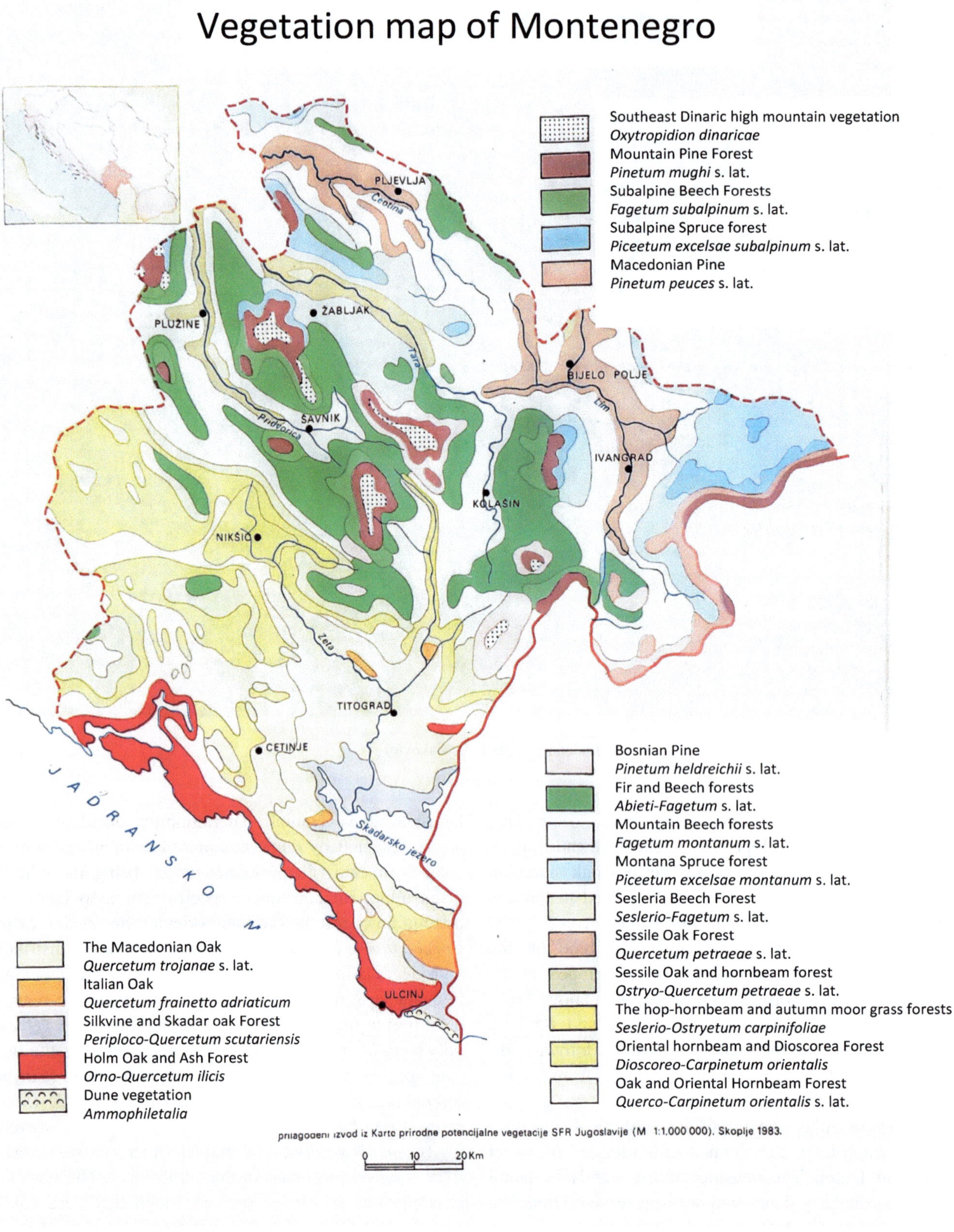

Fig. 3.35 Vegetation map of Montenegro (Lakušić, 1987)

myrtetosum) as the differential species and a poorly developed tree layer.

Above the oriental hornbeam belt, on the southern slopes of the coastal Dinarides (Orjen, Lovćen, Rumija), the belt of hop-hornbeam (*Ostrya carpinifolia*) continues, with forests of Turkey oak (*Quercus cerris*) and Macedonian oak (*Quercus trojana*) also present. These forest types are widely distributed in the subMediterranean and central parts of the country (Petrović 2011).

Among deciduous forest communities, for the Mediterranean and sub-Mediterranean area of Montenegro, the specific natural forests of Skadar Pedunculate oak are found in the floodplains of the Mediterranean and subMediterranean rivers in the south of Montenegro (Fig. 3.36). The Skadar oak (*Quercus robur subsp. scutariensis*) is a thermophilic Mediterranean subspecies of English oak. In Montenegro, it can be found on the northern coast of Lake Skadar (village of Gostilj), where it grows in association with narrow-leaved ash (*Fraxinus angustifolia*), and silkvine (*Periploca graeca*). The distribution region of this subspecies also includes Velika Plaža (Long Beach) near Ulcinj. Natural moist forests of the Skadar oak with oriental hornbeam (*Robureto-Carpinetum orientalis*) can be found along the Bojana River, 200–300 m from the sea (Černjavski et al. 1949; Janković and Bogojević 1965).

Grassland communities phytocoenoses prevail on the flat land of the Coast and Zetsko-Bjelopavlićka Plain. Among them, the lowest positions are occupied by marsh and wet meadow communities from which grassland communities, i.e., meadows and pastures extend, which have to a large extent been transformed into arable land with agricultural vegetation—field crops, vegetables and grape vine. Aquatic and marsh vegetation is represented by several associations that appear in several riparian zones of Lake Skadar, namely: 1. submergent plants, submerged below the water surface (*Potamogeton, Myriophyllum, Ceratophyllum*), 2. aquatic plants with floating leaves (*Nuphar, Nymphaea, Trapa*), 3. partially aquatic plants such as reeds (*Phragmites*), cattails (*Typha*), bulrushes (*Scirpus*), and sedges (*Carex, Cyperus*) (Černjavski 1949).

Wet meadows follow the aquatic and marsh vegetation belt, in which hygrophytes are often found: *Orchis laxiflora, Lythrum salicaria, Galium palustre, Oenanthe sp., Juncus sp*. In marshes, plant remains decompose in anaerobic conditions, leading to the accumulation of organic matter and its transformation into peat. In these conditions, genesis goes in the direction of creating Organogenic Soils—Peats. Other organo-mineral soils are formed from the deposition of fresh alluvial sediments, when the mixing of organic and mineral matter occur. In the zone of marsh and wet meadows, which are, unlike marshes, seasonally flooded, anaerobic decomposition is present during the lasting flood waves, and then aerobic processes arise, guiding soil genesis in another direction (Fig. 3.37). Organic matter, not fully

Fig. 3.36 Skadar oak forest—Gostilj (*Photo* G. Roganović)

Fig. 3.37 Aquatic and marsh vegetation on Lake Skadar (*Photo* M. Knežević)

transformed in anaerobic conditions, decomposes faster and becomes less valuable humus. Lower humus quality results from leaching of base cations from the surface layers of Alluvial Soils, which occurs under the influence of rainwater filtration and a drop in the groundwater level. The oscillation of oxidation–reduction processes cause gleysation of deeper soil layers.

Besides marshes and natural grasslands, i.e., meadows and pastures, which have been largely converted into fields, orchards, and vineyards, all other Mediterranean zone terrain has been overgrown with maquis, garrigue, and forest. Cultivated vegetation was established in the form of smaller oases, mostly on the coastal belt terraces (with olive trees) and the Lake Skadar basin, and then in glens, sinkholes, and smaller karst plateaus.

An interesting type of azonal deciduous forest in the Mediterranean and subMediterranean region is formed by the Hungarian oak (*Quercus frainetto*) forests in the Ulcinj area (Zoganj, Kolonza, Donja Klezna, Gornja Klezna, and Možura), which belong to the *Rusco aculeati—Quercetum frainetto—cerris* association. Important constituents of these forests are the Turkey oak (*Quercus cerris*) and the Hungarian oak, butcher's broom (*Ruscus aculeatus*), asparagus (*Asparagus acutifolius*), oriental hornbeam (*Carpinus orientalis*), terebinth (*Pistacia terebrinthus*), *Cornus sanguineus*, and *Acer monspesulanum*. Forests with these edificators can be found in the valleys of the Morača, Zeta, Ćehotina, Tara and Lim rivers, in the vicinity of Virpazar, Ulcinj, Podgorica (Crnci) and Rožaje, as well as on the mountains Durmitor, Hajla, Prokletije, Rumija.

Along the Montenegrin coast, several vegetation types are present due to the degradation of forests and further maquis degradation. These include garigue of various phytocoenological affiliations, meadows formed by forest clearing, and rocky pastures represented by numerous plant communities. Shrubs (generally up to 1 m in height), scattered and with low coverage, prevail in garrigues. Yet, despite harsh living conditions (summer drought and poor soil) it is known for its high floristic richness. *Erico-Cystetum cretici*, an association of heath and rockrose, is a widely distributed garigue type along the Montenegrin coast. It is dominated by shrubs such as heath (*Erica arborea*), rock rose (*Cistus creticus ssp. eriocephalus*), buckthorn (*Frangula rupestris*), myrtle (*Myrtus communis*), garland thorn (*Paliurus spina christi*), pomegranate (*Punica granatum*) (Topalović et al. 2020), and Phoenitian juniper (*Juniperus phoenicea*) . The coastal zone also hosts a broad distribution of garigue association dominated by thorn species (*Paliuretum adriaticum*), which indirectly connects the oriental hornbeam and butcher's broom community (*Rusco-Carpinetum orientalis*) with the holm oak forest zone. The tree layer in the thorn community is absent; only scattered and individual trees can be found in certain locations. On the other hand, the shrub layer is the best developed, and in some places, it has a coverage of up to 90%. The dominant species in the shrub layer include garland thorn (*Paliurus spina christi*), Spanish broom (*Spartium junceum*), and hawthorn (*Crataegus monogyna*).

The final stage of degradation in the thermophilic Mediterranean and subMediterranean forests is represented by communities of dry grasslands and rocky pastures. This vegetation type covers a range of grassy biotopes, but they are all characterised by a low level of nutrients and precipitation in the summer period. The best stands of this habitat type in the coastal zone have developed between Budva and Tivat, on Luštica and Vrmac, above the Old Town of Bar, and in the Ulcinj area. Recently, the areas under this habitat type, formed and maintained by animal grazing, have experienced a decline due to a decrease in the livestock population in the Mediterranean region of Montenegro.

In the Mediterranean zone, the climatic, soil, and topographical conditions foster the existence of evergreen species and prolonged vegetation periods. This is of particular significance for pedogenesis, as the presence of green leaves and grasses all year round helps protect the soil from erosion. The forest and grass vegetation in karst areas plays a vital role in soil genesis, particularly in the form of the accumulation of organic matter in the soil. The karst, being a dry habitat, is inhospitable to lush vegetation growth, and is instead dominated by xerophytic plant species that can withstand summer heat and drought. Grasses in this region quickly dry out, leaving only trees. Such conditions favour the accumulation of plant residues due to weak mineralisation, while the presence of $CaCO_3$ in these habitats leads to the formation of Calcomelanosols (as a dominant soil type) with a high humus content (up to 25 or 30%). Humus abundance is not the result of lush vegetation but of conditions promoting the accumulation of plant remains and their transformation into humus (Fuštić and Đuretić, 2000).

In the inland zone of Montenegro's Mediterranean region, more than thirty NATURA 2000 habitat types have been registered. Habitats linked to beaches and their hinterland are particularly vulnerable, such as Mudflats and sandflats not covered by seawater at low tide, Annual vegetation of drift lines, Mediterranean salt meadows, Embryonic shifting dunes, Brachypodietalia dune grasslands with annuals, and Mediterranean temporary ponds, etc. Most of these vulnerable (rare and, to a large extent, degraded) NATURA 2000 habitats can be found on Ulcinj's Long Beach and Ada Bojana island. Rich in terms of biodiversity are Mediterranean tall humid grasslands (*Molinio-Holoschoenion*), Pseudo-steppe with grasses and annuals (*Thero-Brachipodietea*), while forest Mediterranean habitats such as Riparian mixed forests of the English oak (*Quercus robur*), narrow-leaved ash (*Fraxinus angustifolia*), and the field elm (*Ulmus campestris*), *Castanea sativa* forests, Coppice *Quercus frainetto* woods, *Quercus ilex* and *Quercus rotundifolia* forest, etc.

3.4.3 Vegetation of the Sub-Mediterranean region and its impact on pedogenesis

In the subMediterranean, or central area, the already discussed community of butcher's broom and oriental hornbeam (*Rusco-Carpinetum orientalis*) is widely distributed and is represented by the subassociation *Rusco-Carpinetum orientalis subas. quercetosum*, where the tree layer is better developed than in the Mediterranean subassociation. Significant areas are occupied by thermophilic oak forests, followed by the hop-hornbeam (*Ostrya carpinifolia*) belt, which shifts into a beech belt at higher elevations.

The Macedonian oak (*Quercus trojana*) can be found in various associations at a broad altitude range, from 50 to 700 m a.s.l. It is a distinctive species of the *Rusco-Carpinetum orientalis subas. quercetosum* Blečić et Lakušić subassociation, is widespread in the subMediterranean region of Montenegro, predominantly in the form of coppice. These oak forests and coppice extend to the belt of evergreen forests of holm oak, the extrazonal forests of the Hungarian oak, or the Skadar Pedunculate oak forests. Other species that play a significant role in the tree layer comprise: *Fraxinus ornus, Acer monspessulanum, Carpinus orientalis, Pistacia terebrinthus*, and *Quercus cerris*. The herb layer is composed of species such as *Primula vulgaris, Anemone apenina, Silene italica, Carex flacca, Symphytum tuberosum*, etc.

The turkey oak forests (*Quercetum cerridis mediterraneo-montanum*) are commonly found in the hop-hornbeam belt, spanning altitudes from 700 to 1300–1500 m a.s.l. These Mediterranean-Montane oak communities often border the *Seslerio-Ostryetum carpinifoliae* association and share a similar floristic composition. On certain sites, they alternate in a mosaic-like pattern, with the Mediterranean-Montane oak forests occupying more thermophilic habitats compared to the hop-hornbeam and autumn moor grass forests. The tree layer in the oak forests consists of the same species as the hop-hornbeam and autumn moor grass forests, but the main difference is that *Quercus cerris* is the dominant tree layer species in oak forests.

The hop-hornbeam and autumn moor grass forests (*Seslerio-Ostryetum carpinifoliae*) occupy a large area in Montenegro between the Oriental hornbeam (and thermophilic oaks) belt and the beech belt. The tree layer has a poor floristic composition, with hop-hornbeam (*Ostrya carpinifolia*) and the Turkey oak (*Quercus cerris*) as the main constituents and beech (*Fagus sylvatica*) occasionally appearing at higher elevations. In well-preserved stands, usually those not easily accessible to humans, the tree layer coverage can exceed 80%, but in degraded stands, it may be less than 50%.

The sub-Mediterranean beech forests belong to the thermophilic variant of beech forests, represented by the *Seslerio autumnalis-Fagetum sylvaticae* association. These forests form a belt on the coastal Dinarides but also extend inland to the Piva and Tara canyons. Stratification is pronounced in this community, with *Fagus sylvatica* as the edificator defining the conditions in the lower layers. Hop-hornbeam (*Ostrya carpinifolia*) can be found in the tree layer at lower elevations of the beech forest belt, up to around 980 m a.s.l. At higher elevations, beech absolutely dominates the tree layer, with *Acer obtusatum* appearing rarely. The shrub layer is more diverse in floristic composition. Beech is the key species, and other frequently recorded species include *Fraxinus ornus, Acer* sp., *Crataegus monogyna*, and *Lonicera alpigena*. Autumn moor grass (*Sesleria autumnalis*) is the typical species of association and dominates the herbs layer.

The highest forest belt in the coastal Dinarides is composed of *Pinetum heldreichii mediterraneo-montanum* and *Fritillario-Pinetum heldreichii* forests, which have unfortunately been extensively degraded (Janković 1965). Balkan endemic plant species are the most prevalent, followed by sub-Mediterranean species. Examples of such species *include Lonicera formanekiana, Viburnum maculatum, Senecio visianianus, Thymus balcanus, Festuca vallesiaca*, and others (Vuksanović 2016).

NATURA 2000 sites recorded in the sub-Mediterranean region include Eastern subMediterranean dry grasslands, Hydrophilous tall herb fringe communities, Illyrian Fagus sylvatica forests, Quercus trojana forests, and Forest vegetation with Castanea sativa.

3.4.4 Vegetation of the continental region and mountain areas

In the continental region, valley meadows are found on the lowest flat valley areas with soil that is occasionally moistened or even flooded, for short periods. Marsh vegetation can be found in small areas (around Plavsko Lake), mainly in the form of grassland communities (*Graminea*) with a high number of associations. Along watercourses, thin forests of willows (*Salix*) and poplars (*Populus*) can be found, with an insignificant effect on soil. The narrow strip along the riverbeds is typically filled with rough gravel-sandy deposits, which is why the forests of these species play a key role in stabilising these deposits, which can be dislodged during high water levels. Moving away from river valleys, forest stand zones can be found with natural grassland (locations where forest vegetation has been cleared and maintained by grazing for centuries) and with limited cultivated vegetation on arable land and orchards.

Forests that are closer to settlements, mostly thinned and depleted through clearcutting and other activities, cannot provide adequate protection from soil erosion (Spalević et al. 2020). The devastation of these forest communities leads to frequent surface runoff and gully erosion, resulting in the loss of organic matter and even entire layers of soil, including the subsoil (Fig. 3.38). In higher, more remote areas away from settlements, the forest stands are better preserved, with a denser structure, higher plant residue production (forest litter), and a canopy that eventually transforms into humus materials (Cagliero et al. 2021).

The profile of forest soil differs greatly in terms of morphology, soil structure, and thickness of individual horizons, including their physical, chemical, and biological properties. The layer of forest litter in dense forest stands helps maintain a favourable water regime. The litter and its gradual decomposition serve as a source of organic matter for raw humus (Roh-humus), which eventually transforms into Moder-humus.

In the central, northern, and eastern parts of Montenegro, beech forests can be found in terrains with a range of slopes and aspects (Čurović et al. 2011; Milošević et al. 2019; Čurović et al. 2020). They are sometimes found in open and exposed habitats, while sometimes, they occupy steep, south-facing slopes and glens, or even gorges and canyons where the local climate is modified with higher relative humidity, reduced fluctuations in moisture, and lower summer daytime temperatures.

Mountain beech forests (*Fagetum moesiacae montanum*) occupy a large area and form a separate forest belt above oak forests, reaching over 1200 m of elevation gradient in some places. They are widespread in the central and continental Dinarides in Montenegro in the belt between mesophilic oak and hornbeam forests on the lower side and beech and fir forests on the upper (Fig. 3.39). They occupy different types of parent material and soil, leading to substrate-based differentiation into neutrophil-alkaliphilic and acidophilic.

Beech and fir forests (*Abieto-Fagetum moesiacae*) are widely distributed in central and northern Montenegro and form an elevation belt that can be up to 600 m wide

Fig. 3.38 Soil degradation—Kraljske Bare (*Photo* G. Roganović)

Fig. 3.39 Komovi, Bindža, beech forest in autumn (*Photo* I. Laković)

Fig. 3.40 Beech forest, Ivanova korita, National Park Lovćen (*Photo* G. Roganović)

(Fig. 3.40). The community edificators are beech and fir trees, with spruce (*Picea abies*) joining in the continental Dinarides.

In the subalpine belt of the Durmitor and Prokletije mountain ranges, mesic beech and mountain maple (*Fageto-Aceretum visianii*) forests can be found at elevations ranging from 1400 to 1800 m a.s.l, on all aspects and slopes, on a Mesozoic limestone parent materials. In the Durmitor Mountain area, this phytocoenosis physiognomy is predominantly determined by beech—its gnarled and bent trunks at the bottom, with significantly less represented Balkan maple. However, in the Bjelasica and Prokletije mountains, and in some sites within the Durmitor area, the situation is different—the Balkan maple determines the phytocoenosis' physiognomy.

Subalpine beech forests (*Asyneumo-Fagetum moesiacae*) occupy the vast area of southeastern Dinarides in the subalpine belt, at elevations ranging from 1500 to 1800 m a.s.l, on all aspects and slopes between 10 and 40°. The physiognomy of this phytocoenosis is determined by the subalpine form of beech, which dominates the lower tree and shrub layer. At the lower limit of this belt, fir may occasionally join beech in the shrub layer. These beech forests belong to the NATURA 2000 habitat Illyrian Fagus sylvatica forests.

In floristic terms, the acidophilic forests (*Luzulo-Fagetum*) found throughout Europe are quite uniform in composition, with the main difference being the presence of some boreal species in Central Europe, and some Balkan and southeastern floristic elements in the former Yugoslavia (particularly in the Western part) and Montenegro. These forests thrive on non-carbonate rocks of variable chemical composition. In Montenegro, two types of these communities can be found in the montane and subalpine zones. The acidophilic forests are recognised as a separate type of NATURA 2000 habitat, Acidophilic Fagus sylvatica forests (*Luzulo-Fagetum*).

Black pine forests are present in Durmitor (Fig. 3.41), the Tara and Piva canyons, in the mountains surrounding Nikšić (Vojnik, Golija, Ledenica, Njegoš), in the area of Pljevlja. These forests belong to the NATURA 2000 habitat (sub)Mediterranean pine forests with endemic black pines.

Spruce forests (*Piceetum abietis*) occupy flat or slightly inclined habitats with a deeper humus-accumulative soil horizon. In Montenegro, spruce forests can be found in almost pure form, or mixed with fir, or more rarely, with white pine or beech. The shrub layer in these forests, often poorly developed, can include juniper (*Juniperus communis*), rowan (*Sorbus aucuparia*), and alpine rose (*Rosa pendulina*) in addition to forest layer saplings. Common elements of the herbaceous plant layer in spruce forests include bilberries (*Vaccinium myrtillus*), great woodrush (*Luzula sylvatica*), *Melampyrum sylvaticum* and others. The moss layer is often well-developed, with *Lephozia lycopodioides* as a typical spruce forest species in this layer. Spruce forests are recognised by the European Union as important for protection, and are assigned the Acidophilous Picea forests of the montane to alpine levels (*Vaccinio-Piceetea*) in the Habitat Directive (Čurović et al. 2011).

Fig. 3.41 Black pine forests—Durmitor (*Photo* G. Roganović)

In mixed deciduous and coniferous forests, a layer of herbaceous plants is commonly present, less frequently of shrubby species, of which junipers (*Juniperus communis*) are widespread, particularly on limestones. The greater presence of junipers is often the result of forest thinning in dry, carbonate rock environments with a shallow soil layer and can also spread over the terrain after wildfires. The dominant presence of junipers is an indicator of strong soil degradation. Various species of herbaceous plants can be found in the ground layer of spruce and mixed forests, but bilberries (*Vaccinium myrtillus*) are particularly significant due to their ability to cover the soil with the branchy above-ground part and protect it from erosion with their root systems.

There are several distinct geographical variants of Bosnian pine forests in Montenegro: Pinetum heldreichii mediterraneo-montanum in the coastal Dinarides and mountain areas around Nikšić, Pinetum heldreichii bertiscum in the Prokletije and Komovi mountains, and Pinetum heldreichii continentale in the Bjelasica and Sinjavina mountains. The Mediterranean-montane Bosnian pine forests are found at altitudes of 1200 to 1800 m a.s.l, while the continental forests occur at altitudes of 1500 to 2000 m a.s.l. Bosnian pine can form pure stands or can be accompanied by Macedonian Pine (*Pinus peuce*) (Fig. 3.42), black pine (*Pinus nigra*), and beech (*Fagus sylvatica*). Macedonian pine forests are found in the Prokletije Mountains area, at an altitude of 1500 to 2000 m a.s.l, and both Bosnian and Macedonian pine forests are protected under the Habitats Directive as High oro-Mediterranean pine forests (Vitali et al. 2019; Čurović and Stijović 2018).

In high mountain areas, vegetation is adapted to harsh climatic conditions, so grass communities prevail, with forest communities much less represented.

Above the beech and coniferous forest belt, mountain pastures can be found (Radonjić et al. 2019). The karst terrain, characterised by pronounced rockiness and very shallow soils, is a dry habitat and therefore, grass vegetation is sparse and incomplete, drying out early in the summer diminishing the erosion protection potential of vegetation. The Calcomelanosols with very fine and fine granular aggregates are particularly vulnerable to wind erosion caused by strong and dry northern wind (Bora). The abundant rainfall in the area exacerbates erosion and runoff from elevated terrain, depositing in widespread depressions in the form of sinkholes and glens. Mountain pastures are the most extensively used agricultural land and extend continuously from the border with Bosnia and Herzegovina to the border with Albania. On plateaus, there is a larger area of levelled terrain, valleys, smaller karst fields, and numerous grass-covered pits known as mountain Rudine (Fig. 3.43).

3.4.5 Role of soil fauna in soil formation

Soil formation begins with detritus resulting from the weathering of the parent material and the process,

Fig. 3.42 Pinus peuce (molika) forest around Hridsko Lake (*Photo* J. Lazarević)

according to Polinov (Vilenski 1946) initiates following the colonisation by the first organisms, lithobionts. Soil living organisms or Edaphon influence the soil by enriching it with organic matter, mixing it with minerals, and forming organo-mineral compounds.

The soil fauna, which includes numerous organisms that inhabit the soil temporarily or permanently, plays an important role in the process of soil genesis. Earthworms (*Lumbricidae*) are abundant pedo-fauna representatives in nearly every region of Montenegro. When feeding, they process organic and mineral matter through their digestive system, creating stable soil aggregates. Earthworms are the best soil tillers, an important contributor to organic matter input in the soil layer and also an indicator of favourable soil conditions, as they thrive in fertile soils.

Rodents (*Rodentia*), such as moles (*Talpidae*), hedgehogs (Erinaceus), lesser blind mole-rats (Nannospalax), hamsters (Cricetus), field mice (Apodemus sylvaticus) and other species (Fig. 3.44) also have an impact on soil formation through digging and burrowing, bringing food into their tunnels, eating plant root systems and some even form mounds from soil pushed on the surface. Such activities contribute to the mixing of soil layers, thus improving water percolation.

Microfauna (*Collembola, Acarina*) are organisms sensitive to chemical contamination and serve as an indicator of soil suitability. They also play a significant role in forming soil aggregates of optimal size and stability.

Microorganisms play a crucial role in soil genesis, as they are the main factor in the transformation of plant remains i.e., in processes of humification and mineralisation, oxidation–reduction, nitrogen fixation and the circulation of matter and energy in nature. In Montenegro, the first surveys of soil biodiversity related to fungi diversity (based on DNA isolation) in high mountain forests were conducted (Lazarević and Menkis 2018; Lazarević et al. 2022; Lazarević et al. 2024).

The activity of soil fauna is optimised when soil temperature, moisture, and aeration are favourable. Specific research activities on the impact of soil fauna on pedogenesis have not been conducted in Montenegro (Fig. 3.45).

Fig. 3.43 Mountainous pastures in the area of Žabljak (*Photo* G. Roganović)

3.5 Anthropogenic Factors' Influence on Soil Formation

The anthropogenic influence is one of the more important factors in soil formation, that is, for its genesis and evolution. Most of Montenegro's territory was inhabited over several thousand years, which is shown by the findings in Crvena Stijena near Petrovići (Nikšić Municipality). In early times, humans were engaged in hunting, crop and livestock farming, and the impact on the soil included cultivation, grazing, and forest clearing to provide arable land (Fig. 3.46). In the Coast and Lake Skadar basin region, the greatest anthropogenic impact on soil resulted from the terracing (Fig. 3.47). Terracing in the area of Montenegro and the broader area of the Adriatic Sea is a long-standing experience among all the nations living on the coast of the Mediterranean Sea. Greek and Roman civilisations covered a large area, from the Adriatic to the Black Sea. At certain points in history, demographic explosions occurred, and wars were waged to conquer new territories and ravage natural and acquired resources. All that disrupted the natural flow of soil genesis and evolution. The ways and intensity of human influence on the soil varied, depending on the activities. Some areas were suitable for agricultural activity; thus, soil cultivation and fertilisation resulted in anthropogenised soil layer. Due to cultivation and faster mineralisation in this surface layer, the content of organic matter was decreasing, while the erosion process was more rapid. Under the forest and preserved grass cover, due to the accumulation of plant residues and its transformation, the share of humus increased, influencing the formation of soil with a good structure, permeability, and other physical, chemical, and biological properties. Intensive logging or overexploitation of the timber for shipbuilding paved the way for faster erosion. Some historical events that took place on the territory of Montenegro indicate that there had been much more forest in the karst area and, therefore, more soil. The Romans and the Venetians had been clearing large areas to get timber to build galleys. Later on, erosion developed in the deforested terrain, causing the occurrence of barren land in the karst area of Montenegro.

The soil could not have been retained in the southern area of Montenegro except in the narrow belt of the Coast with significantly terraced terrain and partly around the Lake Skadar basin. Erosion processes caused the movement of soil particles from the higher ground and deposition in

Fig. 3.44 Pomegranate fruits and a snail (*Photo* P. Đurišić)

the coastal fields. This resulted in the formation of Alluvial and Alluvial-Colluvial Soils, which have undergone certain changes, particularly in recent times, following the implementation of amelioration measures, cultivation, and preparation for intensive agricultural production, but also for other purposes, including urbanisation. In the continental hinterland of Montenegro, erosive losses from mountainous terrain caused the formation of alluvial and colluvial deposits, and accordingly Alluvial and Alluvial–Colluvial Soils mainly used for agricultural production, and with cultivation and other agro-technical measures, these soils were anthropogenised.

In recent years, human activities have contributed to the formation of anthropogenic soil types with completely or partly altered natural soil properties and characteristics. According to the national soil classification (Škorić et al. 1985), the following soils are identified in the area of Montenegro: Gardening soils (Hortisols) (Fig. 3.48), Deeply ploughed soil for preparation of vineyards and orchards (Rigosols), Landfill soils (Deposols), and wastewater deposits (Flotisols). The changes in these soil types are consistent with the technology that has influenced the formation of a particular soil type. Genesis in some soil types is in the initial development stages, particularly Deposols, formed mainly by depositing materials during the exploitation of coal, ore, and other mineral raw materials. Open-pit mining generates vast quantities of tailings disposed of in designated landfills. Also, Flotisols are generated by disposal—depositing of flotation waste material. Tailings and flotation materials often contain a certain percentage of metal extracted by ore processing. They may also contain certain pollutants, so these substrates must be brought to a more favourable state by recultivation, especially if the areas are intended for plant production. In Rigosols, mixing the horizons or the layers by deep ploughing depths (60–70 cm), created the anthropogenic horizon, designated with the symbol–P. The garden soils, Hortisols, usually have a homogenous anthropogenic horizon. Due to the more abundant application of organic and mineral fertilisers year after year, this horizon has increased humus content, rich in nutrients, and highly biologically active (Fig. 3.49). The effect of anthropogenic influence on the soil genesis and properties can be positive and negative (Fig. 3.50).

3.6 Time (Soil Age) as a Genesis Factor

Time is an important but passive factor in soil genesis. The importance of time is reflected in its impact on the duration of other factors' effect on soil formation. Climate can be taken as an example, as it was different in different geological periods. During the Pleistocene only, four glacial and three interglacial periods were established (Radulović 2003.). There are also opinions that terra rossa is a relict soil, originating in the Tertiary, although it has also been formed in contemporary conditions (Pavićević 1958). Most of the soils in Montenegro were probably formed after the last glaciation, which was very intensive. The Quaternary lasted about 1.6–2.0 million years. The Upper Quaternary, or the Pleistocene, lasted about 1.6 million years, and the Holocene, the current epoch, began about 10,000 years ago. It is well known that during the Holocene, there were many significant changes in climate conditions, relief forming, etc. The melting of glaciers in the area of high mountains and the Prokletije resulted in numerous new relief forms (cirques, waves, lakes, moraines, terraces, and others) and sediments as residues from the highest mountain peaks to the river valleys, karst fields, and other areas, where they were transported and deposited by glaciofluvial watercourses.

Many of the soils formed or that existed before the glaciation, the so-called relict soils—in the territory of Montenegro, mainly Terra Rossa and Terra Fusca—were destroyed by glaciation; they either disappeared or were covered by deposits, which were facilitated by the existence of karst fractures and karst permeability. The appearance of layers of the Eocene and aeolian tuffs were also registered

Fig. 3.45 Soil fauna (*Photo* G. Roganović)

in the deposits in the area of Crvena Stijena near Bileća and from the Würm glaciation period. Loess-like material was found in all the Würm layers in the Red Rock Cave in Trebinje (Basler et al. 1966). Many pedologists pointed to a connection between the ferralitic (Fe, Al) and fersilalitic soils (Fe, Si, Al). Moreover, they placed the Bauxite Terra Rossa (ferralitic soil) into the older decomposition cycle, which took place in the warm climate of the Tertiary (Pelišek 1965; Škorić 1981). Geologists also deem that the bauxite was formed from Terra Rossa during its transformation process, i.e., by desilication (Marić 1965). The present-day views on bauxite genesis push aside the Terra Rossa's theory and instead attach importance to the lateritisation process, which means that bauxites are, above all, resedimented fossil laterites (Bušinski 1966, according to citation by Ćirić and Pamić 1975). Some pedologists considered that Terra Rossa is genetically older than bauxite, i.e., that bauxite is the illuvial horizon of Terra Rossa (Gračanin 1947; Nejgebauer 1958). More recent research shows that terra rossa may have been formed not only from limestone residues but also from the Eocene sediments and aeolian material (Šinkovec 1974; Ćirić et al. 1988). Moreover, these and possibly also other relict soils or paleosols covered with Paleogenic sediments served as a basis for a new substratum that the present soils were formed of. On such substratum, like moraine and glaciofluvial deposits, Rendzina and Brown Eutric Soils were formed. Rendzina Soils are more widespread on moraine deposits, formed on the slopes and in depressions of the hilly mountain area and some karst fields. Larger areas of Rendzina can be found at the footslopes of the Durmitor, Sinjavina, Pivska Planina, below Bioč and Maglić, Maganik, Kamenik, Žijovo, Orjen, Vojnik, and Lovćen. Larger areas of Rendzinas can be found in the Grahovsko and Dragalj fields, in Njeguši, Cetinjsko, Radovče, Kopilje, and other smaller karst fields. Brown Eutric Soil is most widespread on the old river terraces in the Lim Valley from its source from Plavsko Lake to Bijelo Polje, the Tara valley from Veruša to Donja Dobrilovina, then in the valleys of the Morača, Cijevna, and Gornja Zeta, which flows through the Nikšićko Field (Fuštić 2000).

The evolution of the fossil soils was disrupted as they were covered by younger sediments or deposited under the ground through karst fractures. Pedological research revealed that two-layered soils are present in Montenegro, where the deeper layer is actually the result of older genesis

Fig. 3.46 Soil tillage in Žabljak (*Photo* G. Roganović)

Fig. 3.47 Land terracing for vineyards—Zagarač area (*Photo* G. Roganović)

Fig. 3.48 Intensive vegetable production (*Photo* G. Roganović)

Fig. 3.49 Setup of an experiment—experimental field of the Biotechnical Faculty (*Photo* M. Knežević)

Fig. 3.50 Intensive urbanisation in the area of the Nikšićko Field (*Photo* G. Roganović)

or old soil covered by colluvial or alluvial deposits and other recent formations, which can be caused by erosion and gravity.

According to the time of formation, soils can be recent, such as soils formed in contemporary conditions, under the influence of contemporary pedogenic factors. Soils formed in earlier periods were changing under the influence of the climate. Such soils are called Fossil Soils, as they are covered with new sediments, which disrupt their further development. Primordial Soils, which remained on the surface and over time were exposed to the effects of contemporary pedogenetic factors and processes, belong to relict soils.

The age of the soil is important for acquiring information about the degree of development, that is, its evolution. The properties of such soil can be explained based on the processes that took place in soil under weather conditions different from those of today.

References

Antonijević R, Pavić A, Karović J, Dimitrijević M, Račić R, Pejović D, Pantić S, Roksandić M (1969) Osnovna geološka karta 1:100.000. Tumač za listove Kotor i Budva. Zavod za geol. i geofizička istraž. Beograd

Antonijević R, Pavić A, Karović J, Menković Lj (1968) Osnovna geološka karta 1:100.000. Tumač za listove Peć i Kukes, Zavod za geološka i geofizička istraživanja. Beograd

Basler Dj, Malez M, Brunnacker K (1966) Die Rote Hohle (Crvena stijena) beo Bileća (Jugoslavien) Eiszeitaler und Gegenwart, Band 17.

Bešić Z (1948).Geotektonska struktura sjeverne Crne Gore. Glasnik Prirodnjački muzej srpske zemlje. Serija A, knjiga 1, Beograd

Bešić Z (1969) Geologija Crne Gore, knj. 1 i 2. Titograd, 1969 i 1975

Blečić V (1982) Crna Gora – Biljni pokrivač Crne Gore. Encikl Jugosl 2:717–720, Zagreb

Burić M, Fuštić B, Bulajić P (2017) Atlas zemljišta Crne Gore, Crnogorska akademija nauka i umjetnosti (CANU), Podgorica

Burić M, Micev B, Mitrović L (2012) Atlas klime Crne Gore, Crnogorska akademija nauka i umjetnosti, Podgorica

Bušinski GJ (1966) Uspehi izučeniaja genezisa boksitov za poslednjeje desetletije (1955–1966)

Cagliero E., Morresi D, Paradis L, Curovic M, Spalevic V, Marchi N, Meloni F, Bentaleb I, Motta R, Garbarino M, Lingua E, Finsinger W (2021) Legacies of past human activities on one of the largest old-growth forests in the south-east European mountains. Veget Hist Archaeobot. https://doi.org/10.1007/s00334-021-00862-x

Cvijić J (1924) Geomorfologija, knjiga I, Beograd

Cvijić J (1926a) Cirkulacija vode i erozija u karstu, Glasnik SGD, br. 12. Beograd

Cvijić J (1926b) Geomorfologija, knjiga II, Beograd

Černjavski P, Grebenščikov O, Pavlović Z (1949) O vegetaciji i flori Skadarskog područja

Čurović M, Medarević M, Pantić D, Spalević V (2011) Major types of mixed forests of spruce, fir and beech in Montenegro. Austrian J Forest Sci 128(Heft 2):93–111, ISSN: 0379-5292

Čurović M, Stijović A (2018): Structural characteristics of white bark pine (Pinus heldreichii) forests in Montenegro; International Conference: Sustainable development of mountain areas—experiences, challenges and perspectives, Žabljak 14–16 septembar 2016; Book of proceedings, CANU, pp 113–120, ISBN: 978-86-7215-445-0

Čurović M, Stijović A, Spalević V, Dudić B, Pajić M (2020) Structural characteristics of the mixed spruce—structural characteristics of the mixed spruce -fir -beech forests on beech forests on Mountain Bjelasica in Montenegro. Notulae Botanicae Horti Agrobotanici Cluj-Napoca 48(3). https://doi.org/10.15835/nbha48311992

Ćirić M, Burlica Č, Martinović J (1988) Geneza i svojstva zemljišnog pokrivača na Karstu. Plenarni referati i izvodi osmog kongresa JDPZ, Cetinje

Ćirić M, Pamić J (1975) Reliktna feralitna zemljišta u okolini Vlasnice. Zemljište i biljka 1–2, Beograd

Fuštić B, Đuretić G (2000) Zemljišta Crne Gore. Univerzitet Crne Gore, Biotehnički institut, Podgorica

Gračanin M (1947) Pedologija, II deo, Zagreb.

Gračanin M (1950) Mjesečni kišni faktori i njihovo značenje u pedološkim istraživanjima, Poljoprivredna znastvena smotra, Sv. 12, p. 51–66, Zagreb

Janković M (1965) Fritillario-Pinetum heldreichii, nova zajednica munike (Pinus heldrelehii) na planini Orjenu iznad Boke kotorske. Arh Biol Nauka (Beograd) 17(3):17p–18p

Janković M (1966) Lauro-Castanetum sativae M. Jank., nova termofilna i eumediteranska zajednica pitomog kestena i lovora u Boki kotorskoj. - Arhiv bioloških nauka 18(1):9p–10p

Janković M, Bogojević R (1965): Robureto-Carpinetum orientalis, nova asocijacija plavnih primorskih šuma kod Ulcinja. Arh Biol Nauka (Beograd) 17(3):15–16

Karaman G and Beeton A (eds) (1981) "The Biota and Limnology of Lake Skadar", University Veljko Vlahović, Institute of Biological and Medicine Research Titograd, Montenegro, Yugoslavia

Karaman V (1997) Flora istočnog dela Bokokotorskog zaliva. Biološki fakultet, Univerzitet u Beogradu, Magistarski rad, pp 1–185

Lakušić R (1987) Šumske zajednice Jugoslavije. Crna Gora. – Šumarska enciklopedija 3:388–395

Lazarević J (2012) Genetic resources in forestry sector of Montenegro, International Conference: Role of research in sustainable development of agriculture and rural areas, May 23–26, 2012, Podgorica, Montenegro- Agriculture & Forestry, Vol 57 (51–71)

Lazarević J, Menkis A (2018) Fungi inhabiting fine roots of Pinus heldreichii in the Montenegrin montane forests. Symbiosis 74:189–197. https://doi.org/10.1007/s13199-017-0504-5

Lazarević J, Topalović A and Menkis A (2022) Patterns of fungal diversity in needles, rootlets and soil of endemic Pinus peuce. Baltic Forestry 28 (1):11–36, article id 58. https://doi.org/10.46490/BF58

Lazarević J, Topalović A, Menkis A (2024) Fungal Diversity in Fire-Affected Pine Forest Soils at the Upper Tree Line. Forests 15, 2012. https://doi.org/10.3390/f15112012

Marić L (1965) Terra Rossa u karstu Jugoslavije. Aauta Geol JAZU Zagreb

Milanović Đ, Caković D, Hadžiablahović S, Vuksanović S, Mačić V, Stešević D, Stanišić-Vujačić M, Biberdžić V, Lakušić D (2021) Priručnik za identifikaciju tipova staništa Crne Gore od značaja za Evropsku uniju sa obrađenim glavnim indikatorskim vrstama. Univerzitet u Banjoj Luci, Agencija za zaštitu životne sredine Crne Gore; Šumarski fakultet

Milošević R, Čurović M, Novaković-Vuković M (2019) Analysis of the floristic composition of mountain beech forests on the territory of Serbia and Montenegro. Fresenius Environ Bull 28(8):5727–5733

Ministry for Spatial Planning and Environment (2008): Spatial Plan of Montenegro until 2020 god. Podgorica

Ministry of Agriculture and Rural Development (MARD), Study of Areas with Natural Constraints in Agriculture and Calculation Design, Final report prepared by OIKON, June 2017

Mirković M (1974) Osnovna geološka karta 1:100000. Tumač za list Gacko. Zavod za geološka istraživanja, Titograd

Mirković M, Kalezić M, Pajović M, Živaljević M, Škuletić D (1968) Osnovna geološka karta 1:100000. Tumač za list Bar i Ulcinj. Zavod za geološka isračivanja Crne Gore, Titograd

Mirković M, Pajović M (1972) Osnovna geološka karta 1:100000. Tumač za list Pljevlja. Zavod za geološka. istraživanja Crne Gore, Titograd.

Nejgebauer V (1958) Prilog problemu crvenice—terra rosse. Zemljište i biljka, 1–3

Odalović A, Prenkić R, Dubak D, Jovančević M, Čizmović M, Radunović Marija (2013) Effect of ecological conditions on expression of biopomological charactersistics of chestnut (Castanea sativa Mill) in natural population of Montenegro, Genetika 45(1):251–260. 0.2298/GENSR1301251O, Beograd

Pavićević N (1956) Buavice na Crnogorskom kršu. Beograd

Pavićević N (1958) Crvenice Crne gore i Hercegovine. Arhiv za poljoprivrerdne nauke, god. XI, sv. 32

Pelišek J (1965) The fossil and recent development of Soils in the Karst region of Czecoslovakia. Problems of the Speleological Research, Prague

Petrović D (2011) Floristička i vegetacijska studija planinskog masiva Rumije. Doktorska disertacija, PMF Studijski program Biologija, Univerzitet Crne Gore

Radonjić D, Djordjević N, Marković B, Marković M, Stešević D, Dajic-Stevanović Z (2019) The effect of phenological phase of dry grazing pasture on fetty acid composition of the cows milk. Chilean J Agricul Res 2(79):278–287. https://doi.org/10.4067/S0718-58392019000200278

Radulović M (2003) Osnovi geologije. Univerzitet Crne Gore, Građevinski fakultet, Podgorica

Radusinović S, Pajović M (2024) Geology of montenegro. In: Barovic G (eds) Speleology of montenegro. Cave and Karst Systems of the World. Springer, Cham. https://doi.org/10.1007/978-3-031-49375-1_3

Spalević V, Barović G, Vujačić D, Čurović M, Behzadfar M, Djurović N, Dudić B, Billi (2020) The impact of land use changes on Soil Erosion in the River Basin of Miocki Potok, Montenegro. Water 12(11):2973. https://doi.org/10.3390/w12112973

Šinkovec B (1974) Porijeklo terra rosse Istre, Geol. Vijesnik 27, Zagreb

Škorić A (1981) Tla Istre. Liburnijske teme, Opatija

Škorić A, Filipovski G, Ćirić M (1985) Klasifikacija zemljišta Jugoslavije. Akademija nauka i umjetnosti Bosne i Hecegovine. Posebna izdanja, urednik: Tihomir Vuković, Sarajevo, knjiga LXXVIII. Odeljenje prirodnih i matematičkih nauka. Međuakademijski odbor za proučavanje zemljišta

Topalović A, Knežević M, Gačnik S, Mikulic-Petkovsek M (2020) Detailed chemical composition of juice from autochthonous pomegranate genotypes (*Punica granatum* L.) grown in different locations in Montenegro. Food Chemistry, 330:127261

Vilenski DG (1946) The role and importance of dokuchaev in soil science. Pochvovedenie No. 6:343–352

Vitali A, Garbarino M, Camarero J, Malandra F, Toromani E, Spalevic V, Čurović M, Urbinati C (2019) Pine recolonization dynamics in Mediterranean human-disturbed treeline ecotones. For Ecol Manage 435:28–37

Vuksanović S (2016) Rasprostranjenje, horološka struktura i centri diverziteta balkanske endemične flore u Crnoj Gori. Biološki fakultet, Univerzitet u Beogradu, Doktorska disertacija

Živaljević M, Pajović M, Djokic V, Škuletić D (1973) Osnovna geološka karta 1:100000. Tumač za list Titograd. Zavod za geološka isračivanja, Titograd

Živaljević M, Vujisić P, Stijović V (1987) Osnovna geološka karta 1:100000. Tumač za list Žabljak. Savezni geološki zavod, Beograd

4 Soil Classification

Abstract

Montenegro inherited the "Genetic school" of pedology, like other Yugoslav Republics, and therefore, genetic principles are introduced into its soil classification system. This chapter briefly describes the intensive and systematic work on the study of soils in the former Yugoslavia and the role of the Yugoslav Soil Science Society, which organised scientific meetings striving to establish better communication, facilitate understanding, and improve cooperation regarding pedological research at the territory of the country. Decisions on soil classification issues were passed at the Congresses of the Soil Science Society of Yugoslavia. This chapter explains the principles of soil classification and its links to FAO and WRB for soil resources. It also gives insight into soil horizon description and designation used in the country and additional symbols. The hierarchical units of the soil classification system (orders, classes, types, subtypes, varieties, and forms), soil classification logic, and differentiation criteria from higher to lower soil categories are explained with an emphasis on soil types as a central soil category. The chapter also presents, in a tabular form, a soil classification system, underlining the soils present in Montenegro and some modifications introduced into the system in the last few decades.

All natural sciences, including pedology, classify different bodies or entities, or organisms, with the objective to better organise the knowledge, facilitate remembering their characteristics, clearly understand the relationships, and make technology transfer and communication easier. Worldwide, soil classification or soil taxonomy was conducted in different periods according to various principles and criteria. In Montenegro, people were living for centuries depending on soil and its products. As such, the soils were sung for centuries in epic and lyric poetry, and the terms of vernacular pedology were developed well, indicating the link of people to soil, land and ground. In those times, the first soil nomenclature and classification of ordinary people included Red Soils—Crvenice, Black Soils—Crnice, Sandy Soils—Pjeskuše, Lithosols—Kamenjari. In the Yugoslav countries belonging to the Austria–Hungarian empire, geological, physical, and chemical classifications started to be used at the end of the nineteenth century, as well as the German system of land evaluation.

After the First World War, with the foundation of the Faculty of Agriculture and Forestry in Belgrade (1919), Kingdome of Serbs, Croats, and Slovenians, the Russian "Genetic school" of soils, whose founder was V.V. Dokuchaev, was accepted in the country. At that time, the Sibircev and Glinka classifications were current achievements that were also used and applied in the research and study of the soils in Serbia and other areas of the formed country. Based on these approaches in soil science, the first results and knowledge about soils appeared, and the findings were reported in the publications of individual authors. Soon, the first soil classification (of national importance) was developed by Aleksandar Stebut (1927), a professor at the Faculty of Agriculture and Forestry in Belgrade. He also published the "Soil Map of the Kingdom of Yugoslavia: Soil Types"', which was published by the Geographical Society in 1931 (Stebut 1931). This work was followed by the classification of Mihovil Gračanin (1951), a professor at the Faculty of Agriculture in Zagreb. A significant contribution to soil knowledge in the territory of the former Yugoslavia was the appearance of pedology textbooks, i.e., Stebut's Scientific Knowledge about Soils in 1927, Agropedology in 1949, 1950 and 1953 (Stebut 1927, 1949, 1950, 1953), and Gračanin's Pedology Part III—Soil Systematics, textbooks of the University of Zagreb (Gračanin 1951).

The second half of the twentieth century was characterised by intensive and systematic work on studying soils in all the Republics and Autonomous Provinces in the former Yugoslavia. At the beginning of that period, the Yugoslav

M. Knežević et al., *The Soils of Montenegro*, World Soils Book Series,
https://doi.org/10.1007/978-3-031-82919-2_4

Fig. 4.1 Eighth Congress of the Yugoslav Society of Soil Science (JDPZ), Cetinje 1988 (Photo-archive CZM)

Soil Science Society (JDPZ) was founded. The first, foundation meeting, of Yugoslav Soil Science Society took place in Belgrade on November 8, 1953 (Životić and Gajić 2023). Thereafter, the JDPZ organised scientific meetings—congresses, symposiums, etc., striving to establish better communication, facilitate understanding, and improve cooperation regarding pedological research at the territory of the country. At the congresses in Ohrid (1963) (Jugoslovensko društvo za proučavanje zemljišta 1963), Belgrade (1972) (Jugoslovensko društvo za proučavanje zemljišta 1972), Novi Sad (1980) (Jugoslovensko društvo za proučavanje zemljišta 1980), and Cetinje (1988) (Jugoslovensko društvo za proučavanje zemljišta 1988), were brought and made decisions on the soil classification, which should be applied in all Republics and Provinces (Fig. 4.1). The so-called "Ohrid classification" was developed by professors Nejgebauer et al. (1963), and it was based on genetic principles. It summarised new knowledge about soils in the world and Yugoslav pedological practice, so it is also the first national classification that critically united the first two versions of classification from 1927 and 1951 made by A. Stebut and M. Gračanin. Nevertheless, between the congresses organised every fourth year, symposia were organised in order to stimulate consultations and meetings of the Committee for Soil Genesis, Cartography, and Classification, which was an organ of the JDPZ. As a result of continuous work on the study of the soil, a general Yugoslav project was adopted under the title: Pedo-genetic Study of Soils and Creation of a Soil Map on a scale of 1:50,000. On that occasion, a committee was formed composed of representatives of all institutions dealing with soil issues, with the task of monitoring and evaluating published works in all republics and provinces. In order to standardise the criteria, field, and laboratory research manuals were prepared. They were edited by the Yugoslav Society of Soil Science in the following years: Kovačević and Jakšić (1964), Bogdanović et al. (1966), and Filipovski et al. (1967). The manuals prescribed unique standards and criteria, as well as a unique methodology for laboratory analyses, which until then were different from institution to institution.

Very extensive research was controlled, coordinated, and evaluated by the Scientific Committee for Soil Map. This enabled very good communication between institutions, both within the Republics and among them. The formed group for the preparation of the pedological map of Yugoslavia, the publication of numerous scientific works and textbooks, as well as the participation in scientific meetings abroad made it possible to follow the world's achievements and implement them so that our needs in that regard were fulfilled and on a modern level.

The results of the joint continuous work of pedo–genetic study of soils with the creation of a soil map on a scale of 1:50,000 (since 1964) enabled the appearance of two innovative versions of earlier classifications: 1. Nejgebauer et al. (1963), under the title: Classification of the Soils of Yugoslavia, and 2. Filipovski et al. (1964), entitled: Soil Classification in Yugoslavia. The newer versions of the classification are from Škorić et al. (1973): Classification of Soils of Yugoslavia and Škorić et al. (1985): Classification of Soils in Yugoslavia. One of the main achievements of JDPZ is its coordinating, consulting, and supporting role in preparation of the soil map of Yugoslavia (Životić and Gajić 2023).

Today, pedology, neither at the European nor at the world level, has a single soil classification system. Other natural sciences, such as botany, zoology, and mineralogy, have had such a system for a longer time. When it comes to soil, which represents a natural body and is subjected to constant changes, difficulties arise due to the very diverse nature of soil individuals as objects of soil classification. The reason for this might be related to the youth of pedology as an independent natural science. Although there is no single system, there have been attempts to make it, and thus significantly facilitate and help scientists to understand each other, especially in public debates and communication and, therefore, overrun obstacles in terminology. Such attempts are the FAO classification system (FAO–UNESCO, 1971–1981; FAO 1988) and the World Reference Base (WRB) for Soil Resources. The first official version of the WRB classification was published at the 16th World Congress of Soil Science in Montpellier, France, in 1998 (FAO). The second edition was published in 2006 (IUSS Working Group WRB 2006), and the third edition in 2014 (IUSS Working Group WRB 2014). There is also an updated version of the classification of 2014, which was published the year after (IUSS Working Group WRB 2015), and the latest version published in 2022 (IUSS Working Group WRB 2022).

There are a number of classification systems in the world that differ from each other. The differences are based on the choice of criteria that are the basis for the classification. Among the more well-known principles in classification are Western European, Russian, and American. In addition to these, many countries have their own national classification systems with their principles.

In Montenegro, pedological research was sporadic until the middle of the twentieth century. They were performed incidentally during excursion tours of some areas (Zloković 1930; Carikov 1930; Stebut 1934). Their works presented the first knowledge about the soils of Montenegro. At the beginning of the second half of the twentieth century, systematic soil surveys began, first at the plains and on the flat terrain, and then throughout the regions. The soils in the plains required to be reclaimed and improved in order to be used for more intensive agricultural production. At that time, Montenegro did not have its own soil classification system, but the Yugoslav system was used. The integration into this system started in 1965, so in accordance with that, its entire territory was mapped on a scale of 1:50,000. The printing of pedological maps was completed in 1988, and thus, Montenegro received a complete inventory of its soils. The monograph on the Soils of Montenegro rounded off the entire work within the aforementioned general Yugoslav project (Fuštić and Đuretić 2000). In the monograph, among other issues, a more detailed classification of the soils distributed across the territory of Montenegro was given, which was in accordance with the latest version of the soil classification in Yugoslavia (Škorić et al. 1985). The basic principles of the first version of the Yugoslav classification (Nejgebauer et al. 1963) were presented at the Second Congress of the JDPZ in Ohrid, and then announced at the 8th World Congress of Soil Science. Over time, the principles were described and developed in more detail by Škorić et al. (1973, 1985). Thus, the last version of the classification from 1985 contains the following principles:

1. The classification is based on the soil properties and not on the external conditions of soil formation. Soil characteristics that are the basis for classification must be morphologically recognisable or easily measurable. Thus, qualitative indicators of the soil characteristics are used for the identification of soils. Also, the soil horizons are precisely defined.
2. The classification is based on soil morphology and genesis (morpho–genetic approach), and it also takes into consideration soil evolution. That is why soil horizon sequence and characteristics of genetic horizons are taken into account when classifying soils into types and at higher categories. Clustering the soil types into higher hierarchical categories is carried out on a genetic and evolutionary basis.
3. The soil type represents the central (basic) unit of soil classification, so the differentiation into soil types and higher systematic categories is done on the basis of unique criteria. However, for categories lower than the type, there are no unique criteria for all soil types, which is a disadvantage of this system.
4. Classification is organised to be suitable for soil productive and ecological assessment and soil evaluation, as well as for the development of technical soil classifications. Therefore, the authors intended to link soil genesis to overall soil quality and functions and soil utilisation and management.
5. All soil orders include the anthropogenic Soils, and various human impacts to soil formation and genesis.
6. The classification system has an open nature and enables easy harmonisation of new knowledge about soils and their classification.

This classification was applied in Serbia and Montenegro even after the break-up of Yugoslavia, with the fact that changes and additions were presented in 1997 at the 9th Congress of the JDPZ in Novi Sad.

Our national classification system is natural or scientific. It has six taxonomic categories, namely: orders, classes, types, subtypes, varieties, and forms. Taxonomic units are hierarchically linked by means of logical principles.

The modern national classification should also include some changes adopted by WRB (IUSS Working Group

2014). Taking into account the solutions given in it, we could design and propose a special project aimed at more detailed research in the field of soil classification. In this way, our classification would be better correlated with WRB and other classifications. Of course, in line with real possibilities because WRB is a diagnostic system, and ours is genetic-evolutionary.

Earlier, when adopting each version of the national classification, our pedologists considered the proposals of the FAO–UNESCO and International Union of Soil Sciences (IUSS) working groups from 1967 to 1998, when the first official version of the WRB classification was adopted. It was preceded by the decision that the revised FAO–UNESCO Soil Map of the World (FAO–UNESCO Soil Map of the World) will be the basis for the further development of the World Reference Base for Soil Resources (WRB). After eight years of testing, the second version of the WRB appeared in 2006, and further improvements were published in 2014 and 2015.

4.1 Soil Horizons Description and Designation

Following the activities related to soil classification at the world level, the Committee for Genesis, Cartography, and Soil Classification of JDPZ (at this time, the name of the country was already Serbia and Montenegro, and not Yugoslavia), but the old name of the Society remained in that period at Congresses in Novi Sad (1997) (Jugoslovensko društvo za proučavanje zemljišta 1997), Vrnjačka Banja (2001), (Jugoslovensko društvo za proučavanje zemljišta 2001) and Budva (2005) (Društvo za proučavanje zemljišta Srbije i Crne Gore 2005) (Fig. 4.2) considered proposals for improvements of the current classification system (from 1985). The principles of classification remained the same, and there were proposals for certain modifications. Those solutions that correspond to our conditions and basic concepts of classification were adopted. It was done in a similar way in the surrounding countries, the former Yugoslav Republics of Croatia, Bosnia and Herzegovina, and Macedonia, and after 2006 in Serbia, whereas in Montenegro, its soil experts presented their announcements at scientific gatherings organised by other countries in the surrounding area and beyond.

Since 1967, the basis for defining and designating soil horizons in Yugoslavia has been the proposals of the FAO–UNESCO and the International Union of Soil Sciences working groups. Other useful materials for Pedology in Yugoslavia were the legend of the soil map of the world (FAO 1968, 1969), and the version of the American Taxonomy (Soil Survey Staff 1975), as well as the finding in reported in other national classifications.

Using these experiences, with certain modifications, our classification contains the following definitions and designations of master horizons and subhorizons (Table 4.1.):

Layers formed by fluvial, aeolian or colluvial deposition, as in alluvial, aeolian and diluvial sediments, are designated with Roman numerals I, II, III, etc.

The absolute soil depth is the total soil depth from the surface of the ground to the parent material. Soil depth is easier to determine in soils formed on hard rock than in soft or loose parent materials. According to our classification, soils are classified into: very shallow (< 10 cm), shallow (10–30 cm), moderately deep (30–60 cm), deep (60–120), and very deep (> 120 cm) soils. The thickness of individual horizons can be observed through a change in colour or some other soil characteristics.

In the national classification system, the gradual boundary between two horizons, which is thicker than 10 cm, is designated with the symbols of the two horizons between which this gradual boundary is located, e.g., AC, EB, (B)C…

If two or more pedogenic processes occur in one soil horizon, the morphological consequences of which are visible, then such horizon is called a mixed or complex horizon and is designated with a slash between the letters, e.g., A/G or A/E.

Designations, descriptions, and definitions of soil horizons are quite aligned with those internationally used. However, in the classification system (A), (B), g, G, T, and P horizons are distinguished, which are not found in the FAO legend. T–horizon is marked as H–horizon in FAO (2006).

There are also additional symbols used to explain the characteristics of soil horizons more precisely—subordinate characteristics. The following symbols are used in the national classification system:

a—hydrogenic form of humus, b—buried horizons, ca—accumulation of calcium-carbonate, cn—concretions of sesqui-oxides and manganese, cs—accumulation of gypsum, fe or s—accumulation of sesqui-oxides, sa—accumulation of salts, v—vertic characteristics, na—increased content of sodium adsorbed on colloidal surfaces, m—cementation of soil horizon, p—ploughed horizons, t—accumulation of clay after migration.

Accordingly, these designations often appear in practice: A_{sa}, A_{vt}, $B_{t,na}$ etc.

Fig. 4.2 Participants of the eleventh congress of JDPZ, Budva 2005 (*Photo-archive* CZM)

4.2 Soil Classification—Systematic Categories

4.2.1 Soil Orders

Soil orders in the national classification system are distinguished based on the character of soil wetting, as well as the origin and chemical composition of the water moisturising the soil.

In the classification system, four soil orders are distinguished:

1. Autogenic[1] (Automorphic, terrestrial) Soils: They are characterised by surface wetting with atmospheric water that freely infiltrates and percolates throughout the entire thickness of the soil profile without stagnating on the impermeable horizon for a longer time.
2. Hydrogenic (Hydromorphic) Soils are characterised by temporary or permanent saturation of soil pores with water that is not saline or alkalised. Because of this, reduction processes are present in the part of the soil profile or over the entire thickness of the profile, and gleysation or pseudogleysation of the soil is observed, with clearly expressed signs of these processes. Soil water saturation can occur due to stagnation of atmospheric precipitation or a rise in the level of groundwater.
3. Halogenic (Halomorphic) Soils are characterised by supplementary wetting by mineralised groundwater and less often by surface saline and alkalised waters.
4. Subaquatic (Subhydric) Soils are formed in underwater environments of shallow, stagnant waters, such as in marshes on the shores of lakes and coastal areas of the sea, and then ponds in some depressions. The processes of pedogenesis and sedimentation alternate in these soils.

[1] According to the latest version of Yugoslav soil classification system (Ćirić et al., 1985) the names of the orders are automorphic, hydromorphic, halomorphic, and subaqueous. In Montenegro, the older names of the orders such as autogenic, hydrogenic, halogenic and subaqautic are still in use and they will be used in further text.

Table 4.1 Designations and definitions of master horizons and subhorizons

Horizon or subhorizon designation	Definition of the horizon or subhorizon
O	Organic surface horizon rich in organic matter having varying degrees of decomposition that overlays the mineral part of the soil and was created in predominantly aerobic conditions. Different subhorizons are designated depending on the degree of the decomposition of organic matter:
O_l	Organic horizon of litter (l stands from litter) represents the accumulation of organic material (litter) with clearly visible parts of plants that permeate the mycelium of fungi. This subhorizon is particularly distributed in forests—forests floors.
O_f	Organic subhorizon of semi-decomposed organic remains (f stands for fermentation). Macroscopically, it is difficult to recognise the primary appearance or the structure of the plant residues (raw humus—mohr humus).
O_h	Organic subhorizon with decomposed (h stands for humified) organic remains. In it, the structure of plant remains is not discernible, rather, there is a slight mixture of organic and mineral part of the soil in the contact zone.
$O_{f,h}$	Organic subhorizon presenting a transition (mixture) of O_f and O_h subhorizons, and denotes transitional humus (moder humus).
(A)	Initial soil horizon of poorly developed soils that represents a zone where the root system of plants is actively developing, but the processes of humification and soil structuring are weakly expressed or completely absent, so its colour is slightly different from the unconsolidated rock from which it was formed.
A	Humus-accumulative horizon in which organic matter is humified and mixed with the mineral part of the soil in the form of organo-mineral complexes, i.e., structural aggregates. It appears in several different forms (modifications):
A_{mo}	Mollic horizon (from Latin mollis—mild, soft) is a modification of humus–accumulative horizon with different thicknesses depending on whether it was formed on hard or loose parent materials, having dark colour, well-developed structure, and base saturation higher than 50%.
A_{um}	Umbric horizon (from Latin umbra—shadow) is similar in colour and thickness to the mollic horizon, but base saturation is less than 50%, soil structure is less pronounced, and when dry, it is harder.
A_{oh}	Ochric horizon (from Greek ochros—pale) is shallower than the previous modifications of A–horizon, lighter in colour, with a weakly developed structure, and might be hard and compact when dry.
E	Eluvial horizon represents the zone from which clay, humus and/or sesquioxides are migrated by water. It is located between O or A, and B–horizons. It is lighter in colour than the horizons between which it is located and usually has a lighter texture.
B	Illuvial horizon underlies the eluvial horizon, and this is the zone of the accumulation of matter removed from the eluvial horizon, i.e., clay, humus, and/or sesquioxide. It appears in three modifications:
B_t	Argiluvic B–horizon (from Latin argilla—clay and luvi—to leach) represents the accumulation zone of clay particles migrated from the E–horizon, and it underlies E–horizon.
B_h	Humus–illuvial B–horizon lies below E–horizon and represents the zone of humus accumulated after migration from the O, A/E and E–horizons. It has black-brown colour and sandy texture, sometimes with grains of ornstein.
B_{fe}	Iron–illuvial or ferro–illuvial B–horizon underlies E or B_h–horizons and represents the zone where sesquioxides are mainly accumulated in acid conditions. It has a rusty or reddish-brown colour with coatings of amorphous Fe and Al–sesquioxides.
(B)	Cambic horizon is the horizon of metamorphosis in situ. It is located between O or A, and C or R–horizons and is formed at the expense of their changes. It differs from the horizons between which it is located by its colour, structure, and texture. It occurs in the form of two modifications depending on the substrates on which it is formed.
$(B)_v$	Modification of cambic horizon resulting from oxidative and hydrolytic weathering of primary minerals and argillo–synthesis in situ (clay synthesis, argillo-genesis), which takes place in soils formed on siliceous and soft calcareous parent materials.
$(B)_{rz}$	Residual cambic horizon is mainly formed by residual accumulation (clay-accumulation) during the chemical dissolution of limestones and dolomites.
C	A loose part of the parent material or unconsolidated rocks in which there are no signs of changes under the influence of pedogenic processes or they are slightly presented.
R	Hard, compacted bedrock.
G	Gleyic (gley) horizon has hydrogenic signs of reduction and secondary oxidation in permanent or temporary anaerobic conditions. The colour is bluish-grey or greenish-grey, with rusty mottles on the surface of the structural aggregates. It has two subhorizons:
G_o or G_{so}	secondary oxidised subhorizon of gleyic horizon located in the zone of the fluctuation of the groundwater, characterised by alterations in oxidising and reducing conditions. It is characterised by the appearance of mottles on the surface of structural aggregates.

(continued)

Table 4.1 (continued)

Horizon or subhorizon designation	Definition of the horizon or subhorizon
G_r	Gleyic subhorizon of reduction is part of the gleyic horizon in which reduction processes constantly take place because the horizon is constantly saturated with water. It is characterised by the appearance of greenish-grey and bluish-grey microzones.
g	Pseudogleyic (pseudogley) horizon is formed by stagnation of surface waters in the soil profile due to the presence of impermeable layers or horizons. The wet, moist, and dry phases alternate during the year, and this causes the appearance of rusty mottles and whitish-grey spots, i.e., marbling or marmoring of the soil in dry conditions.
T	Peat horizon represents the horizon of accumulation of organic matter in anaerobic conditions. Depending on the degree of organic matter transformation, it has three subhorizons:
Tfi	Fibric (fibris lat.—tissue) subhorizon consists of fibrous, weakly humified organic matter (<1/3 is humified).
T_{he}	Hemic subhorizon (hemisus, Greek—half) of the peat horizon where the organic matter is moderately decomposed (1/3–2/3 of organic matter).
T_s	Sapric subhorizon (sapros Greek-rotten) of peat horizon where more than 2/3 of the organic matter is humified.
P	Anthropogenic horizon is created by ploughing, mixing, and homogenisation of several horizons or layers. It is created after deep ploughing during which two or more natural horizons or layers are mixed.

4.2.2 Soil Classes

The differentiation of soils into the soil classes was conducted based on the soil horizon sequence, that is, the order and appearance of soil horizons, without taking into account the genesis or characteristics of the soils. Therefore, all soils belonging to one soil class have the same soil horizon sequence, regardless of the differences in their properties and main pedogenic processes that occur. The soil classes within the soil orders are given below.

A. Classes in the Order of Autogenic Soils

1. Class of Undeveloped or Poorly Developed Soils with soil horizon sequence (A)–C. Undeveloped or poorly developed soils are in the initial stage of development. They can have an (A) or A_p horizon that is formed on different parent materials, except for recent fluvial, lacustrine, or maritime deposits.
2. Class of Humus–Accumulative Soils with soil horizon sequence A–C. This class includes all types of soils with well-developed A or AC–horizons, which overlie hard rock or unconsolidated parent materials.
3. Class of Cambic Soils with soil horizon sequence A–(B)–C. This class is characterised by cambic (brown) horizon, which is located between A or O–horizons and C or R–horizons.
4. Class of Eluvial–Illuvial Soils with soil horizon sequence A–E–B–C. This class of soils is characterised by the presence of eluvial–E and illuvial–B–horizons.
5. Class of Anthropogenic Soils with P–C soil horizon sequence. This class is characterised by the presence of the anthropogenic P–horizon, which was formed under the stronger impact of human activities.
6. Class of Technogenic Soils in which we distinguish layers of sedimentation with sequences including the layers of sedimentation—I, II, III. These soils are created by the deposition of various overburden or mine tailings materials and other types of waste materials.

B. Classes in the Order of Hydrogenic Soils

1. Class of Pseudogleyic Soils with soil horizon sequences A–E/g–B/g–C or also used A–Eg–Bg–C or A–g_1–g_2–C. This class of soils is characterised by the appearance of the pseudogleyic g–horizon, i.e., by the temporary presence of stagnant water at some depth in the soil profile. The "upper groundwater" stagnates in the profile above the clayey horizon or layer because it cannot percolate.
2. Class of Undeveloped Hydrogenic Soils with soil horizon sequence (A)–C or (A)–G. Recent fluvial, lacustrine and maritime deposits belong to this class of soils. The processes of pedogenesis are weakly expressed due to the youth of the deposits with possible formation of (A) or Ap–horizons, but also due to the predominance of sedimentation, which returns pedogenesis to the initial stage.
3. Class of Semigleyic Soils with soil horizon sequence A–C–G. These soils have a humus-accumulative horizon that overlies loose parent material into which groundwater penetrates and forms a gleyic horizon–G at depths below 100 cm. These soils are formed in river valleys.
4. Class of Gleyic Soils with soil horizon sequence A–G. In this class of soils, gleyic horizon is present at depths lower than 100 cm.
5. Class of Peat Soils with soil horizon sequence T–G. This class of soils is characterised by the presence of a peat horizon–T, which is more than 30 cm thick. Peat soils are hydrogenic organic soils.

6. Anthropogenic Hydrogenic Soils with soil horizon sequence P–C–G, or P–G. This class of soils includes those soils on which hydro-meliorative, hydro-reclamation measures have been applied, e.g., groundwater level has been lowered by drainage systems, and the course of soil genesis and soil horizon sequence has been changed.

C. Classes in the Order of Halogenic Soils

1. Class of Acutely Saline Soils with soil horizon sequence Asa–G or Asa–CG. This class of soils is characterised by the accumulation of salts in the soil profile, which most often originate from saline groundwater.
2. Class of Solonetz Soils with soil horizon sequence A–$B_{t,na}$–C. This class of soils is characterised by the presence of argilluvic–sodic illuvial horizon, which is formed by the accumulation of clay migrated from the A-horizon, and contains high sodium content on colloidal surfaces.

D. Classes in the order of Subaquatic (Subhydric) Soils

1. Class of Undeveloped Subhydric Soils with soil horizon sequence (A)–C or (A)–G. This class of soils includes the initial stages of soil formation on clastic sediments under water.
2. Class of Developed Subhydric Soils with soil horizon sequence A–C or A–G. This class of soils includes underwater soils that possess humus–accumulative horizons.

4.2.3 Soil Types and Lower Systematic Categories

In the National soil classification system, the central (basic) unit of soil taxonomy is soil type. The soil types are characteried by: (1) uniform soil horizon sequence, (2) uniform basic processes of transformation and migration of organic and mineral substances, and (3) similar physical and chemical characteristics of individual horizons. Lower categories than soil type are: soil subtypes, soil varieties, and soil forms. Differentiation of lower categories is not conducted on the basis of unique criteria but on the basis of properties that determine the variability between soil types. The criteria for further classification are diverse both in terms of the genesis and soil characteristics, as well as in terms of productive capacity. In soil genetic classification systems, dilemmas might sometimes arise as to which soil type some soil unit belongs to because there are no distinct quantitative indicators used to differentiate between soil units or soil processes (Životić et al. 2022).

The names of certain types of soils in national classification correspond to international names, whereas for soil types that do not appear in the FAO legend, the national name of the soil is given as well as the international name based on Greek-Latin origin of words, e.g., Calcomelanosols (Black Soils formed on limestone rocks).

Soil classification categories used in Montenegro and the criteria for their distinguishment are presented in Table 4.2.

Soil classification (Škorić et al. 1985) up to the level of soil variety, is presented in Table 4.3, with additional marks for the soils that are distributed in Montenegro and modification proposed by Fuštić (2000). Soils distributed in Montenegro are bolded in Table 4.3.

Table 4.3 presents Yugoslav soil classification recently modified for Montenegrin conditions by Budimir Fuštić, who was a leading expert in the preparation of the soil map of Montenegro. Proposed changes in soil classification are explained in the footnotes on the above pages. All the soil units mapped in Montenegro are presented as bolded, whereas the remaining soil units are those that occur in the other parts of former Yugoslavia or occasionally are also distributed in Montenegro but not as mapped soil units.

In the order of Autogenic Soils, the class of Eluvial–Illuvial Soils is not presented as a mapping unit in Montenegro. Luvisols, Podzols, and Brunipodzols are identified only in smaller areas. The main reason for this is the effect of topographic conditions, which do not encourage the migration of mineral and organic matter throughout the thickness of the soil profile, but rather encourage surface runoff and surface migration. In the Order of Autogenic Soils, Chernozems, and Vertisols are not distributed in Montenegro. The formation of Chernozems in former Yugoslavia is related to typical bio-climatic conditions and mainly loess and loess-like sediments, neither of which are present in Montenegro. According to National classification, Vertisols are formed on parent materials rich in montmorillonite clay minerals like tertiary lacustrine sediments, i.e., clays, claystones, marls, marly clays, or argillaceous marls, or are formed on regoliths of basic and intermediate igneous, or metamorphic rocks, such as andesite, gabro, argilloschists, or basalt. Vertisols occur rarely in the conditions of Montenegro, but the appearances of vertic properties are found on Eutric Cambisols in Bjelopavlićka Plain.

Soils from the Order of Halogenic Soils are not present in Montenegro as mapped units. The salinisation processes are present in the coastal belt of the Montenegrin coast under the impact of saline sea water, their evaporation and filtration through the layers of fluvial and maritime sediments. During the duration of high tides, saline sea water is mixed with the water from the Bojana River and consequently filtrated throughout the coastal sandy layers of the river. Saline groundwater temporarily riches the

Table 4.2 Systematic units (categories) of soil taxonomy in Yugoslavia (Škorić et al. 1985)

Soil systematic units	Criteria of the definition of systematic units
Soil orders	Based on the character of soil wetting and the chemical composition of the water, there are four soil orders: Autogenic, Hydrogenic, Halogenic, and Subaqueous Soils
Soil class	Group of soils with uniform soil horizon sequence, which represents a stage of soil evolution
Soil types	Uniform soil horizon sequence, uniform basic pedogenic processes of transformation and migration of mineral and organic matter, similar physical and chemical characteristics of genetic horizons
Subtypes, varieties, and forms	Criteria for systematic units are different for each soil type

Table 4.3 Soil taxonomy of Montenegro (Škorić et al. 1985) modified by Fustic (2000)

A. Order of Autogenic Soils		
Soil type	Soil subtype	Soil variety
Class 1— Weakly Developed Soils: (A)—R or (A)—C		
1. **Lithosols–Litosol (Kamenjar)**	**Based on parent material:** 1.1. **Acid rocks** 1.2. **Intermediate or basic rocks** 1.3. Peridotite 1.4. **Limestones and dolomites**	–
2. **Regosols (Sirozems)–Sirozem (Regosol)**	Based on parent material:	Base saturation/Parent material/Soil depth:
	2.1. **Siliceous**	2.1.1. **Dystric** 2.1.2. **Eutric**
	2.2. **Siliceous–Calcaric**	2.2.1. Loess and loess-like sediments 2.2.2. Marls, marly clays, argillceous limestones 2.2.3. **Calcareous sandstones**
	2.3. **Sandy-Dolomitic**	2.3.1. **Shallow (< 40 cm)** 2.3.2. **Deep (> 40 cm)**
3. **Arenosols (Sandy Soils, Eolian Sands)–Arenosol (Eolski živi pijeskovi)**	Based on parent material: 3.1. Quartz sand (>95% of Quartz) 3.2. Siliceous 3.3. **Siliceous–Calcareous (< 95% SiO_2+CaCO_3)**	Origin of sand (3.1.–3.3.): 1. **Continental** 2. **Maritime**
4. **Colluvial Soils (Colluvium)–Koluvijalna zemljišta (Koluviumi)**	Characteristics of deposited material: 4.1. **Eutric–Siliceous** 4.2. **Dystric–Siliceous** 4.3. **Calcareous** 4.4. Deposited material with buried soils	Variable criteria for varieties (4.1.–4.4.) 1. **Regolithic**[2] 2. **Regolitic, gleyic** 3. **Earthy,**[3] 4. **Earthy gleyic** 5. **Alluvial–Colluvial** 6. Alluvial–Colluvial, gleyic
Class 2—Humus–Accumulative soils: A–C, A–R or A–C–R		
1. Calcomelansols (Black Soils on limestones and dolomites)–Kalkomelanosol (Krečnjačko-dolomitna crnica)	1.1. **Organogenic** 1.2. **Organo–mineral** 1.3. **Cambic**[5] **(Brownised)** 1.4. **Cambic (Rhodic or Rubified–thin red cambic horizon)**[6]	Soil varieties: 1.1.1 **Lithic**[4] 1.1.2 **Skeletal, Colluvial (Skeletic, Colluvic)** 1.2.1. **Lithic** 1.2.2. **Colluvial (Colluvic)**
2. Rendzinas (Rendzina Soils)–Rendzina	Based on parent material: 2.1 Marls, argilliceous and soft limestones 2.2 Loess and loesslike sediments 2.3 **Glacial (moraine) deposits** 2.4 **Dolomitic regolith** 2.5 **Calcareous sand** 2.6 **Calcareous gravel**	Soil varieties (2.1–2.6): 1. **Calcareous** 2. Leached carbonates 3. **Cambic (Brownised)** 4. **Colluvial (Colluvic)**

[2] Pre-dominance of regolithic material over earthy material.

[3] Pre-dominance of earthy material over regolithic material.

[4] Soil formed on hard rocks–lithic contact, soil depth is lower than 15 cm.

[5] Does not have lower systematic units.

[6] Same as above.

Table 4.3 (continued)

A. Order of Autogenic Soils		
Soil type	Soil subtype	Soil variety
3. Rankers (Humus-Siliceous Soils)–Ranker (Humusno-silikatno zemljište)	Based on base saturation: 3.1. **Eutric**	Contact with parent material and brownisation process: 3.1.1. **Lithic** 3.1.2. **Regolithic** 3.1.3. **Cambic (Brownised), Lithic** 3.1.4. **Cambic (Brownised), Regolithic** 3.1.5. Colluvial (Colluvic)
	3.2. **Dystric**	Lithic contact and brownisation/ podzolisation processes: 3.2.1. **Lithic** 3.2.2. **Regolithic** 3.2.3. **Cambic (Brownised), Lithic** 3.2.4. **Cambic (Brownised), Regolithic** 3.2.5. Podzolised, Lithic 3.2.6. Podzolised, Regolithic 3.2.7. Colluvial
4. Chernozems—Černozem	Based on parent material: 4.1 Loess and loess-like sediments	4.1.1. Calcareous (Calcaric) 4.1.2. Leachedcarbonates 4.1.3. Cambic (Brownised) 4.1.4. Calcareous (Calcaric), Gleyic 4.1.5. Leached carbonates, Gleyic 4.1.6. Cambic (Brownised), Gleyic 4.1.7. Salinised (Salic) and Alkalised (Sodic)
	4.2 Calcareous aeolian sand 4.3 Fluvial deposits	(4.2–4.3) 1. Calcareous (Calcaric) 2. Leached carbonates 3. Calcareous (Calcaric), Gleyic 4. Leached carbonates, Gleyic
5. Vertisols (Smonitza)—Smonica	5.1 Calcareous (calcaric)	Based on parent material: 5.1.1. Marls 5.1.2. Clayey (Clay) sediments
	5.1 Non-calcareous 5.2 Cambic (Brownised)	(5.2.–5.3.) 1. Clayey (Clay) sediments 2. Basic and ultrabasic rocks
Class 3: Cambic Soils: A–(B)–C or A–(B)–R soil horizon sequence		
1. **Eutric Cambisols (Eutric Brown Soils)–Eutrični kambisoli (Eutrično-smeđe zemljište)**	Based on parent material: 1.1 Loess and loess-like sediments	Criteria for varieties depend on subtypes 1.1.1 Typical (Haplic) 1.1.2 Illimerised (Luvic) 1.1.3 Typical (Haplic), Gleyic 1.1.4 Illimerised (Luvic), Gleyic
	1.2 Basic and intermediate igneous rocks	1.2.1 Lithic 1.2.2 Regolithic 1.2.3 Vertic
	1.3 Peridotite and serpentinite	1.3.1 Lithic 1.3.2 Vertic
	1.4 **Lacustrine sediments**	1.4.1 **Typical (Haplic)** 1.4.2 **Illimerised (Luvic)** 1.4.3 **Vertic** 1.4.4**Pseudogleyic**
	1.5 **Alluvial, colluvial and aeolian deposits**	1.5.1 **Typical (Haplic)** 1.5.2. **Illimerised (Luvic)** 1.5.3 **Vertic** 1.5.4 Typical, Gleyic 1.5.5 Illimerised (Luvic), Gleyic 15.6Vertic Gleyic

Table 4.3 (continued)

A. Order of Autogenic Soils		
Soil type	Soil subtype	Soil variety
	1.6 Claystone	1.6.1. Typical (Haplic) 1.6.2. Illimerised (Luvic) 1.6.3Vertic
	1.7 Amphibolitic schists	1.7.1Lithic 1.7.2 Regolithic 1.7.3 Illimerised (Luvic)
	1.8 **Flysch Sediments**[7]	1.8.1 **Typical (Haplic)** 1.8.2 **Anthorpogenised**
2. Dystric Cambisols (Dystric Brown Soils)–Distrični kambisol (Distrično smeđe zemljište)	2.1. **Typical (Haplic)** 2.2. **Humic** 2.3. Illimerised (Luvic) 2.4. Pseudogleyic	Based on parent material (2.1.–2.4.) 1. Breccia 2. **Sandstones** 3. Claystones 4. Clays 5. **Sandstones** 6. **Cherts** 7. **Igneous rocks** 8. **Intermediate and basic igneous rocks** 9. **Crystalline schists**
	2.5. Podzolised	2.5.1. Acid siliceous rocks 2.5.2. Substrates rich in Quartz
3. **Calcocambisols (Brown Soils on limestones and dolomites)–Kalkokambisol (Smeđe zemljište na krečnjaku i dolomitu)**	3.1. **Typical (Haplic)** 3.2. **Illimerised (Luvic)**	Depth criteria (3.1.–3.2.): 1. **Shallow (<35 cm)** 2. **Moderately deep (35–50 cm)** 3. **Deep (>50 cm)**
4. **Terra Rossa (Red Soils on limestones and dolomites)—Crvenica (Crvenica na krečnjaku i dolomitu)**	4.1. **Typical (Haplic)** 4.2. **Illimerised (Luvic)** 4.3. Feralithic[8] 4.4. **Colluvial**[9]	Depth criteria (4.1.–4.3.): 1. **Shallow (< 40 cm)** 2. **Moderately deep (40–70 cm)** 3. **Deep (>70 cm)**
Class 4—Illimerised (Luvic) Soils: A–E–B–C or A–E–B–R		
1. Ilimerised Soils (Luvisols)–Ilimerizovano zemljište (Lesivirano zemljište, Luvisol)	1.1 Siliceous and Siliceous–Calcareous substrates	Typical (Haplic) Pseudogleyic Typical, Gleyic Pseudogleyic, Gleyic Podzolised
	1.2 Limestones and dolomites	Typical (Haplic) Acric
2. Podzols–Podzol	2.1 Ferrous 2.2 Humus–Ferrous	Thickness of eluvial horizon: 1. Weak Podzols (E–horizon < 10 cm) 2. Moderate Podzols (10–20 cm) 3. Strong Podzols (>20 cm)
3. Brunipodzols-Brunipodzol	Based on parent material: 3.1 Quartz sandstones 3.2 Quarzites 3.3 Cherts 3.4 Acid igneous rocks 3.5 Phyllite 3.6 Sands	Contact with parent material (3.1.–3.6.): 1. Lithic 2. Regolithic
Class 5—Anthropogenic Soils: P–C soil horizon sequence		
1. **Rigosols–Rigosol (Rigolovano zemljište)**	1.1. Vitisols (Vineyard Soils) 1.2. Soils of intensive orchards 1.3. Deeply–Ploghed Arable Soils (Arisols)	
2. **Hortisols (Vegetable Garden Soils)–Hortisol (Vrtno zemljište)**	–	–

(continued)

[7] This subtype does not exist in ex-Yugoslavia classification system.
[8] The term does not completely correspond to the process of feralithisation.
[9] This subtype does not exist in ex-Yugoslavia classification system.

Table 4.3 (continued)

A. Order of Autogenic Soils		
Soil type	Soil subtype	Soil variety
Class 6—Technogenic Soils: I–II–III soil horizon sequence		
1. **Deposols (Landfill Soils, Soils of Disposal Sites)–Deposoli**	Based on type of material	–
2. Rekultisols[10]		
3. Flotisols (Flotation deposits)–Flotisol	Based on the type of material and thickness	-
4. Aeroprecipitate Soils–Aeroprecipitati	Based on the type of material and quantity	
B Order of Hydrogenic Soils		
Class 1—Pseudogleyic Soils: A–Eg–Bg–C, A–E/g–B/g–C, A–g_1–g_2–C		
1. **Pseudogleys (Pseudogley Soils)—Pseudoglej**	1.1. **Pseudogleys formed on flat terrain** 1.2. Pseudogleys formed on undulated terrain	Based on the depth of impermeable layer (1.1.–1.2.): 1. **Shallow (<25 cm)** 2. **Moderately deep (25–50 cm)** 3. Deep (50–70 cm)
Class 2—Weakly Developed Soils: (A)–C or (A)–G		
1. **Fluvisols (Alluvial or Fluviatile Soils)–Fluvisol (Aluvijalna ili fluvijativna zemljišta)**	1.1. **Calcareous (Calcaric)** 1.2. **Calcareous (Calcaric), Gleyic** 1.3. Calcareous (Calcaric), Gleyic, Salinised (Salic), Alkalised (Sodic) 1.4. **Non-calcareous** 1.5. **Non-calcareous, Gleyic** 1.6. Non-calcareous, Gleyic, Salinised (Salic) and Alkalised (Sodic)	Appearance of the layer of gravel/depth criteria (1.1.–1.6.) 1. **Shallow (<40 cm),** 2. **Moderately deep (40–80 cm),** 3. **Deep (80–120 cm),** 4. **Very deep (>120 cm),** 5. **With buried soil**
Class 3—Semigleyic Soils: A–C–G		
1. Humofluvisols (Fluviatile Meadow Soils)–Humofluvisol (Fluvijatilno livadsko zemljište)	Depth of the appearance of reducing conditions, G–horizon: 1.1 Shallow gleysation (100–150 cm) 1.2 Moderately deep gleysation (150–200 cm) 1.3 Deep gleysation (>200 cm)	1. Calcareous (Calcaric) 2. Calcareous (Calcaric), Salinised (Salic), Alkalised (Sodic) 3. Non-calcareous 4. Non-calcareous, Salinised (Salic) and Alkalised (Sodic)
Class 4—Gleyic Soils: A–G		
1. Pseudogleys–Gleyic Soils—Pseudoglejno-glejno zemljište		
2. Humogleys—Humoglej (Ritska crnica)	2.1 Calcareous (Calcaric) 2.2 Non-calcareous 2.3 Calcareous (Calcaric), Vertic 2.4 Non-calcareous, Vertic	Level of salinisation (2.1.–2.4.) 1. Weakly salinised (0.25–0.5% of salts) 2. Moderately salinised (0.5–0.7% of salts) 3. Strongly salinised (0.7–1.0% of salts) Level of alkalisation (2.1.–2.4.) 1. Weakly alkalised (7–15% of adsorbed sodium) 2. Moderately alkalised (15–40% of adsorbed sodium) 3. Strongly alkalised (>40% of adsorbed sodium)
3. **Eugleys (Marsh–Gleyic Soils, Mineral–Marsh Soils)–Euglej (Močvarno-glejno zemljište, Močvarno mineralno zemljište)**	Origin of soil wetting: 3.1 Hipogleyic (groundwater) 3.2 Epigleyic (Flooded water) 3.3 Amphigleyic (Both)	For 3.1.–3.3 1. Mineral (<10% of humus content) 2. Humic (10–30% of humus content) 3. Peat–Gleyic (T–horizon, but thinner than 30 cm)
Class 5—Peat Soils: T–G		
1. **Acrohistosols–Akroshistosol (Izdignuto tresetište)**	Thickness of T–horizon: 1.1 Thin (30–50 cm) 1.2 Moderately thick (50–100 cm) 1.3 Thick (>100 cm)	Subtype of T–horizon (1.1.–1.3.): 1. Fibric 2. Hemic 3. Sapric

(continued)

[10]This is an addition to ex–Yugoslavia soil classification system.

Table 4.3 (continued)

Soil type	Soil subtype	Soil variety
A. Order of Autogenic Soils		
2. Acro–Plano–Histosols (Intermediate Peat Soils)–Akroplanohistosol (Prelazno tresetište)	Thickness of T–horizon: 2.1 Thin (30–50 cm) 2.2 Moderately thick (50–100 cm) 2.3 Thick (>100 cm)	Subtype of T–horizon (2.1.–2.3.): 1. Fibric 2. Hemic 3. Sapric
3. **Planohistosols (Lowland Peat Soils)–Planohistosols (Nizijsko tresetište)**	Thickness of T–horizon: 3.1 Thin (30–50 cm) 3.2 Moderately thick (50–100 cm) 3.3 Thick (>100 cm)	Subtype of T–horizon (3.1.–3.3.): 1. Fibric 2. Hemic 3. Sapric
Class 6—Anthropogenic Hydrogenic Soils[11]: P–G or Ap–G soil horizon sequence		
1. Deeply–Ploughed Peat Soils—Rigolovano tresetište		
2. Risosols (Rice-Grown Soils)–Rizosol (Zemljišta rižišta)		
3. **Hydro–Meliorated Soils (Hydro-Reclaimed Soils)–Hidromeliorisano zemljište**		
C. Order of Halogenic Soil		
Class 1. Acutely Salinised Soils or Saline Soils: A_{sa}–G or A_{sa}–CG		
Solonchaks–Solončak	Type of Soils: 1.1 Sodaic 1.2 Sulphatic 1.3 Chloridic 1.4 Sodaic-sulphatic 1.5 Sulphatic-sodaic 1.6 Sodaic-chloridic 1.7 Chloridic-sodaic 1.8 Sulphatic-chloridic 1.9 Chloridic-sulphatic 1.10 Mixture of three different types of salts	Based on the upper depth of saline horizon (1.1.–1.10.) 1. Surface salinisation (up to 30 cm depth) 2. Moderately deep salinisation (30–80 cm) 3. Deep salinisation (80–125 cm)
Class 2. Solonetz class, Alkali Soils, Sodic Soils or Eluvial-Illuvial Alkali Soils: A/E–$B_{t,na}$–C		
1. Solonetz - Solonec	1. Solonetz–Solonchak 2. Typical (Haplic) 3.Mollic 4. Luvic 5. Pseudogleyic	For 1.1.–1.4.: 1. Non-Saline 2. Sodaic 3. Sulphatic-sodaic 4. Chloridic-sodaic 5. Chloridic-sulphatic 6. Sulphatic-chloridic 7. Sulphatic 8. Chloridic For 1.5.: 1. Typical (Haplic) 2. Glossic
D. Order of Subaqueous Soils		
Class 1. Weakly developed subaqueous Soils: (A)–C and (A)–CG soil horizon sequences		
1. Protopedon - Protopedon		
Class 2. Sub-hydric Soils: A–C or A–G soil horizon sequence		
1. Gyttja - Gitja		
2. Dy–Daj		
3. Sapropel–Sapropel		
Class 3. Antropogenised Subaqueous Soils: P–C or P–G		
1. Drained Gyttja–Odvodnjena gitja		
2. Drained Sapropel–Odvodnjeni sapropel		

(continued)

[11] This class is present in the area of drainage reclamation projects, not mapped.

effective part of soil solum, whereas in depressions called "knete" overlies the surface of the terrain, creating small swamps. Groundwater, its level, and annual dynamics are affected by the topographic conditions and flooding waves of the Bojana River, other water courses, and Šasko Lake. Groundwater fluctuation causes alteration of oxidising and reducing conditions, gleysation of the soil profile, followed by salinisation. These processes are found in the area around Ulcinj, where there are the largest areas of Marsh-Gleyic Soils and saline soils, but also in Buljarice and parts of Tivatsko and Mrčevo fields.

In the order of Hydrogenic Soils, soils from the classes of Fluviatile–Meadow Soils and Gleyic Soils, like Humofluvisols and Humogleys, are not mapped. Certainly, they are distributed in smaller areas in the Lake Skadar basin. Also, soils from the order of subaqueous soils are distributed to some extent around the Lake Skadar basin, Plavsko, Šasko, and other lakes, but they were neither investigated nor mapped during the formation of the soil map of Montenegro. Their areas are mainly added to Eugley Soils.

References

Bogdanović M, Škorić A, Sertić V, Velikonja N, Cencelj J, Rac Z, Korunović R, Živković M, Miljković N, Frleta Brnetić I, Juras I, Todorović E, Džamić R, Petrović M, Ivović P, Vajnberger A, Nikodijević V, Filipović Đ, Gavrilović V, Anić J, Čaleta Z, Resulović H, Petrović G (1966) Priručnik za ispitivanje zemljišta, Knjiga I Hemijske metode ipistivanja zemljišta, Jugoslovensko društvo za proučavanje zemljišta, Beograd

Burić M, Fuštić B, Bulajić P (2017) Atlas zemljišta Crne Gore, Crnogorska akademija nauka i umjetnosti (CANU), Podgorica

Carikov D (1930) O zemljištu i ekonomskim prilikama područja Skadarskog jezera u granicama naše države, Glasnik ministarstva poljoprivrede, sv. 31., Beograd

Društvo za proučavanje zemljišta Srbije i Crne Gore (2005) Zemljište kao resurs održivog razvoja (Soil as a Resource of Sustainable Development). XI Kongres društva za proučavanje zemljišta Srbije i Crne Gore, Budva 13–16. septembar 2005. Univerzitet Crne Gore – Biotehnički institut, Centar za zemljište i melioracije, Društvo za proučavanje zemljišta Srbije i Crne Gore (DPZSCG). Plenarni referati i apstrakti, p. 210

FAO (1968) World Soil Resources Reports. Definitions of soil units for the Soil Map of the World. Rome

FAO (1969) World Soil Resources Reports. Supplement to Definitions of soil units for the Soil Map of the World. Rome

FAO (1988) Soil map of the world. Revised legend, by FAO–UNESCO–ISRIC. World Soil Resources Report No. 60. Rome

FAO (1998) World reference base for soil resources, by ISSS–ISRIC–FAO. world soil resources report no. 84. Rome

FAO (2006) Guidelines for soil description, Rome

Filipovski G, Nejgebauer V, Ćirić M, Škorić A, Živković M (1964) Soil Classification in Yugoslavia 8th Inter. Congress of Soil Since, Bucharest. Transaction, Volume V. Publishing House of the Academy of the S.R. of Romania, Bucharest

Filipovski G, Škorić A, Racz Z, Kalinić M, Stepančić, D, Kovačević J, Antonović G, Kodrič M, Resulović H, Burlica Č, Altarac-Manuševa L, Tomašegović Z, Juras I, Kovačević P, Pavšer M, Ćirić M (1967) Priručnik za ispitivanje zemljišta, Knjiga IV Metodika terenskog ispitivanja zemljišta i izrada pedoloških karata, Jugoslovensko društvo za proučavanje zemljišta, Beograd

Fuštić B, Đuretić G (2000) Zemljišta Crne Gore. Univerzitet Crne Gore, Biotehnički institut Podgorica

Gračanin M (1951) Pedologija, III deo, Školska knjiga, Zagreb

IUSS Working Group WRB (2006) World Reference Base for Soil Resources 2006. World Soil Resources Report No. 103, FAO, Rome

IUSS Working Group WRB (2014) World Reference Base for Soil Resources 2014. International Soil Classification System for Naming Soils and Creating Legends for Soil Maps. World Soil Resources Reports No 106, Rome: FAO

IUSS Working Group WRB (2015) World reference base for soil resources 2014, update 2015. International soil classification system for naming soils and creating legends for soil maps. In: World Soil Resources Reports No. 106. FAO, Rome

Jugoslovensko društvo za proučavanje zemljišta (1963) Drugi kongres Jugoslovenskog društva za proučavanje zemljišta (Second Congress of Yugoslav Society of Soil Science); Jelenić Đ. (Ed.). Ohrid, 28.V–3.VI 1963. Jugoslovensko društvo za proučavanje zemljišta, p. 714

Jugoslovensko društvo za proučavanje zemljišta (1972) Četvrti kongres Jugoslovenskog društva za proučavanje zemljišta (Forth Congress of Yugoslav Society of Soil Science), Beograd, 4–12.9.1972. Jugoslovensko društvo za proučavanje zemljišta, Unija bioloških naučnih društava Jugoslavije

Jugoslovensko društvo za proučavanje zemljišta (1980) Šesti kongres Jugoslovenskog društva za proučavanje zemljišta (Sixth Congress of Yugoslav Society of Soil Science), Novi Sad, 7– 14. VI 1980. Jugoslovensko društvo za proučavanje zemljišta, p. 407

Jugoslovensko društvo za proučavanje zemljišta (1988) Plenarni referati i izvodi (Introductional lecture and summary). Jubilarni JDPZ Kongres, Cetinje 23–27. maj 1988. Jugoslovensko društvo za proučavanje zemljišta, p. 292

Jugoslovensko društvo za proučavanje zemljišta (1997) Uređenje, korišćenje i očuvanje zemljišta: radovi devetog kongresa Jugoslovenskog društva za proučavanje zemljišta (Papers of the Ninth Congress of Yugoslav Society of Soil Science): Dragović S. (Ed.)., Novi Sad, 23–27.VI 1997., Jugoslovensko društvo za proučavanje zemljišta, p. 803

Jugoslovensko društvo za proučavanje zemljišta (2001) Deseti kongres Jugoslovenskog društva za proučavanje zemljišta (Tenth Congress of Yugoslav Society of Soil Science), Antonović G. (ed.)., Vrnjačka banja, 22–26.X 2001, Apstrakti. Jugoslovensko društvo za proučavanje zemljišta, p. 145

Kovačević P, Jakšić V (1964) Priručnik za terenska pedološka istraživanja, str. 72, Sarajevo

Nejgebauer V, Filipovski Ǵ, Ćirić M, Živković M (1963) Klasifikacija zemljišta Jugoslavije [Yugoslav soil classification]. II kongres JDPZ, Ohrid

Soil Survey Staff (1975) Soil Taxonomy. A Basic System of Soil Classification for Making and Interpreting Soil Surveys. U.S.D.A. Agricultural Handbook, 436

Stebut A (1927) Nauka o poznavanju zemljišta (Pedologija). Izdanje Ministarstva poljoprivrede i voda, Odeljenje za ratarstvo – knjiga 10, Beograd

Stebut A (1931) Pedološka karta Kraljevine Jugoslavije: pregled tipova zemljišta. Geografsko društvo, Beograd, Štamparija Davidović, Pavlovića i druga, p 15

Stebut A (1934) Klimatske, hidrološke i pedološke osobine istočnog dela Crne Gore
Stebut A (1949) Agropedologija. Prvi deo. Naučna knjiga, Beograd
Stebut A (1950) Agropedologija. Drugi deo. Naučna knjiga, Beograd
Stebut A (1953) Agropedologija. Treći deo, Beograd
Škorić A, Filipovski G, Ćirić M (1973) Klasifikacija tala Jugoslavije. Zavod za pedologiju Poljoprivrednog i Šumarskog fakulteta sveučilišta u Zagrebu, Zagreb
Škorić A, Filipovski G, Ćirić M (1985) Klasifikacija zemljišta Jugoslavije. Akademija nauka i umjetnosti Bosne i Hecegovine. Posebna izdanja, urednik: Tihomir Vuković, Sarajevo, knjiga LXXVIII. Odeljenje prirodnih i matematičkih nauka. Međuakademijski odbor za proučavanje zemljišta
USS Working Group WRB (2022) World reference base for soil resources. International soil classification system for naming soils and creating legends for soil maps, 4th edition. International Union of Soil Sciences (IUSS), Vienna, Austria
Zloković Đ (1930) Pedološka ispitivanja na Nikšićkom polju, Geološki anali Balk. poluostrva, knj. X, sv. 1, Beograd
Životić LJ, Mrvić V, Gajić B, Kaluđerović L (2022) Challenges in the Classification of Colluvial Soils Formed on Limestones in the area of Great Field – Foot Slopes of Mountain Vukan. Symposium: "Soils in the era of precision agriculture and informatin technologies" – Soil AgroIT 2022", 16–17. June 2022, Novi Sad. Book of Abstracts, p 54–55
Životić LJ, Gajić B (2023) The history, activities and future perspectives of the Serbian soil science society. Zemljiste i Biljka 72(1):62–79. https://doi.org/10.5937/ZemBilj2301062Z

Soil Types—Properties and Distribution

5

Abstract

This chapter describes the most distributed soil types in Montenegro from Autogenic and Hydrogenic Soil Orders. It starts with the explanation of soils from the class of Weakly Developed Soils: Lithosols, Regosols, Arenosols, and Colluvial Soils, which cover a smaller portion of Montenegrin territory. After that, it gives a deeper insight into the Class of Humus-Accumulative Soils, which are the most distributed soils in Montenegro. The focus is on Ranker Soils, Rendzina Soils, and Calcomelanosols. Calcomelanosols are the most distributed soils in Montenegro, and therefore, the description of Organogenic, Organo-Mineral, and Brownised soil subtypes is given, as well as the Colluvial Calcomelanosols soil variety. Further, there is the description of soils from the Cambic Soil Class: Eutric and Dystric Cambisols, Calcocambisols, and Terra Rossa Soils, which are the second soil class in Montenegro by means of the territory these soils cover. The soil types of Hydrogenic Soil Order are presented through Fluvisols, Eugley, Pseudogleys, and Histosols. The chapter also deals with the soil types from Anthropogenic Classes of Autogenic Soil Orders: Rigosols, Hortisols, Deposols, and Recultisols. For each soil type, an explanation is given for the soil name, the description of soil genesis and soil horizon sequence, position in an evolutionary series, tentative WRB classification, spatial distribution, soil characteristics, and soil management with utilisation. Some examples of soil physical and chemical characteristics are presented in tables, whereas the spatial distribution of soils is shown on the map of Montenegro. This chapter is illustrated with photographs of soil profiles, landscape, vegetation, and soil features.

The soil cover of Montenegro is very heterogeneous, and it is conditioned by the various constellations of pedogenic factors and processes that affect soil genesis and evolution.

Moderately deep and deep soils prevail in the flatter area of the Coastal zone and Zetsko–Bjelopavlićka Plain, as well as in the valleys of the Morača, Zeta, Tara, Lim, Ibar, and Ćehotina rivers, in the karst fields, and on the plateaus. They are quite diverse among each other, as alluvial soils spatially interfere with other soil types found in river valleys. Very shallow and shallow soils are dominantly represented in the hilly-mountainous region and in the extensive karst area, and this was influenced by the geological composition, climate, relief with the strong altitudinal gradients, soil erosion processes and other factors.

A monograph, Soils of Montenegro (Fuštić and Đuretić 2000) and Atlas of Soil maps 1:50,000 (Fuštić 2000) provide a wealth of relevant data on the diversity of the soil cover in Montenegro. Also, the publication Atlas of the Soils of Montenegro (Burić et al. 2017), printed in Serbian and English, contains a representation of the soils with maps at a scale of 1:50,000, with data characterising the soil types on each of the 16 sheets, including numerous photographs of pedological profiles, processes of soil erosion, application of reclamation measures, etc. The data about soils indicate that soils can be classified in several ways, depending on the criteria taken as the basis for classification. Soils are natural formations, so in accordance with the Russian genetic pedological school (founder V. V. Dokuchaev), at first glance, the soils of Montenegro could be classified as genetic (developed) and agenetic (undeveloped) soils. Genetic soils are more distributed in Montenegro compared with agenetic soils. All soils in Montenegro are classified using a strong genetic emphasis on evaluating soil properties linking them to pedogenic processes and soil-forming factors. Also, the action

M. Knežević et al., *The Soils of Montenegro*, World Soils Book Series,
https://doi.org/10.1007/978-3-031-82919-2_5

of soil-forming factors to produce soil properties in kinds of profiles called "soil types", or genetic soil types, was strongly emphasised (Gerasimov and Ivanova 1959). Therefore, soil classification uses a morphological–genetic–evolutionary approach and scientific achievements in the environment. This classification principle has been applied, elaborated, and improved in our country from the very beginning until today, taking into consideration only local conditions.

The dominant impact of some pedogenic factors in soil formation, as well as the predominance of some processes, caused the appearance of different soils. From this point of view, the soils of Montenegro are divided into lithogenic, topogenic, climatogenic, hydrogenic, organogenic, anthropogenic, and technogenic.

The group of climatogenic soils dominantly comprises Red Soils (Terra Rossa Soils) and Black Soils on limestones and dolomites (Calcomelanosols), whereas conditionally, some other soils like Humus–Siliceous Soils (Rankers), Calcocambisols, Eutric Cambisols, Luvic, and Pseudogleyic Eutric Cambisols can also be climatogenic soils. Pedogenic zonation is noted as moving from the southern Mediterranean zone to the mountainous areas. The zonation is properly observed in the case of Red Soils, which are formed on limestones up to 500–600 m a.s.l, over Calcomelanosols that are distributed at higher altitudes, up to the highest mountain peaks, to Lithosols. All these three soil types are formed on limestones, but the climate affects the formation of a particular soil. Taking into consideration other soils, there is no clearly expressed zonation because the relief and other factors limit the occurrence of zonal soils due to their effects and reduce possible zonality to narrower—local areas.

The complete classification of soils in Montenegro is presented in Table 4.2, and it shows the division of soils into four soil orders: Autogenic, Hydrogenic, Halogenic, and Subaqueous Soils according to the old Yugoslav classification system. Within the soil orders, soils are classified into soil classes, whereas within the soil classes, they are classified into soil types. Soil types are further differentiated into soil subtypes, varieties, and forms. Table 5.1 shows the tentative correlation of soil types in the national and WRB classification. In the monograph Soils of Montenegro (Fuštić and Đuretić 2000), all the soils of Montenegro are described in detail, and among the other knowledge and data on soils in Montenegro, this book utilised their valuable achievements.

5.1 Order of Autogenic Soils

The order of Autogenic Soils occupies the largest area of Montenegro. That is understandable, considering the distinct hilly–mountainous terrain over around 95% of the country's territory. The genesis of Autogenic Soils occurs under conditions of wetting only by atmospheric water, which is partially retained in the soil and partially percolates freely throughout the entire thickness of the soil profile without creating any reducing conditions in the soils. More intense wetting of the soil due to abundant precipitation in the autumn–winter period gradually decreases with the distance from the Mediterranean Sea and into the direction of the continental hinterland. During the summer period, the continental part of the country has a more favourable distribution of precipitation compared with the southern part of Montenegro, where aridity during the summer season is pronounced. In recent years, climate changes have caused an increased frequency of long-term droughts and thus triggered the initial processes of desertification in the southern region of Montenegro. Long-term high temperatures without precipitation or insufficient amounts of rainfall interfere with the growth of plants and even cause them to suffer in moisture–lack conditions, often due to wildfires, occurring as side effects of droughts. Naturally, that is reflected in the soils, biodiversity, and soil development processes.

5.1.1 Class of Weakly Developed Soils [(A)–C or (A)–R Sequence of Horizons]

Soils of the class of Weakly Developed Soils are in the initial stage of soil genesis either because soil genesis is very slow or because soil particles are constantly carried away by water and wind, disabling the formation of a well-developed surface horizon. In some situations, the previously formed soil layers are covered with newly deposited soil material, and soil formation rebegins. Because of this, the soils constantly remain undeveloped or are weakly developed and have an initial horizon—(A), which might contain fragments of hard rocks or a disintegrated part of the substrate (regolith or detritus), as well as plant roots.

5.1.1.1 Lithosols

The name of the soil type. The name Lithosols comes from the word lithos—which means rock, and with this soil, the soil's parent material is the detritus of the rocks (gravel and stones) that prevail on the surface of the terrain. The type of substrate gives the main characteristics to Lithosols.

The WRB corresponding RSG (Reference Soil Group): Nudilithic Leptosols. Nudilithic and Lithic Leptosols form a very widespread soil association in Montenegro.

Soil genesis and soil horizon sequence. Lithosols are the "youngest" member in the evolutionary series of soils formed on hard rocks with the following soil horizon

Table 5.1 Tentative correlation of soil types in the national soil classification, WRB classification and soil taxonomy (Soil Survey Staff, 2022)

Soil type – International name	Soil type – National name	Tentative WRB soil group	Tentative USDA great groups etc.
Lithosols	Litosol or Kamenjar	Nudilithic Leptosols	Lithic subgroups of Entisols and Inceptisols, and/or Dystric subgroups of Entisols: Lithic Torri/Xero/Cryo/Udorthents, Dystric Lithic Udorthents, Lithic Xerepts/Cryepts.
Regosols or Sirozems	Regosol or Sirozem	Calcaric Regosols, Dystric Regosols	Inceptisols: Haplustepts, Dystrustepts, Dystroxerepts, Haploxerepts, Dystrudepts, Eutrudepts. Entisols: Ust/ Udo/ Xerorthents, Dystric Lithic Udorthents, Dystro/ Haploxerepts. Lithic subgroup adjective can be used also. Alfisols: Haplustalfs, Hapludalfs. Ultisols: Hapludults, Haplustults, Haploxerults
Arenosols or Aeolian Sands	Arenosol or Eolski ”živi” pijesak	Arenosols	Entisols: Psamments
Colluvial Soils	Koluvium or Koluvijalno zemljište	Colluvic Regosols, Leptosols (Colluvic)	Cumulic subgroup of Entisols, Inceptisols and/or Mollisols: Orthents, Dystro/Haplocryepts, Humustepts, Humixerepts, Humudepts, Dystrudepts, Humaquepts, Haplocryolls, Hapludolls/Haplustolls/ Haploxerolls
Calcomelanosols or Black Soils formed on Limestones and Dolomites	Kalkomelanosol or Krečnjačko–Dolomitna crnica	Lithic Leptosols, Rendzic Leptosols, Dystric Leptosols, Leptic Phaeozems, Leptic Umbrisols, Cambic Umbrisols	Mollisols with lithic subgroup and cumulic: Rendolls, Cryrendolls, Haprendolls, Argixerolls, Haploxerolls/ustolls/udolls. Inceptisols with lithic subgroup on smaller areas: Humicryepts, Humixerepts, Humudepts, Humustepts.
Rendzinas or Rendzina Soils	Rendzina	Rendzic Leptosols (also Skeletic, Rendzic, Cambic, Calcaric and Eutric). Leptic and Rendzic Phaeozems (also Cambic, Skeletic, Dolomitic, and Calcaric). Calcaric Regosols or Eutric Regosols.	Mollisols with lithic subgroup: Rendolls, Cryrendolls, Haprendolls, Haploxerolls/ustolls, Calcixerolls/ustolls. Inceptisols with lithic and/or calcic subgroup: Humixerepts, Humustepts.
Rankers or Humus–Siliceous Soils, Ranker Soils	Ranker or Humusno–silikatno zemljište	Lithic Leptosols (also Skeletic, Umbric, Mollic, Cambic, Dystric, and Eutric). Leptic Phaeozems (also Cambic or Skeletic). Leptic Umbrisols (also Cambic or Skeletic). Leptosols (Colluvic) or Leptic Phaeozems (Colluvic), or Umbrisols (Colluvic).	Mollisols with lithic subgroup: Haploxerolls/cryolls/ustolls/udolls Inceptisols with lithic subgroup: Humixerepts, Humicryepts, Humixerepts.
Eutric Cambisols or Brown Eutric Soils	Eutrični kambisol or Eutrično smeđe zemljište	Eutric Cambisols (also Leptic, Vertic, Stagnic, Skeletic) and Cambic Phaeozems (also Leptic, Vertic, Luvic, Skeletic or Haplic).	Inceptisols: Haplustepts, Humustepts, Humixerepts, Haploxerepts. Alfisols: Hapludalfs. Mollisols: Argiustolls, Argiudolls.
Dystric Cambisols or Brown Acid Soils, Brown Dystric Soils	Districni kambisol or Distrično smeđe zemljište, Smeđe kiselo zemljište	Dystric Cambisols (also Leptic or Skeletic), Cambic Umbrisols (also Leptic, Luvic, Skeletic or Haplic)	Inceptisols with often lithic formative element: Humudepts, Dystrudepts, Dystrustepts, Humustepts, Haplustepts. Alfisols: Haploudalfs.

Table 5.1 (continued)

Soil type – International name	Soil type – National name	Tentative WRB soil group	Tentative USDA great groups etc.
Calcocambisols or Brown Soils on Limestones and Dolomites	Kalkokambisol or Smeđe ili rudo zemljište na krečnjaku i dolomitu	Cambic Leptic Rendzic Phaeozems (Rhodic or Chromic) (also Vertic and Luvic), Cambic Leptic Umbrisols (Chromic or Rhodic). Leptic Cambisols (also Chromic or Rhodic).	Alfisols: Rhodustalfs, Rhodudalfs, Haploxeralfs or utilisation of rhodic/chromic subgroup formative element, and lithic.
Tera Rossa Soils or Red Soils	Crvenica na krečnjaku i dolomitu	Cambic Leptic Rendzic Phaeozems (Rhodic), Cambic Leptic Umbrisols (Rhodic), Leptic Cambisols (also Rhodic, Eutric or Dystric), Leptic Cambisols (Colluvic)	Alfisols: Rhodoxeralfs, Rhodustalfs. Utilisation of lithic subgroup formative element.
Rigosols or Deep¬ly–Ploughed Soils	Rigosol or Rigolovano zemljište	Terric Anthrosols	Anthraltic subgroup adjective used in definition deeply-ploughed vine-yards and orchard soils.
Hortisols or Vegetable Garden Soils	Hortisol or Baštensko or Vrtno zemljište	Hortic Anthrosols	Inceptisols with haploplaggic adjective.
Deposols or Soils of Disposal Sites, Landfill Soils	Deposoli or Zemljišta deponija	Technosols	These are mainly orders of Inceptisols (former Anthrepts) and Entisols, which have anthropic epipedon and anthropic and anthroportic subgroup adjectives.
Recultisols or Recultivated Technogenic Soils	Rekultisol or Rekultivisana zemljišta	Technosols (Novic)	
Fluvisols or Alluvial Soils	Fluvisol or Aluvijalno zemljište	Calcaric Fluvisols (also Eutric, Dystric, Tidalic, Gleyic or Skeletic)	Entisols: Udifluvents, Fluvaquents.
Eugleys or Marsh-Gleyic Soils	Euglej or Močvarno–glejno zemljište	Tidalic Gleysols (also Folic, Histic, Mollic, Stagnic, Calcaric, Eutric or Dystric)	Entisols with humic subgroup formative element: Fluvaquents, Aquents, Epiaquents, Endoaquents. Inceptisols: Humaquepts, Epiaquepts, Endoaquepts Mollisols: Endoaquolls, Epiaquolls.
Histosols or Peat Soils	Tresetište or Tresetno zemljište	Tidalic Histosols (also Fibric, Hemic or Sapric, Dystric or Eutric).	Histosols
Pseudogleys or Pseudogley Soils	Pseudoglej	Stagnosols and Planosols	Albaqualfs, Albaquults, Argiaquolls.

sequences: (A)–C or (A)–R. (A)–horizon represents the parts of unconsolidated or consolidated rocks, which are permeated by plant roots in the form of nests. The most important parent materials for Lithosols in Montenegro are limestones and dolomites, followed by siliceous and eruptive consolidated (compact) rocks. The depth of the soil is insignificant, from a few to a maximum of 10 cm. Soil is formed in situ, mainly by physical weathering, where the smallest particles are prone to the effect of erosion, taken away by water or carried by the wind, or migrated vertically throughout the cracks between the rocks. Due to unfavourable climatic conditions, strong erosion, and constant covering of terrain with new material — detritus, pedogenesis cannot progress. It is rather interrupted—regressive pedogenesis and restarts repeatedly, so the soil is constantly in the initial stage of development (Fig. 5.1).

Position in an evolutionary series. The clear border between Lithosols and other types of soils that are formed on some hard rocks is not easy to distinguish and demarcate. "Kamenjar" or "golet" (Bare rocks or rock outcrops) is an area without vegetation or with very little plant cover. The possibility for certain plants to form and grow on hard rocks is low due to unfavourable climatic (hydro-thermal) conditions and ever-present erosion, and formed earthy material that remains in the cracks of the rocks provides minimal conditions for plants to grow. That is why mostly xerophyte species that have minimal moisture requirements grow and develop on Lithosols. Lithosols, and especially steep rock outcrops along faults and canyon-like gorges, are characterised by specific types of vegetation (Blečić and Lakušić 1976). The influence of vegetation on soil formation is small in terms of intensity but dominant in relation to other pedogenic factors. The vegetation in these locations is sparse and with a small amount of organic remains, but despite this, it is the most dominant pedogenic factor.

When the lithosol is formed on limestones and dolomites, its evolution will go in the direction of the formation of Calcomelanosols and Calcocambisols. If the parent material is acid siliceous rocks, then Lithosols might develop into Ranker Soils or Acid Brown Soils—Dystric Cambisols.

Distribution. The conditions for the formation of Lithosols are quite favourable, considering the large expanse of karst, which has been converted into bare land on a considerable surface. Limestones and dolomites contain only 1–2% of insoluble residual material from which soils are formed. Lithosols are characteristic soils on steep limestone outcrops-cliffs that have frequent occurrence along faults, in gorges and canyons that were intersected by watercourses

Fig. 5.1 Lithosols of Durmitor Mountain (*Photo* G. Roganović)

over a long period. In the karst area, Lithosols also appear on slightly less steep landforms, but often with a high, abundant percentage of rocks on the surface (over 90%), between which a little residual earthy material remains in cracks, scraps, and broken stones. On the mountain peaks, which are covered with rock outcrops, Lithosols are formed due to unfavourable climatic conditions, which cause the absence of vegetation, reduce processes of physical and chemical weathering, and trigger water and wind erosion.

The surfaces of the Lithosols are usually fragmented and do not have a continuous extension but are interrupted by other types of soils. That is why their separation and cartographic display on the 1:50,000 scale pedological map is not completely and clearly demarcated. Most often, Lithosols are mapped as Calcomelanosols or Calcocambisols, with which they are genetically the closest because these soils are formed on limestones and dolomites. Lithosols formed on siliceous parent materials are often classified as Rankers or, less often, as Dystric Cambisols.

In association with other soil types, Lithosols are widespread in Lovćen, Orjen, Rumija, Golija, Moračka Kapa, Sinjavina, Durmitor, Komovi, Prokletije, Pivska Planina, and on other high mountains, especially at altitudes over 2000 m a.s.l. In addition, Lithosols are also widespread on the lower dry limestone-dolomite terrain around the Bjelopavlićka Plain, Zeta, and Morača rivers, and they are also found sporadically along the sea coast. Therefore, Lithosols are practically present throughout Montenegro, but due to the aforementioned fragmentation, they are not presented on pedological maps. This might be corrected by using modern tools, GIS, and remote sensing when correcting the existing pedological map. The area of Lithosols in Montenegro might correspond to the area of bare rocks (Fig. 5.2), which covers 1.53% of Montenegro, according to Corine Land Cover (https://land.copernicus.eu/dashboards/clc-clcc-2000-2018), or even more if we take into account the area of sparse vegetation.

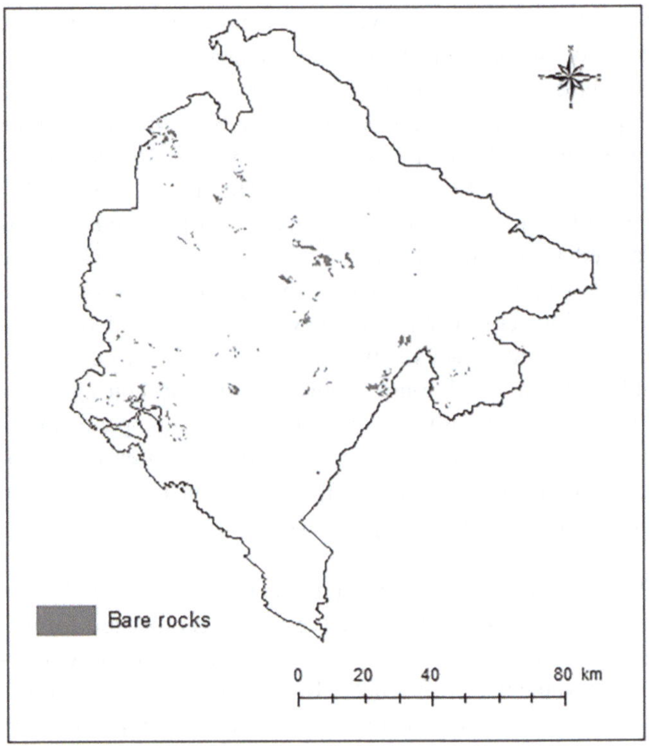

Fig. 5.2 Tentative distribution of Lithosols in Montenegro obtained through information on bare rocks from Corine land cover

Soil management and utilisation. Lithosols are found mostly on inaccessible and difficult-to-reach terrains, and they represent dry, scarce, and unfavourable habitats, except for game and specific plant species. Reclamation measures on Lithosols include afforestation and protection measures. From the environmental impact viewpoint, it is necessary to prevent rockfall around the roads, either by seedlings of trees and bushes or by installing protective nets (Fig. 5.3).

Fig. 5.3 Wild Pomegranate on Lithosols in Rijeka Crnojevića area (*Photo* M. Knežević)

5.1.1.2 Regosols

The name of the soil type. The name Regosols comes from the Greek word *rhegos*, which means cover or blanket. Regosols are formed on unconsolidated—loose parent materials that can be easily physically weathered. This definition of soils excludes alluvial, colluvial, and aeolian sediments. The local name Sirozem means poorly fertile or poorly developed soil.

The WRB corresponding RSG: Calcaric Regosols, Dystric Regosols.

Soil genesis and soil horizon sequence. Regosols are undeveloped soil with the following soil horizon sequence: (A)–C, formed on loose substrates or regoliths of rocks. When the regolith covers the hard rock, then the soil horizon sequence might be (A)–C–R. Regosols represent the initial (first) stage in the development of soils formed on loose parent materials. The reason for weak soil formation and maintenance of soil with the initial horizon—(A) is the erosion processes, which affect the topsoil, causing soil losses and thus disables the formation of the A–horizon. Due to erosion, the initial process of pedogenesis starts again, so the soil is young and possesses a thin (A)–horizon, slightly darker than the parent material it was formed of. The unconsolidated parent material can be different, but it does not include alluvial, colluvial, and aeolian deposits, as well as skeletal detritus, because they are formed in another way and serve as parent materials for soil types other than Regosols. Regosols are usually water-permeable due to the high content of gravel and have low water retention. Also, they have a low content of humus, silt, and clay, light texture, and weak grade of soil structure development (Fig. 5.4). That is why Regosols often represent very dry and warm habitats where only xerophyte plants can survive, i.e., low herbaceous species, weeds, and stunted trees.

Fig. 5.4 Regosols—Initial stage of soil genesis at the eroded terrain and loose substrate (*Photo-archive* CZM)

Position in an evolutionary series. Regosols are soils in the initial stage of soil development because they are constantly prone to soil erosion, which causes regressive pedogenesis and disables soil evolution. Further soil development depends on the type of parent material and the intensity of soil erosion. When the Regosols are formed on siliceous rocks, then the evolution goes in the direction of Rankers or Dystric Cambisols, which depends on the altitude. Regosols on siliceous-carbonate substrates, like on common flysch sediments in Montenegro, which consists of marly limestones, clays, cherts, breccias, and other flysch–series members, might evolve into Eutric Cambisols. This is affected by the presence of $CaCO_3$ in the parent material and various minerals from the feldspar and plagioclase groups. Regosols formed on the weathered dolomite material might evolve in the Rendzina Soils. If the soil erosion process is intensive, then the reverse process can occur, such as when regressive pedogenesis causes the development of Regosols from Rendzina Soils.

Soil characteristics. The physical and chemical properties of Regosols are directly related to the characteristics of the parent material. If the substrate has favourable properties, the soil might be suitable for plant growth. There are three subtypes of Regosols distributed in Montenegro: Siliceous Regosols, Siliceous–Calcareous Regosols, and Sandy-Dolomitic Regosols. The soil varieties of Siliceous Regosols are eutric and dystric, whereas Sandy-Dolomitic Regosols can be thin (< 40 cm) or thick (> 40 cm). Siliceous–Calcareous Regosols are divided into soil forms based on parent material.

Distribution. Siliceous and Siliceous–Calcareous Regosols appear sporadically on small surfaces, without much continuity in appearance, more in the form of patches and lenses on the maps. Siliceous Regosols are found in the zone of Paleozoic schists and sandstones, most widely distributed in the north-eastern part of Montenegro. Terrains with steep slopes and extremely pronounced soil erosion processes are the zones of Regosols formation, as the soils might disappear all the way up to the parent material. Devastation or destruction of the vegetation cover is the cause of severe soil erosion. The total mapped area of Regosols in Montenegro is 615 ha.

Siliceous–Calcareous Regosols on Eocene flysch occur in the coastal belt and in the Lake Skadar basin. These are the terrains with a stunted plant cover, woodland shrubs, and sometimes also a thinned oak forest that poorly protects the soil from erosion. Most of the hills, ridges, and southern aspects of the smaller and larger hills around Ulcinj, from Bar to Budva, in Gornji Grbalj and Boka Kotorska Bay, are overgrown with degraded oak forest and with other poorly developed thin trees.

Sandy–Dolomitic Regosols have the largest distribution from Trebinje to Grahovo and Trešnjevo. Pure dolomites are prone to chemical decomposition and are also physically weathered and crushed in the form of fine silt and regolith. They cover the basins of the Sušička and Nudolska rivers and the area around the source of the Grahovska River. They also appear around Vilusi, Velimlje, and Tupanjeska Ravan (flat), as well as around Rijeka Crnojevića, Dupilo, and Virpazar.

On the soil map 1:50,000, very small, practically insignificant areas of Regosols are mapped and presented as a cartographic unit. Only 615 ha. This is because Regosols appear as small and scattered spots and are mostly associated with Rendzina Soils. Similarly, Dystric Regosols are most often found in association with Dystric Cambisols and Rankers, whereas Eutric Regosols are in association with much more distributed Eutric Cambisols.

Soil management and utilisation. Regosols often have low ecological value. Where possible, these soils should be afforested to protect the soil cover from erosion. In the current stage of soil development, they are habitats for wild animals and are potentially used for grazing domestic animals. Regosols should be reclaimed by soil conservation measures and afforestation.

5.1.1.3 Arenosols

The name of the soil type. The name is derived from the Latin word *arena*—sand, which means sandy soil.

Soil genesis and soil horizon sequence. Arenosols are undeveloped soils with soil horizon sequence (A)–C. These soils are formed on aeolian sand that is constantly moved by the wind. In Montenegro, sands as a parent material for soil formation are distributed only in the coastal belt, and they have two origins, fluvial and maritime (Fig. 5.5). The watercourses, the largest of which is the Bojana River, deposit sand at their mouths into the sea. These sandy deposits are then disturbed by sea waves and currents, dragged away from the mouths, and thrown back onto the shore. Along some beaches from Igalo to Ulcinj, sandy drift covers small areas, concentrated only in smaller bays, inlets, and gulfs (Igalo, Tivat, Jaz, Budva, Buljarica, Čanj, Sutomore, and Bar), and they are of maritime origin. The area of all these Arenosols together is around 50 ha. The largest area of sandy sediment extends from Ulcinj to the mouth of Bojana, including the island of Ada Bojana. "Shifting" sands, which are continuously moved by winds, cover the area of Velika Plaža (Long Beach) (12.5 km long and 250–500 m wide) and part of the island of Ada Bojana. On larger areas of Gornji and Donji Štoj and on the shores of Bojana River, sands are stabilised, overgrown with vegetation, and thus protected, so the wind cannot transport it.

From a morphological point of view, sandy sediments are laid down in layers (layered) with various thicknesses of individual layers. The thickness of sandy layers near Ulcinj is over 25 m (Mirković et al. 1968). Arenosols and sands of Štoj have specific topography from the sea coast to the former Lake Zoganje, where a salt pan was built on an area of 1500 ha. From the seashore, the terrain rises slightly to a height of 3.2–4.4 m a.s.l. in Brijeg Mora and Špatula, but there are also smaller depressions filled with water and swampy vegetation. The local population calls these depressions "knete". This first Štoj mound spread into the lower terrain of the road Ulcinj—Ada Bojana, which is 2.2–2.6 m a.s.l. Beyond this depression, the terrain rises again and reaches an elevation of 4.3 m a.s.l. in Gornji Štoj, up to 4.9 m in Bregvia, and 5.5 m in Donji Štoj. Such a wavy relief indicates that it is the result of aeolian erosion and probably also the transgression of the Adriatic Sea. The wind shifted sand particles, creating and shaping the landforms of dunes and interdunes in the same way it still does today. Nowadays, on a larger area in the hinterland of Long Beach and Ada Bojana Island, the sand is stabilised, overgrown with vegetation, which enables the soil genesis. The accumulation of organic matter and other processes, primarily through the impact of groundwater, contributed

Fig. 5.5 Arenosols, Lučica beach, Petrovac (*Photo* M. Knežević)

to the formation of the "grassy" layer with the initiation of humus–accumulative horizon. Under natural vegetation, the humus content reaches 2% and above, while it is significantly lower on arable land because it decomposes and mineralises more quickly. Arenosols have a layered structure, and individual layers have different thicknesses and sometimes particle sizes, which is particularly pronounced in the area of Štoj mound, where the occurrence of fine to coarse sandy fractions has been identified.

The WRB corresponding RSG: Arenosols.

Position in an evolutionary series. Arenosols are soils in the initial stage of soil development prone to aeolian erosion. Development of these soils is possible only after the stabilisation of sands which can be enabled by the cessation of the wind effect or its reduction through the design of windbreaks (shelterbelts). The evolution of the soil is very slow, taking the direction of the creation of a humus–accumulative horizon. Therefore, the measures applied led to a change in soil horizon sequence, forming Anthropogenised Sands with the same soil horizon sequences A–C, as Rankers or Rendzina Soils. The Arenosol subtypes are differentiated, based on mineral composition, on Quartzic Arenosols, Siliceous, and Siliceous-Calcareous Arenosols, whereas the origin of sand is used for the differentiation of soil varieties into Continental and Maritime Arenosols.

Distribution. In the description of Arenosols genesis, we pointed out in which area this soil type is distributed. Arenosols are spread in the area between Ulcinj and Ada Bojana island, as well as in the small gulfs and bays of Montenegro's shore. These are places with sandy beaches and are used for tourism. Apart from tourism, a significant area around Ulcinj is used in agriculture for the production of vegetables, especially for growing watermelons. Vegetable cultivation is supported by the possibility of using groundwater for irrigation. In addition to vegetables, the soil and climatic conditions are very favourable for citrus cultivation. Arenosols cover an area of 983 ha in Montenegro.

Soil characteristics. Arenosols of the Ulcinj area were examined in more detail. Particle size distribution indicates the dominant presence of fine and coarse sand particles (> 0.02 mm), while silt and clay particles are less present. Therefore, sands and sandy soils have mostly loose, soft, or very friable consistence and weakly developed single-grain soil structure. Individual sandy particles are easily moved by the wind if the land is not overgrown and protected by vegetation. Vegetation protects the sand in two ways: it directly prevents the impact of wind and the movement of the soil particles, and through the formation of humus–accumulative horizon and development of the root system, it improves soil structure and strengthens the soil so that the wind cannot blow it away. The sandy texture has almost always rapid water permeability through macropores, low water and nutrient storage capacities. Therefore, Arenosols are dry habitats and agricultural production is limited without irrigation.

Arenosols are characterised by heterogeneous chemical characteristics. The surface layer is usually non-calcareous, so the pH is weakly acid or neutral, whereas the $CaCO_3$ content increases with depth, and the pH in water ranges from 7 to 8. In addition to $CaCO_3$, the reaction is also affected by a high level of saline groundwater. Arenosols are poor in humus, nitrogen, phosphorus, and potassium.

Characteristics of some Arenosols are presented in Table 5.2.

The shifting sands of Ulcinj's Long Beach (Fig. 5.6) contain in the heavy fraction 93–98% of minerals from the pyroxenes and amphiboles group, followed by carbonates, olivine, and epidote, whereas the remaining 2–7% are magnetite, rutile, chromite, leucoxene, and zircon. These minerals have some medicinal properties.

The content of $CaCO_3$ in Arenosols near Ulcinj, as well as the mentioned minerals, indicates that it is a Siliceous–Calcareous subtype of Arenosols, as it contains less than 95% of quartz (Mirković et al. 1968; Fuštić and Đuretić 2000).

Soil management and utilisation. Reclamation measures on Arenosols include sand consolidation and design of windbreaks, as well as the application of clay to improve soil texture, whereas regular agro-technical measures include humification, fertilisation, and irrigation. Vegetable and fruit production on Arenosols can be quite successful with good management inputs.

5.1.1.4 Colluvial Soils

The name of the soil type. The name Colluvial Soils comes from the Latin word *colluvio*, which means heterogeneous mass or mixture of materials, and it denotes the process of carrying out soil particles by surface runoff, during which soil particles are deposited on lower terrain. The surface water might concentrate into streamlets, brooks or creeks, or larger streams, which may be intermittent, and in the last case, the torrents move and transport not only small soil particles and mud but also boulders. Once the transitional power of the waterstream weakens, sediments are gradually deposited in the foot and toeslopes, and colluvial material is accumulated.

Soil genesis and soil horizon sequence. Colluvial Soils are undeveloped soils with an (A) or A_p–horizon. Soil genesis begins once the deposited material is stabilised, and vegetation develops. Filatov (1945) was the first to describe

Table 5.2 Particle size distribution and chemical characteristics of Arenosols (adopted from Fuštić and Đuretić 2000)[1]

Location	Soil horizon	Soil depth (cm)	Sand (%)	Silt (%)	Clay (%)	pH in water	pH in KCl	$CaCO_3$ (%)	Humus (%)	P_2O_5 (mg/100 g)	K_2O (mg/100 g)
			2–0.02 mm	0.02–0.002 mm	<0.002 mm						
Ada	(A)	0–26	91.8	6.0	2.2	6.38	6.60	0.0	0.97	0.9	8.2
	I	30–50	93.2	4.3	2.5	7.28	7.15	23.3	0.37	1.4	3.7
Gornji Štoj	(A)	0–23	76.3	6.9	16.8	7.71	6.70	11.4	0.69	0.3	4.6
	I	25–50	97.5	1.5	1.0	7.64	7.30	25.4	0.44	0.1	2.5
	II	60–80	98.9	0.6	0.5	7.55	7.00	23.3	0.20	0.4	2.7
Donji Štoj	Ap	0–25	92.4	7.4	0.2	7.44	6.14	0.0	2.00	0.9	3.2
	I	25–40	88.0	4.0	8.0	7.37	6.46	11.1	0.95	0.4	2.5
	II	60–80	98.8	0.7	0.5	7.09	6.79	17.9	0.71	1.0	2.8
	III	100–120	85.5	14.2	0.3	7.07	6.68	19.2	0.62	1.0	2.4
Štoj–Port Milena	Ap_1	0–20	61.4	23.5	15.1	7.68	5.56	0.0	2.45	2.5	6.2
	Ap_2	20–40	70.5	20.0	9.5	7.56	6.46	2.1	2.31	1.6	5.1
	I	45–58	89.5	6.7	3.8	7.81	6.98	3.0	1.31	1.2	2.9
	II	60–80	77.6	21.5	0.9	8.25	7.44	15.0	0.94	1.4	2.7
	III	100–120	81.2	18.8	0.0	8.24	7.61	21.4	0.86	2.0	2.5

Fig. 5.6 Arenosols at the Long Beach (Velika plaža) of Ulcinj (*Photo* A. Topalović)

[1] Sand (%), Silt (%), Clay (%)—Percentage of sand, silt and clay sized particles; pH in water – Active acidity; pH in KCl—Potential activity; $CaCO_3$ (%)—Content of total carbonates in the soil; Humus (%)—Percentage of humified organic matter in the soil; P_2O_5 (mg/100 g)—Available Phosphorus in the soil (expressed as phosphorus pentoxide content); K_2O (mg/100 g)—Available Potassium in the soil (expressed as potassium oxide content).

Fig. 5.7 Colluvial Soils with developed Ap-horizon, Godinje (*Photo* J. Vasin)

the process by which surface runoff detaches and removes soil particles down the slope, calling the products of that process diluvium (Fig. 5.7). Surface runoff is caused by rainwater or snowmelt, and the material that is moved and transported can be of varying composition and size. Larger particles, even larger fragments, and pieces of rocks, are also sometimes moved by high surface runoff. As a result, transported material is deposited at the footslopes, and even on levelled terrain, at a certain shorter distance from slopy terrain. The colluvial process intensity depends on the runoff strength and speed, the slope gradient, vegetation cover on sideslopes, the properties of the transported material, and its erodibility. All these factors affect the thickness of the deposited material, its composition, and size, as well as the shape of the deposited material. Groundwater sometimes influences the genesis of soils formed on colluvial deposits. In addition, colluvial and alluvial deposits are often mixed at the contact of toeslopes and river valleys, which makes it difficult to separate one from another. Such deposits in earlier versions of soil classification were called Alluvial–Deluvial Soils, whereas in actual soil classification, these soils are a variety of Colluvial Soils.

The material forming colluvial sediments is diverse and heterogeneous. The depth and composition of the deposited layer along the sideslopes, where part of the colluvial material is retained and deposited, vary. At the beginning or upper part of the catenas, the deposited material is typically shallower and coarser in texture, while at the footslopes its thickness increases and soil texture varies from heavy to light, with better-sorted particles but without the distinct layering typical for alluvial deposits. Nevertheless, layering can still occur. Larger components of colluvial deposits usually have sharp edges, sometimes are subangular, and are not rounded as in alluvial deposits. Larger components are represented by soil skeleton, and its content varies depending on the characteristics of soils from which the deposited material originates, the strength and speed of the surface runoff, and the slope length. In torrential catchments, a lot of coarse material is transported and further deposited in the form of fans.

The arrival of new, fresh colluvial sediments disrupts the soil genesis, returning the soil formation at the initial stage by preventing the formation of the humus–accumulative horizons. The differentiation of Colluvial Soils into lower systematic units is based on the properties of the deposited material. Therefore, the following subtypes appear in the area of Montenegro: Eutric, Dystric, and Calcareous. The presence of skeletons in the profile is the basis for classifying Colluvial Soils into soil varieties, but the appearance of the pseudogleysation process is also used as a criterion.

The WRB corresponding RSG: Colluvic Regosols and Leptosols (Colluvic).

Position in an evolutionary series. Colluvial Soils are true topogenic soils. The mineralogical, textural, and chemical composition of colluvial deposits is similar to the soil and parent material it originates from. Therefore, the colluvial deposits are quite heterogeneous and diverse, such as Colluvial Soils. In the case of stabilised colluvial deposits, soil genesis moves in the direction of the formation of humus–accumulative horizons or A–horizons, if the soil is cultivated and used in agriculture. An important morphological feature of colluvium is its colour. In the zone of Calcomelanosols, Rendzina Soils, and Ranker Soils, the dark colour of Colluvial Soils predominates, and the soil evolution goes towards the formation of these soil types with A–C soil horizon sequence. Colluvial sediments from the area of schists, sandstones, flysch sediments, eruptive rocks, and Neogene sediments are mostly brown in colour or with higher colour values. The evolution of these Colluvial Soils moves in the direction of the Cambic Soils. Similarly, colluvial deposits from Red Soils and Calcocambisols have a characteristic reddish colour with a lower colour value.

Fig. 5.8 A zone of colluvial processes and colluvial soils, Kuti (*Photo* G. Roganović)

Distribution. The conditions for the formation of Colluvial Soils in Montenegro are favourable, considering the predominantly hilly and mountainous relief (Fig. 5.8). In its north-eastern area, siliceous rocks prevail, and in this area, Cambisols are dominant soils. These rocks are water impermeable, so there are pronounced conditions for surface runoff. Soil and parent material are carried away after every rain, especially from arable fields and places with devastated or destroyed vegetation cover. The production of colluvial deposits depends on the amount and intensity of precipitation, the slope gradient, land use practices and susceptibility to soil erosion, the state of the vegetation cover, and other factors, including human activities. On the soil map on a scale of 1:50,000, Colluvial Soils are presented to the extent allowed by such map scale, and they are not insignificant in this part of Montenegro. A large part of the Colluvial Soils at the footslopes and on gentle slopes is used as arable land. However, in the valleys of streams and smaller rivers, mostly coarse sediments are deposited, which are usually overgrown with various types of trees, shrubs, weeds, but there are also parts without vegetation. Colluvial Soils are mapped on 4061 ha in Montenegro (Fig. 5.9).

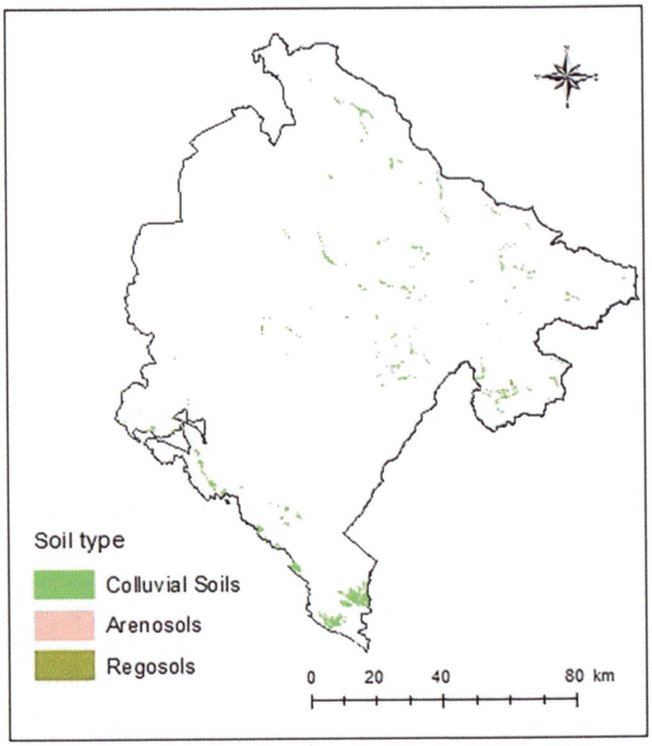

Fig. 5.9 The distribution of weakly developed soils in Montenegro

In the coastal and central part of Montenegro, siliceous rocks are slightly represented, but the soils and substrates are subject to a stronger development of the colluvial processes, which are affected by the abundance of precipitation of high intensity and duration (several days) in the autumn and winter period. Flysch sediments, sandstones, shists, and other siliceous and calcareous rocks and the soils formed on them are very susceptible to soil erosion. As a result, eroded material is produced in larger quantities. This is particularly relevant in deforested areas, with thinned and open oak forests, rare thickets, and grasses that dry out quickly during the hot and very dry period from June to October. The result of intensive colluvial processes are larger areas of Alluvial–Colluvial Soils in the coastal fields from Igalo to Bar, in Crmničko and Orahovsko fields, then in Brajići and Mrkovsko field. On the 1:50,000 scale soil map, separating Alluvial from Colluvial Soils was sometimes difficult. In the narrow coastal zone, almost all bays and fields with Colluvial and Alluvial–Colluvial Soils have been urbanised and converted into urban and sub-urban areas.

In the coastal zone, some large coastal field complexes partially covered with colluvial deposits are utilised for agricultural production, where land reclamation and watercourse regulation have been carried out, like in Sutorina, Tivatsko and Mrčevo fields, and parts of the Barsko Field, and Anamalsko Field near Ulcinj.

In the karst area, the transport of eroded material down the slopes occurs over short distances. The abundance of karst landforms such as sinkholes, depressions, karst fields, caves, and fissures represent local erosional bases where the material is deposited. A part of the eroded material is lost underground, and only the deposits in sinkholes, dolines and footslopes above karst fields are subjected to recent sedimentation when also pedogenesis begins. Typically, Colluvial Soil formed from Calcomelanosols and Terra Rossa Soils does not contain many skeletal particles, as is the case with colluvial sediments originating from siliceous rocks. That is because the karst is very permeable to water, so it does not allow the formation of furrows, ravines, channels, and gullies. In the karst areas, there are no surface watercourses as carriers of torrential activity and causes of greater production of mostly coarse sediments. Colluvial Soils formed from the Rendzina Soils, widespread in the karst fields, also do not contain a large share of skeletons. The current national classification does not distinguish Colluvial Soils formed from these soil types as Colluvial Soils, but as a subtype or a variety of Calcomelanosols, Red Soils, or Rendzinas.

Soil characteristics. Particle size distribution of Colluvial Soils is uneven, that is, heterogeneous. Colluvial Soils from the areas of shists, sandstones, and eruptive rocks, have predominantly sandy-gravelly soil particles. Older stabilised colluvial deposits contain slightly more silt and clay. Colluvial Soils formed from Neogene sediments have mostly loamy or clayey texture. The situation is similar to the colluvial deposits from soils formed on flysch sediments, which sometimes have more skeletal particles. The colluvial (and alluvial) deposits along the coastal area, in Crmnica, and around the Bjelopavlićka Plain are characterised by greater silt and clay content.

Colluvial soils are most often unstructured or with a weak structure development, although the surface layer is sometimes crumbly or cloddy. This is especially noticeable in the soils cultivated and/or soils covered by vegetation, where organic matter is formed and positively affects the soil structure. Colluvial Soils have uneven chemical properties. They can be calcareous or non-calcareous, and depending on that, there is a wide range of pH values, from acid reaction to alkaline. The content of humus varies, which also affects the content of nitrogen, phosphorus, and potassium. Crop production is affected by physical and chemical properties, soil texture, depth of active soil layer, skeletal content, and level of available nutrients.

Characteristics of some Colluvial Soils are presented in Table 5.3. Colluvial Soils from Martinovići and Liverovići are calcareous, whereas from Pelev Brijeg is non-calcareous. The soil profile excavated at the Židovići location represents Alluvial–Colluvial Soil (not Gleyic).

Soil management and utilisation. Colluvial Soils used in agricultural production require regular agrotechnical measures such as fertilisation, manuring, and irrigation. In acid Colluvial Soils, it is necessary to apply calcification with humification. Colluvial Soils are often in sub-urban zones, and there are many kitchen gardens and gardens on these soils. Also, some areas under weakly developed Colluvial Soils need to be afforested in order to increase vegetation density, foster soil genesis and stabilise terrain. Conservation protection measures in upper parts of watersheds reduce surface runoff, soil erosion processes, and production of colluvial material.

5.1.2 Class of Humus–Accumulative Soils [A–C, A–R, or A–C–R Sequence of Horizons]

The class of Humus–Accumulative Soils includes evolutionarily younger soils that represent the next stage of development, after the initial one, in soil evolution. Their pedological profile has the following soil horizon sequences: A–C or A–R, but also A–C–R and A–AC–C. The humus-accumulative horizon is characterised by the accumulation of humus, which is the dominant process in the genesis of these soils. The assessment of the impact of this process is based on the thickness of the soil horizon and the type of the humus-accumulative A–horizon. Soils of this class are known by the quite common name "black soil", and they are characterised by the synthesis and transformation of organic matter, which intensively mixes with the mineral soil matter and forms organo-mineral complexes, i.e., soil structure. Therefore, soils from this class are enriched with humus and have a good grade of soil

Table 5.3 Particle size distribution and chemical characteristics of some Colluvial Soils (adopted from Fuštić and Đuretić 2000)

Location	Soil horizon	Soil depth (cm)	Gravel (%)	Sand (%)	Silt (%)	Clay (%)	pH in water	pH in KCl	$CaCO_3$ (%)	Humus (%)	T (meq/100 g)	BS (%)	P_2O_5 (mg/100 g)	K_2O (mg/100 g)
			>2 mm	2–0.02 mm	0.02–0.002 mm	<0.002 mm								
Martinovići	Ap	0–22	38.3	66.9	24.1	9.0	7.85	7.35	5.9	5.45	–	–	–	–
	Ah	22–50	36.0	64.5	25.9	9.6	7.55	6.90	7.1	4.97	–	–	>50	20.1
Liverovići	Ap	0–25	33.0	76.4	19.1	4.5	7.40	6.52	50.8	6.72		–	>50	15.3
	Ah	25–45	14.5	77.9	26.9	5.2	7.50	6.82	50.5	4.61	–	–	42.8	10.0
	I	60–80	36.0	77.7	14.4	7.9	7.40	6.82	56.3	1.95	–	–	12.2	6.2
Pelev Brijeg	Ap1	0–11	15.7	41.3	43.8	15.9	5.58	4.50	0.0	5.52	26.7	75.5	10.0	15.7
	Ap2	11–38	10.7	41.2	36.2	22.6	5.82	4.57	0.0	4.29	25.1	84.4	6.8	12.9
	I	38–58	5.3	45.1	40.0	18.0	6.08	5.46	0.0	4.45	23.5	87.6	3.0	11.0
Židovići	Ap	0–25	13.0	44.7	35.5	19.8	6.70	5.70	0.0	4.31	17.4	87.5	6.8	11.5
	Ah	25–50	7.0	44.8	25.9	29.3	7.00	5.70	0.0	2.62	19.6	92.0	1.6	10.5
	I	50–70	0.0	45.0	24.2	30.8	6.80	5.70	0.0	1.96	20.0	95.1	1.5	10.5

T Cation exchange capacity; *BS* base saturation

structure development. Accumulation of humus occurs entirely in the A–horizon, which overlies C or R–horizons, which are soil parent materials, but humus accumulation can also occur in the transitional AC–horizon overlying unconsolidated parent materials.

The humus-accumulative horizon of these soils most often contains well-humified organic matter mixed with the mineral part of the soil in the form of organo–mineral complexes, with mollic modification of humus–accumulative horizon—A_{mo}. Such is the case with Black Soil formed on limestones and dolomites, Rendzinas and Ranker Soils when they are in an advanced development stage, but also with Chernozems and Vertisols, which are not distributed across Montenegro. In this soil class in Montenegro, an organic O–horizon often appears, formed in conditions of slower mineralisation of organic matter due to the colder climate, especially under well-preserved natural plant cover.

In the case of soils of this class, the main changes in the pedological profile occur through the biological circulation of substances, where the most pronounced processes are the accumulation of humus and biogenic elements. This led to other significant changes, such as the formation of a favourable structure, good porosity, and other mostly good physical and chemical characteristics.

In Montenegro, the following soil types are represented within this class: (1) Calcomelanosols: Black Soil formed on limestones and dolomites, (2) Rendzina Soils and (3) Ranker Soils: Humus–Siliceous Soils.

5.1.2.1 Calcomelanosols

The name of the soil type. The name Calcomelanosols comes from the German word *kalk* which means limestone, and Greek word *melas*—black, internationalised in English as melano, which means dark, black, and explains that these soils are formed on hard calcareous rocks—limestones and dolomites and that they have the dark colour of the A–horizon, because of the high humus content. These soils are also called Black Soils on limestones and dolomites, Black Mountainous Soils, or simply Black Soils, which is the term not to be confused with Chernozems, which also means Black Soils.

The WRB corresponding RSG—Lithic Leptosols and Rendzic Leptosols, but also Eutric and Dystric can be used as principle qualifiers. Also, Calcomelanosols can correspond to Leptic Phaeozems, and principle qualifiers such as Rendzic, Chernic, and Cambic can be used. If they have an umbric horizon they can correspond to Leptic Umbrisols or Cambic Umbrisols.

Soil genesis and soil horizon sequence. Calcomelanosols in Montenegro are formed exclusively on hard and compact limestones and dolomites. Such rocks are those from the Mesozoic, i.e., of the Jurassic and Cretaceous periods, and they contain around 1% of insoluble non-calcareous material in the rocks. Also, Triassic limestones and dolomites with 1–2% of insoluble residual material serve as a parent material for the development of these soils. Other soil types are also formed on hard limestones and dolomites—Calcocambisols and Terra Rossa Soils, but they represent the further stage in soil genesis. The formation of these soil types is influenced by different climatic conditions, to some extent the relief, but it is also affected by the nature of calcareous rocks, which affect microclimate and hydrological conditions. The territory of Montenegro, from the coast of the Adriatic Sea to the highest peaks of the Dinaric Mountains, is characterised by great variations and oscillations both in terms of temperature and precipitation. Therefore, the climate affects, to a very large extent, the formation of different soil types on limestones and dolomites. The warmer climate in the southern part of Montenegro favours the processes of brownisation and rubification in the subsoil, leading the soil genesis in the direction of the formation of Red Soils and Calcocambisols, whereas in the colder climate, humus mineralisation is slower and organic matter accumulates in the soil resulting in the formation of Calcomelansols.

A specific feature of these calcareous rocks is their uniform mineralogical composition, as they consist of 98–99% of $CaCO_3$ and $MgCO_3$. This implies that these rocks are completely different from other types of parent materials. Soils on calcareous rocks are not formed from the basic dominant rock mass because the rocks are prone to chemical weathering, and $CaCO_3$ and $MgCO_3$ are dissolved and leached out of in the form of calcium bicarbonate—$Ca(HCO_3)_2$. Therefore, the earthy soil material is formed from the insoluble residual material of these rocks, which accounts for only 1–2% of the rock mass. This is one of the dominant processes in the formation of all soil types on karst, with the fact that in the case of Calcomelansols, the other dominant process is the accumulation of organic matter. This accumulation of humus is higher than in Red Soils and Calcocambisols due to different climatic conditions. The small content of residual material in the rocks implies that soil formation on limestones and dolomites is a very slow and long-term process that can take several thousands of years just to create a soil layer of several centimetres.

In addition to the soil and climate, relief and vegetation also play a significant role in the formation of soils on limestones and dolomites. The relief has a strong impact because on steep slopes, especially when there are bare rocks and coarse fragments on topsoil, the soils are often found in the initial stage of formation and are shallow. On undulating terrain with gentle slopes, shallow soils are also formed, whereas, in depressions such as sinkholes, dolines, levelled plateaus, and smaller fields, the soil is moderately deep to deep. In addition to the direct influence on the soil thickness, the relief also affects it indirectly through surface runoff and pronounced erosion that occurs in the karst area as a colluvial process, and the results of such action are Buried or Colluvial soils. They fill the depressions in the karst and reach a depth of several metres in sinkholes.

Calcomelanosols are soils of uneven solum depth that depends on the intensity of chemical weathering of limestone rocks, which has a localised character. The border between the humus–accumulative horizon and solid rock is not flat, but interrupted, broken, and it often happens that in one location, the soil is only a few centimetres thick, and 5–10 m from there, it is 50–60 cm thick. This characteristic of Calcomelanosols poses an obstacle to more intensive agricultural production in good thermo–hydric conditions.

The vegetation cover of the karst area consists of various types of forest trees, bushes, shrubs, and grasses. Forests are usually sparse due to the high percentage of bare rocks on the ground; trees are often stunted and devastated by clearing, fires, etc. The dryness of the karst habitats affects the natural vegetation, represented by xerophytes that can adapt to such conditions. Vegetation cover in the karst area affects the soil genesis and evolution in a specific way.

Position in an evolutionary series. Soils on hard limestones are very old because chemical dissolution is very slow, and soil formation takes place over a very long period of time. For example, it takes 8000–10,000 years to form a 1 cm soil layer from an insoluble residual material (1–2%) (Duschafour 1982). Pedogenesis occurs in different bioclimatic conditions; therefore, soils on limestones and dolomites are polygenetic, with different characteristics. The dissolution of carbonates is a process that takes place with variable intensity in certain climatic conditions, which is why soil formation is the slowest in the initial stage of soil development. The characteristics of the soil surface and the relief of the karst area, rich in cracks, sinkholes, and other morphological forms, indicate the complexity of the karst landscape, which manifests itself in various ways and in different directions, including the genesis and evolution of the soils of karst. A complete picture of the vast karst area of Montenegro can be gained after taking into account the state of the vegetation cover, the intensity of wind and water erosion processes, and the mosaic-like appearance of the soils encountered there.

Soil evolution on limestones and dolomites is related to the thickness of the solum. The higher the soil thickness, the greater the water retention, and accordingly, the biological activity and the processes of mineralisation of organic matter accumulated in the soils in the initial soil development stage increase. The evolution of soil on limestones, in general, follows this pathway: Lithosols (initial stages of humus accumulation from physical weathering of rocks)—Calcomelanosols: Organogenic Calcomelanosols (accumulation of humus and presence of a smaller portion of mineral material in the soil—insoluble residual material of limestones or dolomites)—Organo-mineral Calcomelanosols (in addition to the accumulation of humus, organo-mineral complexes are formed)—Brownised or Cambic Calcomelanosols (increased mineralisation of humus and increased formation of the mineral components)—Calcocambisols (decrease in the content of humus due to more intensive mineralisation, and increased synthesis of secondary minerals in the soil subsurface) or Red Soils (also increased portion of mineral matter in the soils, reduction of humus due to intensive mineralisation, and increased clay content in the profile, with the typical red soil colour due to hematite formation). The final stage of evolution does not appear in Montenegro, which is the formation of Ilimerised (Luvic) Calcocambisols and Red Soils, and finally, the formation of Luvisols on limestones. These soils are formed after further reduction in the humus content, increased clay content in the profile, clay migration due to acidification, and formation of E and B–horizons.

According to the stage of soil development, Limestone–dolomite Black Soils are differentiated into several subtypes based on the stage of soil genesis. The following subtypes are distinguished: Organogenic Calcomelanosols, Organo–Mineral Calcomelanosols, Brownised (Cambic) and Rubified (Rubificated, Rhodic) Calcomelanosols, whereas Lithic Calcomelanosols and Colluvial (Buried) Calcomelanosols are distinguished at the level of soil varieties.

Distribution. Calcomelanosols are soils of the Karst Region of Montenegro, and are distributed on around 651,249 ha, which means that these soils are the most widespread soils in the country, spreading on 47.1% of its territory. Nevertheless, this area practically needs to be revised as Calcomelanosols occur in association with Lithosols on limestones, which are not mapped. However, the share of Lithosols is not so significant, so we can say the first ranked among soils in Montenegro are Calcomelanosols. They are distributed on all high karst mountains of Montenegro, namely Orjen, Lovćen, Durmitor (Fig. 5.10), Sinjavina, Komovi, Žijovo, Maganik, Moračka Kapa, Lola, Vojnik, Golija, Pivska Planina, Volujak, Maglić, Ljubišnja, Hajla, Prokletije and Mokra mountains, also in the karst fields and sinkholes and dolines, as well as in the area of Banjani, Vilusi, Grahovo, around Rijeka Crnojevića, on the side-slopes of mountains above Bjelopavlićka Plain, and border

Fig. 5.10 Brownised and Colluvial Calcomelanosols in association with Lithosols and Regosols (Sirozems), Prutaš—Durmitor (*Photo* G. Roganović)

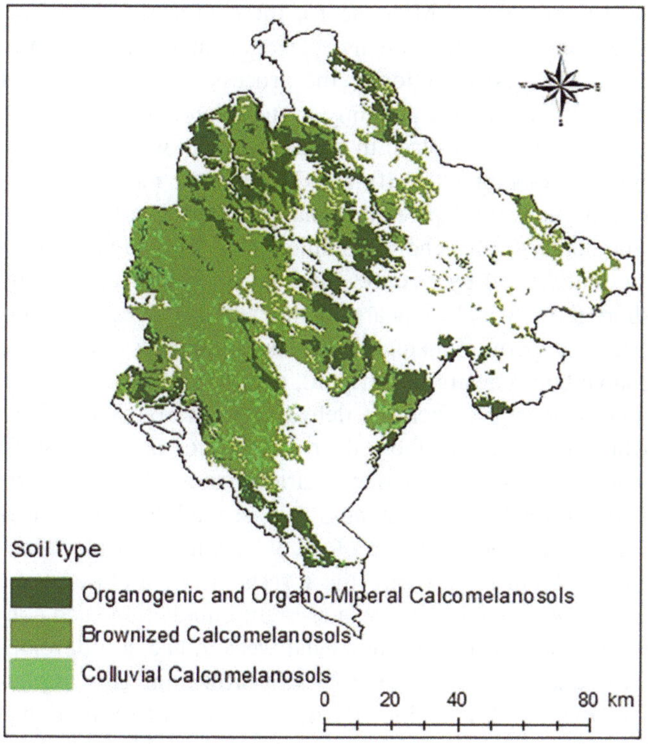

Fig. 5.11 The distribution of Brownised and Organo-Mineral and Organogenic subtypes of Calcomelanosols in Montenegro, and Colluvial Calcomelanosols soil variety

of Montenegro and Serbia in the area near the town of Pljevlja. To better depict the distribution of hard limestones and dolomites in Montenegro, it should be mentioned that the area of old Montenegrin Katunska Nahija, and even the broader area between towns of Cetinje, Nikšić, and Trebinje is called the "Rocky Sea"—"Kameno more". The area under Calcomelanosols is shown in Fig. 5.11.

Soil characteristics. As Calcomelanosols are the most widespread soils over the territory of Montenegro, the characteristics of three Calcomelanosols subtypes and one Calcomelanosols variety will be presented in the text below.

Organogenic Calcomelanosols are very shallow soils because they consist entirely of humus–rich A–horizon that directly overlies hard rock. These soils have an A–R soil horizon sequence. Organogenic Calcomelanosols contain more than 25% of humus, which is the factor responsible for the black or dark colour (chroma < 1, value < 2 according to Munsell) of these soils. Calcomelanosols are characterised by the uneven depth of the occurrence of hard rock, i.e., a broken boundary between earthy material and rocks. Rocks can cover the surface of the terrain, and their distribution can range from none and very few to abundant or even dominant (>80% of the terrain covered with bare rocks). Earthy material, i.e., formed soil, lies between the rocks, and therefore, this phenomenon represents an association of Lithosols on limestones and Calcomelanosols.

The high content of humus means that the structure of these soils is very stable, very fine to fine granular, and formed from young organo-mineral complexes made mainly of calcium-humates. Calcium–humates are stable compounds, difficult to disintegrate, and are present in the soil in larger quantities. That is why these soils have much better aggregation and water permeability of their aggregates than some other soil types (Životić et al. 2019). The soil texture of these soils is loamy sand, silty loam, or sandy loam because silt predominates the clay content. That is why these soils are porous and water-permeable, providing dry and shallow habitats most suitable for the growth of grasses.

Calcomelanosols are non-calcareous soils, although they are formed on calcareous rocks, and they have an acid to neutral soil reaction. In terms of other chemical characteristics, they have very high cation exchange capacity and moderate to high base saturation.

Organo–Mineral Calcomelanosols. In terms of main morphological characteristics, soil horizon sequence, and up to some extent of colour, Organo–Mineral Calcomelanosols are almost indistinguishable from Organogenic. Here, too, the entire profile consists of a humus–accumulative horizon overlying hard rock. The soil profile sequence is the same: A–R. The soil is thicker compared with organogenic Calcomelanosols, although it is essentially shallow, up to 25 cm deep, and sometimes deeper on milder forms of relief (Fig. 5.12). They represent a further stage in the Organogenic Black Soils development, formed of organo-mineral complexes with lower organic matter content than Organogenic Calcomelanosols, and hence, higher mineral portion(). The process begins with the mineralisation of organic matter and the creation of a certain amount of clay, which is responsible for the formation of the so-called organo–clay cores, which are the most important part of organo–mineral complexes. These blends of organic and mineral components resulted in the formation of a strongly developed soil structure and fine to medium granular soil aggregates. Such small structural aggregates are formed by enveloping the small mineral particles with thin humus membranes, and in the presence of sufficient calcium, the organo-mineral complexes become stable. Humus, which the soil contains up to 25%, plays a crucial role in the formation of organo–mineral complexes, whereas the role of clays is even more emphasised in the next stage of soil development when the Calcomelanosols are prone to the processes of brownisation, i.e., the formation of (B)–horizon.

The content of the mineral part, along with the lower content of humus, causes the colour of A–horizon to change

Fig. 5.12 Organo–Mineral Calcomelanosols at Lovćen Mountain (*Photo* Lj. Životić)

from extremely black to very dark brown and dark brown (chroma 1–2, value 2–3). Also, this is reflected in the particle size distribution because there is a greater portion of fine sand and silt, and the soil texture of the mineral part is heavier compared to Organogenic Calcomelanosols. Lighter loamy texture and high humus content make these soils friable, loose, water-permeable, and easy to cultivate.

Cation exchange capacity (CEC) of Organo–Mineral Calcomelanosols have slightly larger oscillations, as well as the sum of adsorbed bases and base saturation. The humus content has the greatest impact on CEC here, which is followed by the clay content and mineralogy. Clay minerals in Organo–Mineral Calcomelanosols have a fairly wide SiO_2:R_2O_3 ratio (Pavićević 1956). Organo–Mineral Calcomelanosols are moderately to well supplied with potassium (between 10 and 40 mg/100 g), while they are poor with phosphorus.

Characteristics of some Organogenic an Organo–Mineral Calcomelanosols are presented in Table 5.4.

Cambic (Brownised) Calcomelanosols. This subtype represents the next higher stage of Calcomelanosols evolution (Fig. 5.13). Black Mountainous Soils are prone to the processes of stronger or weaker rubification during evolution. In the territory of Montenegro, these processes have two directions depending on the elevation and climate conditions. The first direction is the process of rubification[2] of the soil, which occurs mainly up to about 500–600 m a.s.l. This process takes place in a warm climate with an average annual temperature of 10–15 °C. Up to the mentioned elevation, the natural vegetation mainly consists of transitional woodland shrubs and xerophile grass formations, which are represented by pastures, and are often accompanied by short flora, which appears in thinned and devastated forests. The warm climate, open forests, and grassy vegetation enable better heating of the soil, and thus faster mineralisation of organic matter and dehydration of iron–hydroxides, which gives rise to the red or reddish colour of the soils, which is the most important characteristic of Red or Brown Soils formed on limestones. The second direction takes place under the influence of a colder climate at elevations above 600 m a.s.l, where the forests are of a closed type, and the grasslands get dry later in the summer. In such conditions, soil heating is somewhat weaker, and soil development goes in the direction of soil browning. This implies the creation of a cambic (B)–horizon that is less thick than the humus–accumulative A–horizon. Also, this horizon is less red in colour, and has more brown, sometimes yellowish-brown or reddish-brown colour.

In addition to climate and vegetation, other factors, such as relief and its forms, soil erosion processes, a lower percentage of bare rocks or coarse fragments, and other processes, affect the browning of Calcomelansols. As a result, Cambic Calcomelanosols occupy less steep slopes, their foot and toeslopes, plains, valleys, and the sides of sinkholes, the bottom of which is filled with Colluvial Calcomelanosols. These conditions enable the formation of deeper soils, and with the greater soil depth, soil morphology changes, as well as soil water, physical and chemical characteristics.

Being a higher stage in soil development, Cambic Calcomelanosols represent the transition to Red Soils or Brown Soils on Limestones. This is a consequence of the stronger mineralisation of organic matter and the release of clay particles enveloped by organic colloids. The increased content of clay is caused by the destruction of the organo–mineral complexes, and agillo-accumulation, accumulation of residual clays, and not by illimerisation–lessivage from the overlying horizons. This affects both the soil texture

[2] According to international literature, soil rubification is defined as a pedogenesis stage in which iron is released from primary minerals to form free iron oxides that coat quartz particles in soils with a thin reddish film (Buol et al., 1973; Ben-Dor et al., 2006). In our national practice, formation of red colour is related to the formation of mineral hematite, and Red Soils, whereas the formation of reddish or brownish–red colour is mainly related to Calcocambisols, and in that case the mineral goethite is formed.

Table 5.4 Particle size distribution and chemical characteristics of Organogenic and Organo–Mineral Calcomelanosols with A–R soil horizon sequences (adopted from Fuštić and Đuretić 2000)

Location	Soil depth (cm)	Gravel (%)	Sand (%)	Silt (%)	Clay (%)	pH in water	pH in KCl	$CaCO_3$ (%)	Humus (%)	T (meq/100 g)	BS (%)	P_2O_5 (mg/100 g)	K_2O (mg/100 g)
		>2 mm	2–0.02 mm	0.02–0.002 mm	<0.002 mm								
Luke Bojovića	0–18	6.8	75.7	22.5	1.8	6.05	5.20	0.0	27.48	57.4	71.7	2.6	14.2
Crljena greda	0–10	19.9	66.6	29.4	3.0	6.85	6.02	0.0	27.19	70.1	97.9	2.1	23.2
Orjen	0–12	8.9	68.1	28.3	4.6	6.98	6.34	0.0	24.68	57.6	97.7	2.3	22.6
Pišče–Brvno	0–20	1.2	51.2	32.2	16.8	5.51	4.72	0.0	14.26	49.0	61.0	2.7	42.1
Seljkovac	0–18	8.9	71.6	22.7	5.7	5.77	4.54	0.0	13.26	45.0	71.5	4.0	19.8
	20–40	5.6	74.0	20.7	5.3	5.85	4.76	0.0	10.64	40.1	75.3	2.1	12.2
Belić–Vusanje	0–16	0.0	53.1	36.6	10.3	5.96	4.91	0.0	16.28	70.8	71.9	5.9	32.0
	16–30	0.0	55.7	33.1	11.1	5.19	4.31	0.0	14.70	63.7	80.8	4.4	8.9
Vratlo (katun)	0–17	24.2	63.8	29.7	6.6	6.90	5.85	0.0	18.67	78.8	99.2	2.0	9.4
Stravče	0–18	0.0	54.2	42.0	3.8	5.93	4.52	0.0	15.62	34.1	60.5	2.5	13.3
Žijevo Mountain	0–17	12.1	51.2	35.50	13.3	5.98	4.32	0.0	7.56	38.6	63.3	1.9	9.4
Kupić	0–12	0.5	60.8	33.7	5.5	5.90	4.90	0.0	16.08	57.6	69.5	1.4	44
Pločnik	0–35	0.0	52.6	36.9	10.5	5.53	4.63	0.0	15.56	58.0	53.4	2.1	9.7
Rašov Kiljan	0–18	6.7	46.5	53.3	0.2	6.72	5.87	0.0	15.64	66.5	95.9	1.6	26

Fig. 5.13 Brownised Calcomelanosols in the lessivage process, Ivanova Korita (*Photo* M. Knežević)

and soil structure. The formation of clay by the destruction of aggregates of the organo–mineral complexes, Pavićević (1956) called the "oglinavanje"—" residual clay rise" (increase in residual clay), of Calcareous–Dolomite Black Soils, for which he then used the term "Buavica".

Cambic Calcomelanosols have similar properties to Organo–Mineral Calcomelanosols, because they have a high humus content (8–15%), a crumbly or medium granular structure, and a dark brown soil colour (chroma from 1 to 2, value 2–3). Their thicknesses are, on average, 15–30 cm, they have lighter texture since the predominant fractions are fine sand and silt, and the content of clay is 10–20% (Fuštić and Đuretić 2000), which is higher compared to the less developed, previously described Calcomelanosols.

The properties of the (B)–horizon are often quite different. In addition to the brown or reddish colour, there is a significantly lower content of humus which is transformed, and an increased portion of clay originating from residual clay accumulation, which both contribute to the formation of granular and polyhedral structural aggregates or fine blocky subangular aggregates. This horizon often occurs with more clay, which means that its soil texture is significantly different from the A–horizon. The greater thickness of the soil profile, texture of (B)–horizon, and the lower water permeability enable increased water retention in the soil layers, which affects the process of pedogenesis, and thus physical, chemical, and biological properties, and enhanced microbiological activity.

Cambic Calcomelanosols have slightly more pronounced acidity, which is a consequence of the lower presence of base cations in the adsorptive complex. Like the previous subtypes of Calcomelanosols, Cambic Calcomelanosols are poor in phosphorus, but moderately to well supplied with potassium.

Characteristics of some Cambic (Brownised) Calcomelanosols are presented in Table 5.5.

Colluvial Calcomelanosols. These soils are a variety of Calcomelanosols. Its origin is influenced by the same pedogenic factors as in all the previously described Black Soils it originates from. However, the process of colluviation plays a crucial role since, in the karst area, numerous landforms interchange very dynamically over short distances (Fig. 5.14). Because of this, eroded soil material from the surrounding higher terrain travels a short distance to sinkholes, dolines, gullies and smaller fields where the eroded material is deposited. Colluvic material is a heterogeneous mixture of material that has moved down a slope under gravitational action (Životić et al. 2017).

The character of the karst relief and the abundance of precipitation (2000–3500 mm per year) in the area of the so-called "hot karst" of the southwestern parts of Montenegro play a crucial role in the formation of Colluvial Calcomelanosols. A large amount of precipitation contributes to the formation of various relief forms, of which there are numerous craggy hills, ridges, heights, and mounds, from which the material is eroded and then deposited in nearby sinkholes, dolines, gullies, and smaller karst fields. The dissolution of limestones under the influence of rainwater and CO_2 is carried out not only on the surface but also in depth. Because of this, there are frequent cracks, pits, and sinkholes in the karst, and a part of the eroded and transported material is carried underground and lost forever.

The relief depressions with formed Colluvial Calcomelanosols are shadowless, flat, or slightly inclined and resemble small oases in a vast rocky desert. Although they are fragmented and scattered, they are the only arable areas in the karst. They are characterised by a thicker layer of soil, which in some sinkholes reaches a thickness of several metres. Fresh colluvial sediment is constantly deposited into these relief forms, which slows down the pedogenesis process or returns it to the initial stage. The surrounding rocky terrain is constantly exposed to erosion and forever remains with a thin soil layer and is often turned into rock outcrops.

Table 5.5 Particle size distribution and chemical characteristics of Cambic Calcomelanosols (adopted from Fuštić and Đuretić 2000)

Location	Soil horizon	Soil depth (cm)	Gravel (%)	Sand (%)	Silt (%)	Clay (%)	pH in water	pH in KCl	$CaCO_3$ (%)	Humus (%)	T (meq/100 g)	BS (%)	P_2O_5 (mg/100 g)	K_2O (mg/100 g)
			>2 mm	2–0.02 mm	0.02–0.002 mm	<0.002 mm								
Pišče	A	0–20	1.0	55.1	29.7	15.3	5.71	4.82	0.0	13.11	53.2	47.6	1.2	11.3
	A	22–36	4.2	47.7	22.4	30	6.48	5.83	0.0	10.43	38.5	74.5	0.9	8.4
	(B)	36–50	1.4	32.6	22.2	45.2	6.58	5.93	0.0	7.01	45.5	94.7	2.4	13.6
Rasovi vrh	A	0–20	0.0	65.3	26.4	8.3	6.00	5.10	0.0	16.38	62.2	67.4	4.3	16.4
	A(B)	20–35	8.4	64.7	21.7	13.7	6.50	5.60	0.0	9.12	36.2	88.4	3.2	8.1
	(B)	40–55	1.8	48.9	27.2	23.9	6.90	5.90	0.0	2.66	19.3	92.2	8.4	6.7
Katun Potrk	A	0–4	0.3	60.8	31.5	7.7	5.61	4.34	0.0	11.55	24.0	50.4	2.4	11.8
	A	4–32	0.3	59.8	29.9	10.3	5.61	4.30	0.0	11.26	25.7	51.7	20.0	11.4
	(B)	32–62	0.2	55.8	31.5	12.7	5.98	4.84	0.0	7.04	29.6	70.8	1.4	9.4
Kučka Krajina	A	0–20	0.0	56.1	32.5	11.4	5.90	4.80	0.0	10.73	46.4	65.4	3.5	15.7
	(B)	22–35	0.0	38.2	33.5	28.3	6.20	5.23	0.0	5.87	44.5	81.9	1.8	9.7
Brezojevica	A	0–18	0.0	62.7	32.8	4.5	6.46	5.54	0.0	15.74	34.7	67.2	1.1	14.2
	A	18–31	0.0	51	43.8	5.2	6.44	5.44	0.0	10.03	40.9	68.9	0.9	10.0
	(B)	31–46	0.4	54	32.9	13.1	6.55	5.50	0.0	5.90	41.1	89.6	0.9	10.5
Dubrave	A	0–25	0.0	59.4	34.7	5.9	5.20	4.09	0.0	12.71	48.6	42.3	3.0	19.8
	(B)	25–50	0.0	25.8	37.3	36.9	5.42	4.19	0.0	3.97	31.4	46.0	1.6	14.4
Praga	A	0–25	15.8	47.9	36.6	15.5	6.90	5.60	0.0	9.25	36.0	91.7	–	–
	(B)	25–50	18.2	51.4	27.2	21.4	6.50	5.40	0.0	6.65	30.6	70.2	–	–
Brestice	A	0–15	0.0	50.2	24.5	25.4	6.60	5.85	0.0	11.08	38.7	91.8	1.0	40.2
	(B)	15–30	0.0	50.8	12.8	36.5	6.30	5.50	0.0	4.94	39.0	89.2	0.6	26.6

Fig. 5.14 Colluvial Calcomelanosol used for grapevine production—Gornji Ceklin (*Photo* J. Vasin)

In addition to precipitation and parent material, the colluvial process is accelerated by scarce, thinned, or completely destroyed vegetation cover, then by grazing, fires, or some other human activities. The most important human impact is reflected in the building of terraces. In an effort to increase the area of arable land as much as possible, humans quarried stone and built karst dry–stone walls, behind which soils remained. In this way, humans shortened or redirected the paths of eroded material and, at the same time, shaped the natural forms of relief and appearance of dolines, sides of sinkholes, and valleys, converting them into terraces. Erosion of material from the surrounding higher terrain or the addition of earthy material by humans caused the fulfilment of the space above the dry boundary wall, so they could have different heights, sometimes up to 2 m, which also depended on the slope. The depth of the soil above the walls depends on the height of the wall, which also affects the width and slope of the plateau of the terraces. Calcomelanosols and other soils on terraces are not classified as Anthropogenic Soils in the national classification system, although there are serious anthropogenic impacts on the formation of these soils.

Due to the weaker chemical dissolution of dolomites and somewhat more pronounced physical weathering, the occurrence of sinkholes, pits, and dolines is far less frequent, and as a result, there are less Colluvial Calcomelanosols in the area of dolomites.

Also, in the area of Triassic limestones, which cover a large area in the high mountainous region of Montenegro, the sinkholes are not deep but much wider, predominantly saucer-shaped or have the shape of a pelvis, and often pass into small valleys and fields. The surrounding terrain has a smaller percentage of rocks on the surface and often gentle slopes, so the better-developed vegetation cover is a factor that mitigates the more intensive soil erosion.

Colluvial Calcomelanosols have many characteristics similar to that of the surrounding terrain, which is understandable, knowing that they were formed from the surrounding earthy material by the process of colluviation. The similarity is reflected in terms of colour, which is black or dark, then in terms of soil structure, soil texture, etc. There is no differentiation of horizons in the solum, and the soil profile is uniform and homogenous throughout the thickness of the soil, with a predominant loamy texture, which makes these soils easy to cultivate and enables normal plant rooting. They do not contain gravel except for rare limestones fragments. The exceptions are Colluvial Calcomelanosols of the terraces, which indicates that the presence of gravel is a consequence of clearing and quarrying. During terracing, the man stacked large stones in the walls, while the smaller fragments remained on topsoil and were mixed with the soil over time during cultivation.

Due to its greater depth and its position in the depressions, Colluvial Calcomelanosols have significantly higher water retention capacity, and the ability to retain and accumulate water. This entails changes in the direction of humus mineralisation, residual argillo-accumulation, browning and/or rubification. Such conditions can foster a change in the soil profile morphology, and the colour of the subsurface layers gradually changes from dark to brown or reddish; the structure changes in the direction of the formation of larger aggregates, soil compaction occurs, soil stickiness increases, etc.

In terms of chemical characteristics of Colluvial Calcomelanosols, there are no noticeable differences compared to Organo–Mineral and Cambic (Brownised) Calcomelanosols. A–horizon contains 8–12% of humus, while its content decreases with depth. The soil reaction is weakly acid to neutral, but the unevenness of certain parameters of the adsorptive complex is noticeable, particularly those related to the base saturation.

Characteristics of some Colluvial Calcomelanosols are presented in Table 5.6.

Soil management and utilisation. Calcomelanosols are mostly used in animal husbandry, but deeper Organo–Mineral and Cambic (Brownised) subtypes, as well as

Table 5.6 Particle size distribution and chemical characteristics of Colluvial Calcomelanosols with A–R soil horizon sequence (adopted from Fuštić and Đuretić 2000)

Location	Soil depth (cm)	Gravel (%)	Sand (%)	Silt (%)	Clay (%)	pH in water	pH in KCl	$CaCO_3$ (%)	Humus (%)	T (meq/100 g)	BS (%)	P_2O_5 (mg/100 g)	K_2O (mg/100 g)
		>2 mm	2–0.02	0.02–0.002 mm	<0.002 mm								
Gornji Zagarač	0–25	0.3	33	46.4	20.6	5.83	5.03	0.0	8.57	39.1	78.4	5.2	>50
	25–50	0.0	33.6	43.6	22.8	6.04	5.33	0.0	6.67	34.0	79.9	3.0	>50
	60–80	0.0	45.5	36.9	17.6	5.63	4.87	0.0	8.80	32.6	73.6	2.8	>50
Gradina	0–23	0.0	29.7	44.4	25.9	6.14	5.43	0.0	6.20	34.9	76.7	3.6	>50
	23–48	0.0	39.5	45.3	15.2	5.68	4.87	0.0	7.47	34.4	72.6	2.8	40.0
Lastva	0–15	0.0	48.2	35.1	16.7	6.24	5.47	0.0	13.18	–	–	1.4	28.6
	20–40	0.0	51.7	21.1	27.3	6.84	6.01	0.0	6.48	–	–	1.3	19.4
Bajov Kamen	0–20	8.0	45.6	40.5	14	6.05	5.17	0.0	6.59	–	–	3.0	>50
	20–40	6.8	41.7	36.6	21.7	6.25	5.27	0.0	5.86	–	–	2.2	>50
Petrovići	0–20	0.0	45.1	38.1	16.8	6.43	5.39	0.0	9.41	39.0	77.9	1.8	50.0
	20–45	0.0	35.1	37.4	27.1	6.28	5.39	0.0	8.23	39.1	73.4	1.5	23.2
Velji Do	0–18	2.9	38.3	38.9	22.8	6.20	5.25	0.0	11.83	41.2	66.4	1.8	>50
	18–62	0.0	42.7	33.2	24.1	6.40	5.45	0.0	9.88	38.5	74.2	1.7	33.4
Vukov Do	0–20	0.1	36.9	41.4	21.7	6.70	5.94	0.0	8.92	39.4	93.9	2.8	>50
	20–50	9.2	37.2	30.5	32.3	6.98	5.94	0.0	7.53	36.2	95.4	1.3	20.0
Lipova Glava	0–19	7.9	47.7	38.9	13.4	7.03	6.10	0.0	12.65	52.7	98.6	2.1	16.4
	19–42	5.1	61.4	22.9	15.7	6.98	6.10	0.0	12.82	60.7	98.8	1.9	14
Cerini	0–25	0.0	37.3	42.5	20.2	6.31	5.42	0.0	8.80	45.9	85.2	2.8	19.6
	30–60	0.0	36.3	38.4	25.3	6.40	4.92	0.0	8.19	46.9	92.6	1.6	15.3

Colluvial varieties, are used in agricultural and vegetable production, primarily for growing potatoes, but also leafy vegetables, under suitable climate conditions. For good yields, it is necessary to apply regular agrotechnical measures such as fertilisation with mineral fertilsers and irrigation and remove coarse fragments from the surface of the field and topsoil in the case of Colluvial Black Soils.

5.1.2.2 Rendzinas

The name of the soil type. The name Rendzina comes from the Polish language and is based on the characteristic sound heard during ploughing—rzendzic! which occurs when there is gravel in the soil that grates in contact with a plough blade. This term entered the world literature on soils, and was widely accepted. Rendzina Soils are humus-rich soils, rich in gravel, formed on soft calcareous rocks.

Soil genesis and soil horizon sequence. Rendzina Soils have the following soil horizon sequence: A–C or A–AC–C. They are characterised by a mollic horizon—$A_{mo,}$ which has a gradual border with unconsolidated C–horizon. In the territory of Montenegro, moraine and glaciofluvial deposits are the most widespread parent materials for Rendzina Soils (Fig. 5.15), followed by soft limestones and dolomitic materials such as broken angular rock fragments, rock fall material, screes, and talus deposits. The first ones were created during the intense glaciation of the high Montenegrin mountains in the postglacial period when the glaciers melted and numerous glacial streams were formed. Glaciers and watercourses removed huge amounts of mostly coarse material, of different sizes and composition, with various routes and moving directions. Part of this material remained on the sideslopes, or was deposited in certain dolines, sinkholes, and karst fields. The other part of the material was deposited by watercourses in the valleys of Lim, Tara, Piva, Morača, Zeta and Cijevna rivers, as well as in the valleys of some of their tributaries. Old river terraces were built from this material. The remaining materials are a product of modern times, and they were created mainly by physical weathering (disintegration) of limestones and dolomites, by erosion and colluvial sedimentation.

All these materials, deposits, are poorly sorted, both vertically and spatially. Their common characteristic is that their origin is calcareous rocks, so as a parent material for soil formation, they contain a high percentage of $CaCO_3$. The main process of pedogenesis in Rendzina Soils is the accumulation of neutral humus (saturated with bases), because the presence of the carbonate gravel and $CaCO_3$ in the soil matrix of A–horizon provides enough Ca^{2+}–ions to bind with humic acids, forming organo-mineral complexes of the calcium–humate type. The humus–accumulative (A_{mo}–horizon) formed in this way is quite similar to the A–horizon of Calcomelanosols, in terms of colour, structure, lighter texture, and good water permeability. The difference is that it does not directly overlie the parent material, but gradually, or over the transitional AC–horizon, it passes into the loose C–horizon, which is the soil parent material, and in terms of gravel, makes this horizon very different from A–horizon of Calcomelanosols.

The thickness of the A–horizon is different and mostly depends on the position, i.e., slope gradients and recent soil erosion processes, and it ranges from a few to 20–30 cm. Sometimes, the A–horizon is very shallow and on the flat terrain of some karst fields such as Cetinjsko, Dragalj, Kopilje (Fig. 5.16), Radovče, etc. The gradual transition of the A–horizon into the parent material is especially noticeable in Rendzina Soils formed on moraine and glaciofluvial deposits. The transition is either gradual, or the AC horizon is slightly larger than 10 cm and represents a mixture of fine soil of the A–horizon and loose parent material that are diffusely mixed, with broken, wedge-shaped, or irregular boundaries.

Rendzinas have a stable, very fine and fine granular structure formed by neutralisation of humic acids with calcium cations and their further coagulation, which occur especially in deeper varieties of Rendzina Soils. The content of rock fragments in Rendzinas of Montenegro is a regular phenomenon, but it is often found in the A–horizon,

Fig. 5.15 Rendzina Soils, Kopilje (*Photo* M. Knežević)

Fig. 5.16 Rendzina Soils formed on calcareous gravel, Kopilje (*Photo* M. Knežević)

while the transitional AC and C–horizons are strongly skeletal, and less often skeletoidal.[3]

The WRB corresponding RSG: Rendzina Soils might correspond to Rendzic Leptosols, but also principal qualifiers such as Skeletic, Rendzic, Cambic, Calcaric, and Eutric can be used, depending on the soil morphology and characteristics. Also, Rendzina Soils with well-developed and thicker mollic horizon can correspond to Leptic and Rendzic Phaeozems (Radmanović et al. 2017), but also principal qualifiers like Cambic, Skeletic, Dolomitic, and Calcaric can be used. In some rare situations, Rendzina Soils can correspond to Calcaric Regosols, and principle qualifiers like Leptic, Skeletic, Dolomitic, or Eutric can also be used.

[3]According to the classification of Gračanin (1945) soils containing rock fragments are classified as slightly skeletoidal soils (<10% of coarse fragments), skeletoidal soils (10–30%), strongly skeletoidal soils (30–50%), skeletal soils (50–70%), strongly skeletal soils (70–90%), and absolutely skeletal soils (>90%).

Position in an evolutionary series. Rendzinas are formed as a gradual transition (stage of soil development) from Regosols formed on unconsolidated calcareous parent materials. Depending on the substrate and topographic conditions, Rendzina Soils have several subtypes, but the following are represented in Montenegro: (1) Rendzina Soils formed on moraine deposits, (2) Rendzina Soils formed on calcareous broken coarse fragments (coarse regolith), and (3) Rendzina Soils formed on calcareous gravel. All three subtypes have varieties of Calcareous, Brownised (Cambic), and Colluvial Rendzic Soils. Deepening of the soil solum enables more favourable wetting conditions, so the soil water characteristics positively change, which is particularly reflected in the greater water holding capacity, longer water retention in soil layers, and less rapid water percolation. Such conditions enable the gradual leaching of carbonates, cause changes in the soil reaction and synthesis of clay, thus accelerating the process of brownisation and formation of cambic (B)–horizon that is less thick than the A–horizon. Therefore, the intensity of all these processes in Rendzina Soils depends on the environmental conditions, primarily on topographic position, aspect, slope, and climatic conditions, i.e., pronounced humidity or aridity. Deeper Rendzina Soils mostly appear in dolines, sinkholes, and karst fields. In these relief forms, as well as on northern aspects, soil moistening favours the activity of microorganisms and faster mineralisation of organic matter, at the expense of which the clay content increases. After the carbonates are leached, the process of brownisation occurs, that is, the formation of the cambic (B)–horizon. This pathway leads to the evolution of Rendzina Soils into the formation of Eutric Cambisols.

From Regosols to Eutric Cambisols, typical processes in the evolution of Rendzina Soils occur through the synthesis of humus and formation of a humus–accumulative horizon, decarbonisation or leaching of Ca^{2+}-ions, weak acidification, and increased weathering of primary minerals, resulting in the formation of cambic horizons and Cambic Rendzina Soils. In the final stage, when the cambic horizon becomes thicker than the A–horizon, according to the classification, Eutric Cambisols are formed:

Regosols—Rendzina Soils: Calcareous → Leached → Br ownised or Cambic—Eutric Cambisols.

Rendzinas soil varieties that exist in Montenegro are Calcareous Rendzina Soils with calcium carbonate in the humus-accumulative horizon, Leached Rendzina Soils with carbonates leached from the A–horizon, Brownised Rendzina Soils where the cambic horizon starts to develop, but is less thick compared with the A–horizon, and Colluvial Rendzina Soils which are the next stage in the genesis of Colluvial Soils.

Distribution. Rendzina Soils cover an area of around 32,859 ha and are the sixth most widely distributed soil in Montenegro. The spatial distribution of Rendzina Soils is presented in Fig. 5.17. Moraine deposits and Rendzina Soils formed on them are spread over a relatively large area on the sideslopes of Durmitor, Sinjavina, Prokletije, Žijovo, Prekornica, Vojnik, Golija, Orjen, and Lovćen mountains. Glaciofluvial sediments reached almost all karst fields and the valleys of Lim, Tara, Piva, Morača, and Zeta rivers, also occurring in the valleys of some of their tributaries. The thickness of moraine sediments on the sideslopes is usually lower, whereas in the depressions, foot and toeslopes, it is significantly higher. Deposits of terminal moraines reach tens of metres, as well as deposits on old river terraces and in some karst fields. This caused that, depending on the climate, Rendzina Soils alternates with Brown Eutric Soils, e.g., in Grahovsko and Nikšićko fields, Bakovići and Drijenak near Kolašin, Skić near the town of Plav, Jezera below Durmitor, etc.

Rendzina Soils formed on calcareous irregular rock fragments (coarse regolith) cover a much smaller area

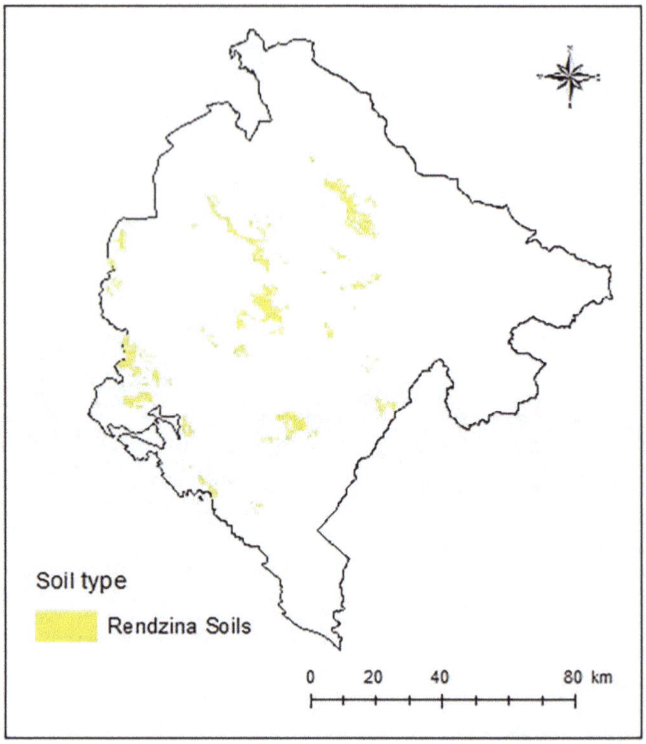

Fig. 5.17 The distribution of Rendzina soils in Montenegro

compared with the Rendzina Soils formed Moraine Deposits. They are formed of friable dolomitic coarse fragments created by the slow weathering of dolomite rocks. Such material is mainly produced by the physical weathering of dolomites, which has a predominantly silty and sandy texture, but it is pervaded by rock fragments. In the initial stage of soil genesis, Regosols are formed, and the next stage is the formation of Rendzinas. That is why these Rendzinas subtypes are most widespread in the area where dolomites and dolomitic–limestones appear, such as in the area from Trebinje to Grahovo, i.e., in the basin of the Nudolska River, and further in the direction of Vilusi and Velimlje, then from Trebinje towards Gacko, and around Rijeka Crnojevića and Virpazar.

Calcareous regoliths like terminal ("osuline", talus deposits) and lateral rock falls—scree ("sipari") are most often encountered in the canyon of Tara, Piva, and Morača rivers and other watercourses, as well as in the canyons of their tributaries. Beneath the limestone cliffs, which are mostly formed after faulting and have an almost vertical appearance, lateral rock falls on the escarpments appear as patches and most often in the form of narrow strips that, like the cliffs, extend longitudinally to the slope of the terrain. Terminal rock falls (screes with talus deposits) occur along the limestone sideslopes and cover a larger area than the lateral rock falls—screes, because they extend below the peaks down the slopes to their base (footslope) forming the talus deposit, body of "rock fall—"sipar", whereas when two bodies of different "screes" are linked, then it is called Plaz—body of scree. Unlike lateral rock falls ("sipari"—scree), which consist of piles of rock fragments, talus deposits are composed of a lot of fine silty, sandy, and gravely material, which can cause the appearance of furrow and gully erosion, often accompanied by landslides in conditions of heavy rainfall, snow avalanches, the movement of game or livestock, and human activities. Screes ("sipari") are mostly without vegetation, but the talus deposits ("osuline") are also sparsely overgrown with stunted bushes and grasses. Because of this, erosion by water and wind constantly advances and prevents further soil genesis. Contemporary talus deposits are more widespread on the southern and southwestern slopes of Durmitor, Sinjavina, Visitor, and Komovi mountains, as well as on other mountains, including Orjen, Rumija, and Lisinj which gravitate towards the coastal area.

Moraine and glaciofluvial deposits, in addition to calcareous fragments, often contain also siliceous components, and this, along with the formation of Rendzina Soils, causes the formation of Brown Eutric Soils in lower areas under the influence of a warmer climate. This is particularly notable in the Nikšićko Field, the fields around Podgorica, part of the Grahovsko Field, and the old river terraces of Lim, Tara, and Morača. Typical Rendzina Soils appear in all other karst fields but also in the parts of Grahovsko and Ćemovsko fields (Karabuš), with the latter located at only 50 m a.s.l.

Soil characteristics. An important characteristic of Rendzina Soils is coarse fragments, present in lower or higher content in all three Rendzina Soils subtypes distributed in Montenegro. They also contain gravel in A–horizon, which varies from a few to 50%. Transitional AC and C–horizons are usually strongly skeletal. Regarding particle size distribution, predominant fractions are fine and coarse sand, and silt content is higher than clay. Such a soil texture makes Rendzina Soils vividly "sandy" soil, but this is valid only for Rendzinas formed on dolomite detritus, while on other parent materials, they mostly have the loamy texture of fine earth. The texture of A–horizon and unconsolidated parent material have a crucial effect on water permeability and low water retention, which are only favourable with Colluvial Rendzina Soils deeper than 40 cm. Shallow Rendzinas are dry habitats, even on the flat terrain of karst fields (Dragalj, Radovče, Kopilje, Lukovo, Kržanja, Trmanje, and others). The process of brownisation, which appears in Rendzina Soils, leads to a change in soil structure, which in the (B)–horizon turns into medium granular, gradually affecting the A–horizon as well. The process is accompanied by a lower loss of water by percolation, which contributes to its longer retention in the soil layer, triggering the processes of weathering and the creation of clay minerals.

When the parent material contains high content of calcareous fragments, then soil reaction ranges from neutral to slightly acid. When, in addition to carbonates, there are siliceous rock fragments, the pH has lower values. Rendzina Soils formed on strongly calcareous parent materials are characterised by high cation exchange capacity. According to the content of basic cations and base saturation, they are similar to Calcomelanosols, and the parent material rich in $CaCO_3$ and organic matter content contributes to that similarity. Shallow Rendzinas, except with humus, are poorly supplied with nutrients and, therefore, have limited agricultural significance except for use as lawns and pastures. Only Colluvial Rendzina Soils deeper than 30– 40 cm are used as good arable soils (Fig. 5.18). Rendzina Soils and other shallow and skeletal soils of the Ćemovsko Field, after the application of reclamation measures, irrigation, and more abundant fertilisation, are successfully used for plantation orchards and vineyards. Plant rooting is facilitated by the formation of loose AC–horizon, formed by mixing of A and

Fig. 5.18 Rendzina soils of Krnovo plateu (*Photo* G. Roganović)

C–horizons during deep ploughing and regular cultivation. The transitional AC–horizon and loose C–horizon practically represent a physiologically deepened soil layer. The Ćemovsko Field serves as a good example of how reclamation measures can convert marginal lands into cultivated ones, which is particularly important for Montenegro, given its small area of arable land per capita.

Characteristics of some Rendzina Soils are presented in Table 5.7. Soils in Velje Duboko, Bršno, and Pirni Do are formed on moraine deposits, whereas soils in the Ćemovsko Field and Karabuš are formed on calcareous sand and gravel.

Soil management and utilisation. The natural vegetation growing on Rendzina Soils consists of xerophytic low, degraded forests, transitional woodland shrubs, and thickets, followed by pastures. In the higher mountainous areas on this soil, good coniferous forests are found in the area between Durmitor and Sinjavina, as well as meadows. Rendzina Soils of karst fields (where they are buried) are deeper and richer in nutrients and biologically more active, and are used as arable land, suitable for growing potatoes and vegetables (carrots, beets and onions), and other agricultural crops (Fig. 5.19).

Rendzina Soils of karst fields, which are shallow and have low nutrient content, are dry and warm habitats. Therefore, their use is only possible under conditions that permit the application of irrigation. Depending on the parent material, the application of intensive fertilisation must be carried out in a sustainable manner to prevent nutrient leaching, not only because they would be lost for plants but also to protect the environment and prevent groundwater contamination with nitrates and phosphates. When the humus content is below 5%, the soil should be fertilised with organic fertilisers, that is, manure and siderites (green fertilisation). It is important to pay attention to the active lime content in the AC or C–horizons and, based on that, select less sensitive cultures that can tolerate a higher content of active lime (Fig. 5.20). In this way, the risk of chlorosis due to the inactivation of accessible iron, which is needed by plants, will be eliminated.

Deeper Rendzina Soils are soils that can be used in agricultural production depending on the climatic conditions. To enhance cultivation and production, the removal of coarse fragments from the soil is useful as it increases the soil retention capacity and reduces water permeability. Regular agrotechnical measures of rational fertilisation are extremely important for high and stable yields. Orchards and vineyards on lower terrains with a Mediterranean and modified Mediterranean climate are commonly found on Rendzina Soils, and irrigation is necessary. The removal of stones, boulders, and gravel from the soil surface and topsoil is also a very effective measure to improve cultivation and agricultural production. Collected coarse material is used to build karst dry-walls, which often form the borders between the parcels and can also affect surface runoff and soil water balance in the area.

5.1.2.3 Rankers

The name of the soil type. Ranker Soils were first mentioned in Austrian literature under the name Humus–Siliceous Soils and subsequently in Germany under the name Rankers (Kubiena 1953). The name of the soils is derived from the German word *Rank,* which means a steep slope since these soils are located mainly in such topographical positions. Rankers are also mentioned in the older literature as Black Mountainous Soils, or simply Black Soils, like Calcomelanosols, and the reader could get confused with the same name for two different soil types.

Soil genesis and soil horizon sequence. Rankers are formed exclusively on siliceous parent materials and thus differ from other soils from the humus–accumulative class, formed on calcareous parent materials. Ranker Soils could

Table 5.7 Particle size distribution and chemical characteristics of Rendzina Soils (adopted from Fuštić and Đuretić 2000)

Location	Soil horizon	Soil depth (cm)	Gravel (%)	Sand (%)	Silt (%)	Clay (%)	pH in water	pH in KCl	$CaCO_3$ (%)	Humus (%)	P_2O_5 (mg/100 g)	K_2O (mg/100 g)
			>2 mm	2–0.02 mm	0.02–0.002 mm	<0.002 mm						
Velje Duboko	A	0–12	22.6	66.5	27.8	5.7	7.40	6.40	7.2	14.27	4.2	17.5
Bršno	A	0–13	33.5	64.1	25.3	10.7	7.23	6.64	19.7	5.46	3.6	41.0
	A	13–39	39.5	62.8	23.7	13.6	7.48	6.69	26.5	4.43	3.6	22.2
Pirni do-brdo	A	0–22	48.8	60.6	31.1	8.4	7.29	6.24	5.3	13.46	3.5	26.0
	A	22–48	34.6	68.4	25.7	5.9	7.44	6.44	4.8	13.39	3.3	17.2
Ćemovsko Field	A	0–25	27.2	45.4	42.1	12.5	7.72	7.02	1.2	6.05	1.2	8.7
Karabuš	A	0–25	7.4	46.7	38.8	14.5	7.22	6.47	0.0	6.04	1.8	7.4
	A	27–40	73.0	70.6	26.1	3.3	7.39	6.72	12.1	7.59	1.2	3.7
	AC	45–58	88.2	91.3	7.7	1.0	8.21	6.67	68.9	1.46	3.6	0.5

Fig. 5.19 Potato production in the Northern Montenegro (*Photo* M. Knežević)

have A–C–R, A–R or A–C soil horizon sequences. The soil has mollic, umbric or organic horizons, which are most often overly hard compacted rocks (R–horizon), and sometimes developed on the products of mechanical weathering of rocks, so A–AC–C–R soil horizon sequence can also be identified. In addition to the differences in soil horizon sequences, depending on parent material, there are certain differences in the characteristics of humus that Ranker Soils are rich in, but also in other soil characteristics. When siliceous rocks are physically weathered, detritus of various dimensions is formed, and this affects soil texture and the content of coarse fragments (Fig. 5.21). Since the parent material is often rich in quartz and poor in basic ions, Rankers can be very acidic soils, which is reflected in other chemical properties. In addition to the parent material, which is the most important decisive factor of soil genesis, the genesis is affected also by climate, relief, and vegetation. That is why the Ranker Soils in Montenegro appear in hilly–mountainous areas, mostly above 1000 m a.s.l on more prominent relief forms and at greater slopes. Due to soil erosion processes, the soil is often in the initial (younger) stage of development and has a thin solum. If the soil is covered with vegetation, then the solum deepens faster. Climatic conditions and topographic factors have an impact on Rankers and trigger the process of brownisation and formation of cambic (B)–horizon that is thinner than the humus–accumulative A–horizon.

The WRB corresponding RSG: Rankers can correspond to Leptosols if they are shallower than 25 cm, and rich in coarse fragments, and then Lithic Leptosols are often found, but also Skeletic, Umbric, Mollic, Cambic, Dystric, and Eutric can be used as qualifiers. Deeper Rankers with mollic horizons can correspond to Leptic Phaeozems, and additional principal qualifiers can also be Cambic or Skeletic. Also, they can correspond to Leptic Umbrisols, and Cambic or Skeletic qualifiers can be used. Colluvic supplementary qualifier can also be used for Colluvial Rankers, which corresponds to Leptosols (Colluvic) or Leptic Phaeozems (Colluvic), or Umbrisols (Colluvic) (Životić et al. 2021).

Fig. 5.20 Immortelle plantation grown on Rendzina Soils, Kopilje (*Photo* M. Knežević)

Position in an evolutionary series. The genesis of Ranker Soils begins with the humus accumulation process, followed by more intensive physical weathering and partial chemical weathering of the rocks. The process occurs in Montenegro in cold climatic conditions and with higher precipitation. The parent materials of Ranker Soils in Montenegro have a very heterogeneous composition. Practically, Rankers are developed on all non-calcareous rocks such as schists and sandstones, quartzites, cherts, and igneous rocks with high SiO_2 content. In the aforementioned climatic conditions, the chemical weathering of rocks and primary minerals is weak, and the processes of mineralisation of organic matter and synthesis of clay are also slow. In addition, at higher altitudes and in more prominent landforms, the soils are shallow, well-drained, and with coarse fragments, so they are dry, which allows them to heat up quickly during the day time and in the summer period, but also cool quickly during nights and in the winter time. Such habitats favour the development of xerophytic vegetation of grasses and other species, mostly of a bushy appearance, such as juniper, blueberry, dwarf mountain pine, and others. Grasses are the most important source of organic matter from which the A–horizon is formed. The decomposition of organic matter and its slow mineralisation are affected by hydrothermal conditions because low temperatures, freezing, and lack of water in the soils prevent the active role of microorganisms. That is why humus accumulates in the soil.

Siliceous rocks can be poor in bases, and when the bases are leached from the soils in given climatic conditions, complete acidification might occur, expressed by low pH values and the presence of acid humus. The absence of bases and low base saturation causes the adsorption of H^+, Al^{3+} and Fe^{3+} ions on colloidal surfaces, which, along with good water permeability and base leaching, contribute to the creation of cambic (B)–horizon, thus the formation of Cambic (Brownised) Rankers and Dystric Cambisols.

The classification of Ranker Soils in Montenegro to the subtype level is based on base saturation, and we distinguish Eutric and Dystric Rankers. Within both subtypes,

Fig. 5.21 Ranker Soil, Lubnice near Berane (*Photo* G. Roganović)

there are varieties with Lithic and Regolithic contact, Cambic (Brownised) Ranker Soils, which have cambic horizon thinner than the A–horizon, and Colluvial Ranker Soils, which have the A–horizon thicker than 40 cm. Ranker Soils can be formed from Lithosols and Regosols on siliceous rocks, as well as with further development of Colluvial Soils. Soils that can be formed with further evolution of Ranker Soils are Eutric and Dystric Cambisols, but in the process of higher acidification, these soils can also evolve into Luvisols and Podzols, which are sporadically present in Montenegro and are not mapped.

Distribution. Rankers cover an area of 6693 ha. Both Eutric Rankers and Dystric Rankers are represented. Eutric Rankers occupy a smaller area since diabase, serpentinite, spilite, and diorite as basic eruptive rocks are less encountered. Dystric Rankers occupy larger areas of Ljubišnja, Bjelasica (Fig. 5.22) distribution of Rankers is presented in Fig. 5.23.

Soil characteristics. Rankers are usually sandy loam soils in Montenegro, with a small content of clay and a larger content of coarse fragments. These soils are well aerated and water-permeable, with an overall low water-holding capacity. The chemical properties depend on the properties of the parent material. This is particularly related to the soil reaction, the sum of bases and base saturation, which often have lower values. The content of humus varies from 5 to 15%, which is closely related to the environmental conditions which do not trigger the mineralisation of organic matter. These conditions are especially unfavourable at higher altitudes and on steep terrain and somewhat more favourable at lower elevations and on flat terrain. As with other soil types, the humus content decreases with depth, especially with the appearance of a cambic (B)–horizon. Ranker Soils are poor in phosphorus and have a low to medium content of potassium.

Characteristics of some Ranker Soils are presented in Table 5.8.

Soil management and utilisation. The productive capacity of the Rankers depends mostly on the soil depth, chemical composition, altitude, and characteristics of the parent material. Ranker Soils are mostly used for pastures and forests, less often as meadows on lower terrains. Pastures with acid reaction are often overgrown with bilberry (*Vaccinium myrtillus*), juniper (*Juniperus communis*), and dwarf mountain pine (*Pinus mugo*).

In order to use Rankers as arable land, it is necessary to improve soil chemical properties in dystric subtypes. This only applies to deeper soils. Therefore, calcification with humisation is the key reclamation measure since the parent material contains small amounts of Ca^{2+} and Mg^{2+}–ions. Improvement of soil reaction and increase in the content of

Fig. 5.22 Ranker soil type formed on igneous rocks, Bjelasica Mountain (*Photo-archive* CZM)

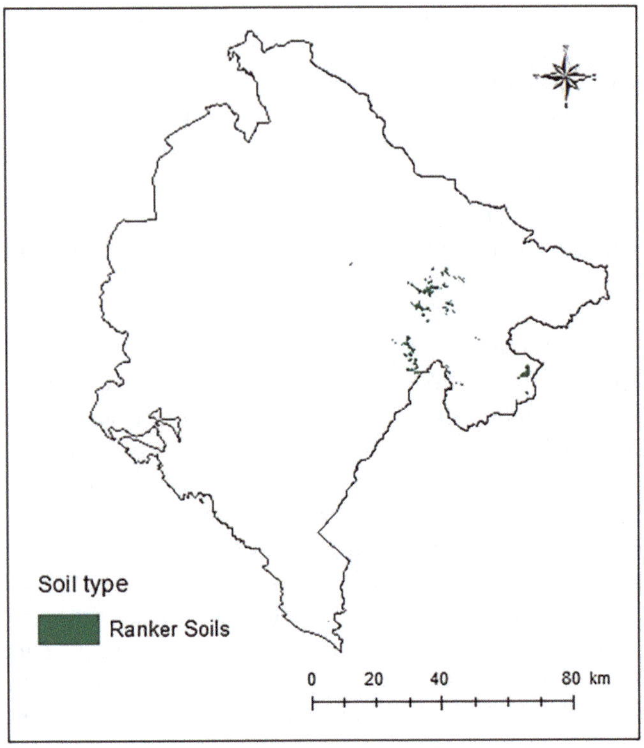

Fig. 5.23 The distribution of Ranker soils in Montenegro

Table 5.8 Particle size distribution and chemical characteristics of one Dystric and one Eutric Ranker Soils (adopted from Fuštić and Đuretić 2000)

Location	Soil horizon	Soil depth (cm)	Gravel (%)	Sand (%)	Silt (%)	Clay (%)	pH in water	pH in KCl	CaCO3 (%)	Humus (%)	T (meq/100 g)	BS (%)	P_2O_5 (mg/100 g)	K_2O (mg/100 g)
			>2 mm	2–0.02	0.02–0.002	<0.002								
Dolsko Brdo	A	0–16	19.6	59.3	29.2	11.5	4.70	3.47	0.0	14.71	53.0	23.9	2.3	14.4
	A	16–34	42.3	50.2	37.4	12.4	4.65	3.42	0.0	13.30	53.1	14.8	1.0	9.4
	AC	40–60	69.3	56.2	31.1	12.7	4.65	3.62	0.0	6.74	58.8	32.0	1.3	6.4
Kuti	A	0–17	64.9	69.6	23.1	7.3	5.71	4.70	0.0	11.89	32.3	67.9	4.6	33.5

Ca and Mg, which could replace H, Al, and Fe–ions on colloidal surfaces, can enable, with adequate fertilisation, the cultivation of potatoes, carrots, and beets. Calcification is also recommended for the alteration of the floristic composition of grasses and the quality of hay.

5.1.3 Class of Cambic Soils [A–(B)–C or A–(B)–R Sequence of Horizons]

The class of Cambic Soils is characterised by the presence of a cambic horizon located between the humus–accumulative horizon and parent material (C and R–horizons). Cambic Soils include four types of soil in the National Classification: Eutric Cambisols, Dystric Cambisols, Calcocambisols, and Terra Rossa Soils. The division of soils within this class is based on parent material. Calcocambisols and Red Soils are formed on hard limestones and dolomites, whereas Eutric and Dystric Cambisols are formed on silicate substrates and soft limestones. The differentiation between Eutric and Dystric Cambisols is based on base saturation. Thus, Eutric Cambisols are most often formed on parent material rich in bases, whether they are sedimentary, metamorphic, or igneous rocks, whereas Dystric Cambisols are formed on bases–poor parent materials, i.e., more acid substrates. Cambic Soils are formed by further development of soils from the Class of Humus–accumulative Soils, through the development of cambic horizon—(B). Cambic horizon is formed at the contact and expense of A and C or A and R–horizons. This horizon has characteristics that differentiate it from the horizons below and above. It differs in terms of soil structure, texture, colour, thickness, and other physical and chemical characteristics. The cambic horizon is the zone in which the formation or synthesis of clay (Eutric and Dystric Cambisols) or accumulation of clay (Calcocambisols, Red Soils) occurs.

The cambic horizon is formed in two different ways and includes several basic pedogenesis processes. Mainly, two modifications of cambic horizons are present. The first way of formation is after leaching $CaCO_3$ in calcareous soils and decalcification of the upper part of the profile, which is conditioned by greater precipitation. These processes are followed by a gradual decrease in pH and mild acidification, as well as the mineralisation of humus, resulting in a reduction of its content. The roots of plants release CO_2, which, in contact with water, forms carbonic acid. After decalcification and leaching of carbonates, H^+– ions dissociated from carbonic acid trigger the processes of hydrolytic and oxidative chemical weathering of primary minerals and formation of secondary minerals. Therefore, the roots of the plants and H^+– ions play an important role in the formation of the cambic horizon, which is formed at the contact of two existing horizons, at the depth where the rooting system of the plants is the most active. During chemical weathering, secondary minerals, i.e., clay minerals, are formed, and this formation is followed by the release of iron oxides that affect the colour of this horizon, which can range from yellowish, light brown, ochre, brown, ore–brown, ore, reddish–brown to red. This clay formation process is called argillo–genesis, argillo–synthesis, or metamorphosis in situ. This process further causes a gradual change in the texture of the soil profile because the clay content in the (B)–horizon increases compared to the content in the A–horizon. This modification of the cambic horizon is designated as $(B)_v$.

Another way of a cambic horizon formation occurs on hard limestones and dolomites where the cambic horizon is formed in the process of argilo–accumulation, the accumulation of clay from the residual material of limestone rocks which remains after decalcification and carbonate leaching from the rocks. These rocks contain more than 98% of $CaCO_3$, which, after chemical weathering, is leached in the form of $Ca(HCO_3)_2$. The cambic horizon is formed by the accumulation of clay from the insoluble residual material and as a result of increased mineralisation of humus under conditions of increased microbiological activity. This modification of the cambic horizon is designated as $(B)_{rz}$, where the suffix rz indicates the name of the process—residual accumulation.

The presence of cambic (B)–horizon is the most important diagnostic sign of all Brown Soils. Nevertheless, this horizon is not the same in all types of brown soils but appears in the form of two modifications: as $(B)_v$, which is formed during the chemical decomposition of primary minerals in situ, and argillo–genesis in the soils formed on silicate and siliceous–calcareous parent materials, and as $(B)_{rz}$, which is formed from the residual material which remained after the chemical weathering of limestones and dolomites in which $CaCO_3$ and $MgCO_3$ contents are 98–99%, whereas the residual material form only 1–2% of the rock.

5.1.3.1 Eutric Cambisols

The name of the soil type. The name Eutric Cambisol is derived from the Italian word *cambiare*—change, and actually denotes the character of the process resulting in the formation of cambic (B)–horizon and the English word eutric, which indicates that the soil has base saturation higher than 50%. In many classifications around the Globe, the traditional name for these soils is Brown Soils, and this term is used also in Montenegro.

Soil genesis and soil horizon sequence. Brown Eutric Soils have the following basic soil horizon sequences: A_{mo}–(B)–C or A_{mo}–(B)–R. Eutric Brown Soils are formed on soft carbonatic parent materials or on parent materials rich in bases, either there are sedimentary, intermediate and basic eruptive rocks or basic metamorphic rocks. Lacustrine sediments and glacio–fluvial deposits represent the most important parent materials for the development of Eutric Cambisols in Montenegro, while Eutric Cambisols formed on neutral and basic igneous and metamorphic rocks, as well as on colluvial deposits, are locally distributed. The parent material of Eutric Cambisols is most often loose, unconsolidated, and less often hard and consolidated or densely compacted like lacustrine loams.

The climatic conditions in which the Brown Eutric Soils are formed are very diverse. In the southern part of Montenegro, the climate is semi-arid to semi-humid (Coastal Zone, Zetsko-Bjelopavlićka Plain, Nikšićko, and Grahovsko fields). Average annual temperatures are 9–15 °C, average annual precipitation range is very broad, between 1300 and 3500 mm, but the summer period is dry and warm, the autumn–winter period with abundant precipitation, and with strong monthly and yearly fluctuations. The north-eastern area of Montenegro is characterised by a semi-humid climate with average temperatures of 7–8 °C and annual precipitation of 700–1000 mm, evenly distributed throughout the year. Therefore, Eutric Cambisols are formed in all above-mentioned climatic conditions.

The relief on which the Eutric Brown Soils are formed is flat in the area of the Zetsko-Bjelopavlićka Plain (Fig. 5.24), Nikšićko and Grahovsko fields, and on the old terraces of the river valleys, whereas in the zone of lacustrine sediments in the Berane basin and around the town of Pljevlja the relief is partly flat and undulating or slightly undulating. These soils are formed at various heights from 10 m a.s.l in the Ćemovsko Field to 1200 m a.s.l on Veruša around the source of Tara River.

The corresponding WRB RSG—Brown Eutric Soils mostly correspond to Cambisols and Phaeozems RSG's of WRB. If the soils have a well-developed mollic horizon, they can correlate to Cambic Phaeozems (Životić et al. 2017), with a possibility to use some other principal qualifiers like Leptic, Vertic, Luvic, Skeletic or Haplic. They can correlate to Eutric Cambisols, but also Leptic, Vertic, Stagnic, and Skeletic principle qualifiers can potentially be used.

Position in an evolutionary series. In relation to the Class of Humus-accumulative Soils, Cambic Soils represent a more developed stage of soil formation. They are relatively older soils. This is affected by good or favourable hydro–thermal conditions that enable a more intensive occurrence of pedogenic processes. As a result of constant moistening and good aeration, intensive processes of chemical and biological weathering of the mineral matter occur. In these processes, the original primary structure of the parent material and minerals are radically changed, and new secondary minerals are formed, whereas iron oxides are released, which are responsible for the colour of (B)–horizon, which

Fig. 5.24 Eutric Cambisol formed on alluvial and colluvial deposits—the experimental field of the Biotechnical Faculty (*Photo* B. Vrščaj)

can vary in a wide range from yellowish and ochre to be brown and reddish. The synthesis of clay minerals (argillo–synthesis) takes place simultaneously with the iron oxides formation. In this manner, (B)–horizon starts to develop at the place of this metamorphosis in situ, when the clay content starts to increase, and the horizon begins to get thicker over time.

The intensity of the development of Eutric Cambisols depends on the parent material, which can be with or without $CaCO_3$, with a lower or higher content of base cations. The parent material, in combination with climate and relief, affects the thickness of the solum and characteristics of the soils. In such variable conditions, certain differences appear among Eutric Cambisols in Montenegro. The following subtypes of Eutric Cambisols are distinguished in the country: (1) Eutric Cambisols formed on lacustrine sediments; (2) Eutric Cambisols formed on colluvial deposits, and (3) Eutric Cambisols formed on flysch, not present in Ex-Yugoslav classification but represent additions to the Montenegrin soil classification system.

Fig. 5.25 Eutric Cambisol formed on lacustrine sediments with water stagnation in deeper layers, Lješkopoljski Lug (*Photo* J. Vasin)

In the case of Eutric Cambisols formed on lacustrine sediments (Fig. 5.25), based on soil evolution, Typical (Haplic), Luvic (Leached), Vertic, and Pseudogleyic (Pseudogleyed) varieties can be distinguished. The following soil varieties of Eutric Cambisols on colluvial deposits are distinguished: Typical (Haplic), Luvic (Leached), and Vertic. Regarding Eutric Cambisols formed on Eocene flysch —Typical and Anthropogenised varieties.

Subtypes of Eutric Cambisols on loess and loess-like sediments do not exist in Montenegro because these substrate types are not distributed across the country. Also, the subtype of Eutric Cambisols formed on neutral and basic eruptive rocks is not presented on the 1:50,000 soil map due to the insignificant presence of these soil subtypes in Montenegro and because they appear together with other more widely distributed soils formed on these rocks.

The differentiation into soil varieties is based on the distinguishment of dominant pedogenic processes, that is, on the evolutionary principle. The main features of the Eutric Cambisols varieties are as follows:

(a) Typical (Haplic)—there are no signs of processes other than those typical for the soil type.
(b) Luvic (Leached)—migration of clay particles from the surface to deeper layers, most often from the A to (B)–horizon with the appearance of colloidal membranes (clay coatings and nodules) in the (B)–horizon and the beginning of the formation of E–horizon (below the O or A–horizons) which has a more pale colour, lower content of humus, clay particles or R_2O_3;
(c) Vertic—contains more than 30% of clay in the (B)–horizon with the appearance of cracks in the dry period, which is particularly pronounced in the summer time in the Zetsko-Bjelopavlićka Plain.
(d) Pseudogleyic — stagnation of surface waters in the profile during the wet part of the year, which results in small changes in the colour of A and (B)–horizons. The change of the dry and wet conditions causes alteration of oxidation–reduction processes, and, consequently, the accumulation of reduced Mn and Fe compounds. This results in the appearance of mottling, with rusty, grey or bluish spots, which are of short duration; otherwise, it would be a Pseudogley Soil type.

Distribution. Eutric Brown Soils cover an area of around 81,240 ha. On lacustrine sediments, this type of soil is found in a larger area in the Berane basin (Budimlje, Dapsići, Jasikovac, Goražde, and Polica) and Polimlje basin, then in the Pljevlja area (Pljevljsko field, Otilovići, Mataruge, Glisnica, Sađe, and Maoče), Bjelopavlićka Plain (Fig. 5.26) and Nikšićko Field. The presence of marls and clays, with interlayers of gravelly–sandy or marly materials, the continental climate, and the slightly undulating to flat relief, are the conditions for the formation of a Typical Eutric Cambisols variety around Pljevlja and Berane towns. In the Bjelopavlićka Plain and parts of the Nikšićko Field, soil with heavy texture was formed on diluvial loams, and on the flat terrain due to abundant wetting from autumn to spring, varieties of Luvic (Leached) and Pseudogleyic Eutric Cambisols prevail, sometimes in succession with Pseudogleys and Typical and Vertic Eutric Cambisols varieties.

The subtype of Eutric Cambisols on alluvial–colluvial sediments is most widespread in the valleys of Lim (Fig. 5.27), Tara, Piva, Morača and other watercourses, on old river terraces formed entirely of gravel and sand, i.e., glacio–fluvial sediments, which is found also in karstic fields. On the banks of Morača River and on Ćemovsko Field, the deposited material is cemented to a considerable extent into conglomerate, which is also the case in the Bjelopavlićka Plain (Donji Martinići, etc.), where glacio–fluvial sediments reached by the Brestica, Rimanić, Ljutotuk, Morava, Suvodola and Smrdana streams. Gravelly–sandy sediments reached by these streams and

Fig. 5.26 The area of Eutric Cambisols formed on lacustrine sediments, Viško Field (*Photo* G. Roganović)

Fig. 5.27 Eutric Cambisols formed on alluvial and colluvial deposits, Berane (*Photo* G. Roganović)

spilled into the former lake, usually in the form of a fan. The developed soil covers a larger area in the fields of Pržine and Janjića Kopita near Spuž, in the area of Donji Martinići, Donje Selo, Jelenak and Brajovići, and then in parts of Viško and Dobro fields, Zagorak and Frutak.

The Eutric Cambisols subtype formed on flysch sediments appears in the coastal area and the bordering (peripheral) part of the Bjelopavići Plain, Kuči and Crmnica. The flysch of Eocene age in this area of Montenegro is dominated by marls, clays, calcareous (carbonate) sandstones, sandy and clayey limestones. The presence of a calcareous component is also significant in the Eocene flysch from Gatačko Field to Krstac, and further from the Duga gorge to Nikšićko field, as in the Piva area, i.e., from Gacko to Plužine. The Eutric Cambisols of the flysch sediments zone of the coastal area, Crmnica and the peripheral part of Bjelopavlićka Plain and Kuči area are found on a significant area on terraces and represents an important arable land, while the soil under natural vegetation is prone to soil erosion due to heavy rainfalls from autumn to spring. The spatial distribution of Eutric Cambisols is presented in Fig. 5.28.

Soil characteristics. Eutric Cambisols have different depths depending on the parent material, landscape characteristics, and soil management (Pavićević 1959). The lowest soil thickness has the soil formed on marls that build more prominent relief forms, such as Polica and Radunovića hill near Berane, and Maoča and Mataruge in the Pljevlja area. Eutric Cambisols also have small depth when they are formed on a gravel of glacio–fluvial origin, especially when the conglomerate rocks are formed, such as on the banks of the Morača and in the Ćemovsko Field. In these cases, the depth of the soil ranges from 25 to 50 cm. Eutric Cambisols have the greatest thickness on the flat terrain of the Bjelopavlićka Plain, Nikšićko and Pljevljsko fields, where the parent materials are loams (with and without marls), gravel, and less often sand and other types of materials.

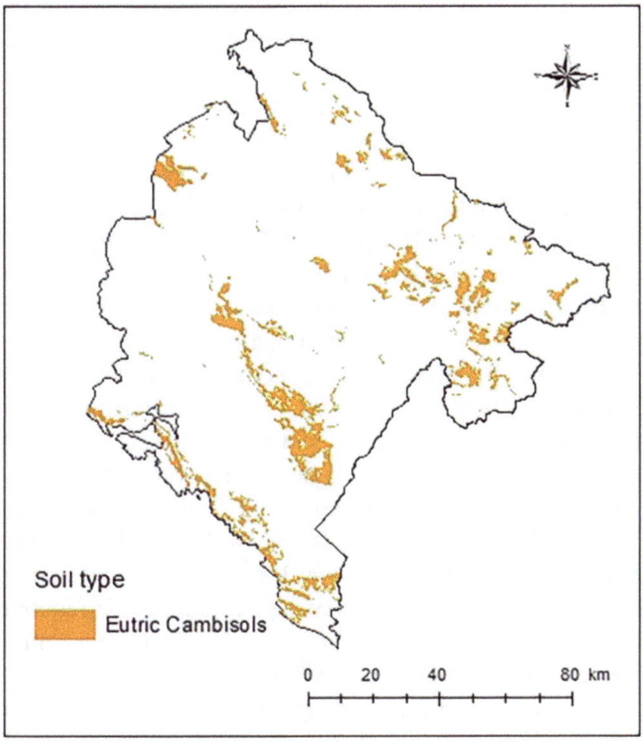

Fig. 5.28 The distribution of Eutric Cambisols in Montenegro

The most important physical and chemical properties of Eutric Cambisols depend on the characteristics of the parent material and its chemical properties, and especially on the type of clay minerals and the duration of the process of argillo-genesis. Therefore, soil texture varies between the horizons. Huge differences are observed between subtypes, but also within the varieties of the same subtype. The smallest differences are present in the subtype of Eutric Cambisols formed on flysch sediments. A–horizon has mostly loamy texture, except when the portion of gravel and a sandy fraction has increased due to soil erosion processes. The cambic horizon shows greater variability in soil texture, as it ranges from loamy to clayey. The clayey texture of the cambic horizon is particularly pronounced in varieties of Luvic (Leached) and Pseudogleyic Eutric Cambisols, and such a soil texture might lead to the deterioration of water permeability and soil water–physical properties, frequent alteration of oxidation–reduction processes and other processes in soil, which might result in the pseudogleysation process with the formation of Pseudogley Soils.

The soil structure in the A–horizon is generally favourable, which is conditioned by the content of humus, but the structure is also affected by cultivation, the wetting regime of the soil surface and other factors. The very thin, crumbly structure is most often encountered in shallow Eutric Cambisols, medium to thick granular structure in soils with a predominantly loamy composition and under natural vegetation. (B)–horizon has mostly unfavourable soil structure, which is especially the case with the Luvic (Leached) and Pseudogleyic variety of Eutric Cambisols. The seal to the formation of structural aggregates in the cambic horizon gives an increased content of clay and a lower content of humus in (B)–horizon of these varieties, so due to alternate wetting and drying cycles, blocky subangular (polyhedral), prismatic or lumpy structural aggregates prevail. Soil drainage is good in Eutric Cambisols varieties on loose materials such as gravel and sand, except when cemented conglomerates are below. Eutric Cambisols formed on clays and loams are the most poorly drained among subtypes, and this is affected by the clay content of (B)–horizon, clay substrates and flat terrain.

The chemical properties of Eutric Cambisols are also fairly variable. This refers both to the soil reaction, which ranges from weakly acid to neutral, as well as to the content of humus, nutrients, cation exchange capacity, etc.

The humus content is influenced by land use and altitude. Arable soil has a humus content of 3 to 5%, and under natural vegetation, the content of humus is from 5 to 10%. The content of humus increases with the increase in altitude, and this is a consequence of slower mineralisation in the conditions of the colder continental climate in the area of Berane and Pljevlja. The exceptions are the soils on the more prominent landforms of the Polički basin, Maoče and Potkrajci, where the humus content is low due to pronounced erosion under natural pasture vegetation (1.5–3%).

Although formed on calcareous parent materials, Eutric Cambisols do not contain $CaCO_3$. Because of this or due to the pronounced process of acidification of the soil, its reaction varies from acid to neutral. This is in correlation with the characteristics of the adsorptive complex, primarily with base saturation (%V), which ranges from 50 to 95%, but sometimes in certain horizons in the soil profile it can be lower than 50%.[4] The lower values of base saturation are certainly the result of the leaching of bases, which is reflected in the increased acidity, i.e., soil acidification. In that case, Ca^{2+} and Mg^{2+}—ions are not more dominant in the adsorptive complex, but are present together with H^+, Al^{3+} and Fe^{3+}– ions. Eutric Cambisols are generally poor in phosphorus (less than 10 mg P_2O_5/100 g), while they are usually moderately supplied with potassium (10–20 mg K_2O/100 g).

Characteristics of some Eutric Cambisols are presented in Table 5.9. The soil at Potkrajci location is formed on lacustrine sediments. The next three profiles are formed on alluvial and colluvial deposits, whereas in Odžina Poljana and Petrovići, soils are formed on flysch sediments.

Soil management and utilisation. Eutric Cambisols are mainly used as arable land, which is conditioned not only by their properties and fertility, but also by the flat or mildly undulating terrain over which they are distributed. The soils of the old river terraces in the valleys of Lim, Tara, Piva and Morača, as well as the soils of Neogene sediments around Berane and Pljevlja, are easy to cultivate, and they are used for growing various agricultural crops. The choice of agricultural crops is limited by the climate, but the soil and the climate conditions are favourable for the cultivation of the majority of field crops and fruits of temperate continental climate. In addition to these, this land is used for natural meadows, while the forest vegetation is almost negligible, appearing only along watercourses that have cut their beds into old river terraces and as forest remnants on the sides of certain ridges and tops of the wavy relief of Neogene sediments. The yields of cultivated agricultural crops are good but not stable and high without irrigation and appropriate fertilisation. Hence, loamy Eutric Cambisols, moderately deep to deep, often have very suitable characteristics for agricultural production, maybe the best among the soil types in Montenegro.

In the coastal zone and in the area surrounding the Lake Skadar basin, Eutric Brown Soils are distributed on the terraces of the flysch terrain. They are easy to cultivate. Vegetables and other types of field crops, citrus (in the coastal zone), and other types of fruits, including olives(Fig. 5.29), figs, pomegranates and vines, are grown on terraces. Good and stable yields of agricultural crops are achieved on these soils, but with irrigation, they are even more stable and higher.

In the basin of Lake Skadar, shallow and skeletal Eutric Cambisols in the Ćemovsko Field and other fields around the city of Podgorica and on the Morača banks, up to Bioče, are less favourable for cultivation. Most of these soils have low fertility, and furthermore, the cultivation of agricultural crops without irrigation is risky due to pronounced dry spells during the summer season. Although these Eutric Cambisols have low fertility, their productive capacity can be significantly increased by carrying out appropriate reclamation measures. This is convincingly confirmed by the example of the Ćemovsko Field with its complexes of vineyards, peaches and other species, which are grown on a surface of about 2300 ha.

Unfavourable physical, chemical and biological properties and water–physical properties of soils formed on the diluvial loams make them less suitable for growing agricultural crops without reclamation measures. Nevertheless, most of these lands in the Bjelopavlićka Plain and parts of the Nikšićko Field (Glibavac, Mokra Njiva, Straševina and Slivlje) are used for growing arable crops and orchards (Fig. 5.30). Forest vegetation has been preserved in fragments only in the Bjelopavlićka Plain, and natural grass vegetation in the Lješkopoljski Lug. As of recently, even in the Bjelopavlićka Plain, arable land use has almost ceased, so the land is used as meadows or pastures.

The low potential of the shallow and skeletal Eutric Cambisols should be improved by land reclamation, as it was done in the Ćemovsko Field near Podgorica. Varieties of Luvic (Leached), Pseudogleyic and Vertic Cambisols require appropriate agrotechnical and hydrotechnical amelioration. In the case of soils with a heavy clay texture, especially those distributed over leached and poorly permeable parent materials, horizontal pipe drainage is required to eliminate the "stagnant water" of abundant precipitation from autumn to spring. Eutric Cambisols are poor in phosphorus, so for high yields of agricultural crops, fertilisation with increased doses of phosphorus fertilisers are required, which also applies to potassium when its content is lower

[4] It is important that the average base saturation over the entire thickness of soil profile is higher than 50%. Otherwise, those are Dystric Cambisols.

Table 5.9 Particle size distribution and chemical characteristics of Eutric Cambisols (adopted from Fuštić and Đuretić 2000)

Location	Soil horizon	Soil depth (cm)	Gravel (%)	Sand (%)	Silt (%)	Clay (%)	pH in water	pH in KCl	$CaCO_3$ (%)	Humus (%)	T (meq/100 g)	BS (%)	P_2O_5 (mg/100 g)	K_2O (mg/100 g)
			>2 mm	2–0.02 mm	0.02–0.002 mm	<0.002 mm								
Potkrajci	Ap	0–25	0.0	35.1	43.1	21.8	6.60	5.70	0.0	7.79	34.1	92.3	22.0	>50
	(B)	25–50	0.0	38.7	36.9	24.4	6.70	5.70	0.0	5.14	27.7	92.7	7.5	28.5
Rudo Field	A	0–13	24.5	52.9	37.6	9.5	5.77	5.03	0.0	5.83	25.5	79.8	9.7	14.4
	(B)	13–37	48.1	53.5	35.7	10.8	6.02	5.18	0.0	3.75	20.6	80.9	8.0	9.8
Donja Gorica	A	0–20	1.6	49.2	36.8	14.1	6.14	5.40	0.0	3.32	17.8	59.4	1.2	8.7
	(B)	30–50	30.0	47.8	24.1	28.2	6.40	5.20	0.0	2.36	19.4	69.8	2.1	9.4
	C	90–120	75.5	69.1	13.8	17.2	7.70	6.70	29.7	2.66	–	–	1.4	6.6
Lekići	A	0–19	2.0	61.8	26.6	11.6	6.85	6.05	0.0	2.01	12.3	79.7	4.5	4.8
	(B)	20–50	2.1	42.4	31.5	26.1	6.55	5.90	0.0	2.65	15.9	79.0	8.2	7.2
	(B)C	60–90	0.8	33.1	25.6	31.3	6.60	5.76	0.0	2.72	17.3	75.6	8.5	7.8
	C	100–120	20.4	37.1	18.9	44.0	6.16	5.70	0.0	2.10	19.8	73.7	10.7	8.7
Odžina poljana	A	0–20	0.0	37	36.8	26.2	5.75	4.49	0.0	9.44	52.3	71.9	4.8	23.4
	(B)	20–50	0.0	38.3	31.4	30.3	6.05	4.89	0.0	2.32	44.1	93.0	2.2	17.8
Petrovići	A	0–20	0.3	36.3	45.9	17.8	6.70	5.63	0.0	5.09	40.9	92.6	3.5	48
	(B)	20–40	0.1	28.4	39.1	32.5	6.65	5.48	0.0	3.84	41.0	87.2	3.0	35.1

Fig. 5.29 'Stara maslina' Mirovica, Bar (*Photo* B. Lazović)

Fig. 5.30 Wild pomegranate groves on Eutric Cambisols formed on alluvial and colluvial deposits (*Photo* P. Đurišić)

than 10 mg/100 g. When Eutric Cambisols are used for growing field crops, orchards and vines, irrigation is unavoidable (Fig. 5.31). In the southern part of Montenegro, with very pronounced droughts during the growing season, successful agricultural production of any crop is impossible without irrigation. Examples from practice indicate that the best method of applying water is through drip emitters. Eutric Cambisols of the flysch zone in the coastal area, as well as those in the continental part of Montenegro, should be irrigated in order to achieve stable yields of vegetables, field crops and orchards. In these areas, agricultural production without irrigation during dry years is risky, as they do not occur rarely. The complete valorisation of Eutric Cambisols for use in agriculture can be obtained after, sometimes and inevitably, regulation of its water regime. Hilly terrain around Berane and Pljevlja towns should be protected from soil water erosion, and those surfaces used for coal exploitation should be recultivated.

Fig. 5.31 Foliar fertilisation at the experimental field of the Biotechnical Faculty (*Photo* A. Topalović)

5.1.3.2 Dystric Cambisols

The name of the soil type. The name Dystric Cambisols is derived from the Italian word *cambiare*—change, indicating the formation of cambic (B)–horizon, and the English word dystric, which refers to the base saturation lower than 50%. In the national classification, these soils are also called Brown Acid Soils.

Soil genesis and soil horizon sequence. The genesis of Dystric Cambisols is highly related to the characteristics of the parent material, which is poor in bases. The soil is formed on acid rocks such as sandstones, schists (Fig. 5.32), acid igneous, and other quartz–siliceous rocks that are commonly found, especially in the north-eastern part of Montenegro. The impact of parent material on genesis and soil characteristics often denotes a lithogenic character of this soil type formed on parent materials represented in Montenegro. These soils are formed in all altitude zones from river valleys to mountain peaks (350–2500 m a.s.l) and climatic conditions characterised by average annual temperatures of 3–8 °C and precipitation between 700 and 2500 mm.

In conditions of such parent material, climate, and relief, Dystric Cambisols are characterised by an ochric humus–accumulative horizon—A_{oh} or umbric humus–accumulative horizon—A_{um}, which overlies cambic $(B)_v$–horizon. The umbric horizon is a characteristic of soils at higher altitudes. The cambic $(B)_v$–horizon of Dystric Cambisols differs from the A and C–horizons in colour, structure, but not always in clay content (Fuštić 2000). The dominant process of pedogenesis is the chemical weathering of primary minerals and argillo-genesis, which is due to more extreme climatic conditions and acid substrates, less pronounced than in Eutric Cambisols. The lack of bases is thus reflected in the lack of formed clay minerals. Unfavourable climatic conditions and high rainfall cause the accumulation of humus, due to poor mineralisation, as well as leaching of bases and acidification. The result of these processes are the following soil horizon sequences: A_{oh}–(B)–C or A_{oh}–(B)–R, and in higher areas, soil horizon sequences A_{um}–(B)–C or

Fig. 5.32 Dystric Cambisols formed on schists, Brskovo (*Photo* M. Knežević)

A_{um}–(B)–R. The soil has a low base saturation (< 50%) and low pH values, lower than 5.5 in water.

A constellation of pedogenic factors determines the rate of weathering of siliceous rocks. The physical weathering of these rocks is intense, but their chemical weathering is more restricted. Nevertheless, a considerable amount of weathered material is formed, which serves as a substrate for soil formation. In milder climatic conditions, these processes are more pronounced. The parent silicate rocks contain SiO_2, and this silicon–dioxide is the dominant component of all the mentioned rocks represented in Montenegro. An acid substrate affects the formation of acid soils in the initial stage of development. During the weathering of silicate rocks in milder climatic conditions, sesquioxides (R_2O_3) are released as products of weathering, which gives the characteristic brown, yellowish or ochre colour to the cambic horizon—$(B)_v$.

With the increase in altitude, due to the colder climate, the chemical weathering is weaker, so the formation of the clay is small, especially in siliceous rocks with a high portion of SiO_2, which do not contain enough of those minerals from which the clay minerals are formed. In conditions of a colder climate, the mineralisation of organic matter is weaker, and the content of humus that is not saturated with bases increases, so the soil becomes even more acidic. The colour of the cambic horizon of Dystric Cambisols is light–brown, light–grey, or ochre. The depth and properties of the humus–accumulative horizon depends on the character of the humus, topographic conditions, slope and altitude, and type of plant cover, in which forest and, less often, grasses are dominant. It can be ochric or umbric, and under forest vegetation, the appearance of an organic horizon—O is found on the topsoil, below which is found A–horizon or directly $(B)_v$–horizon.

The WRB corresponding RSG's—Dystric Cambisols can correlate with Cambisols and Umbrisols RSG's of WRB. Cambic Umbrisols can correspond to Dystric Cambisols if the soils have an umbric horizon. Additional principle qualifiers can be Leptic, Luvic, Skeletic or Haplic. In the case of RSG of Cambisols, Dystric Cambisols correspond to Cambisols with Dystric qualifier, but also Leptic or Skeletic can be used.

Position in an evolutionary series. In the area where siliceous rocks are spread in north-eastern Montenegro, there are dynamic changes in terms of relief, climate and vegetation. In addition to the parent material, these conditions mostly affect the humus formation and content in the soil. This is the main reason for the presence of two subtypes of Dystric Cambisols in Montenegro: Typical (Haplic) Acid Brown Soil—Typical (Haplic) Dystric Cambisols and Humic Acid Brown Soils—Humic Dystric Cambisols. The first subtype is characterised by the presence of A_{oh}–horizon, and it appears in the lower regions, whereas the second subtype with A_{um}–horizon prevails in the highlands and along mountainous areas.

In certain places with a humid or perhumid climate, the conditions are favourable for the processes of clay migration, pseudogleysation and podzolisation, and the appearance of other subtypes of Dystric Cambisols, which are less common in Montenegro. The leaching of certain soil components directs further development of Dystric Cambisols into a subtype of Leached (Luvic) Dystric Cambisols. It is characterised by the appearance of colloidal membranes (clay coatings) in the (B)–horizon, but without clear signs of the presence of E–horizon. On flatter terrain or on slightly undulating relief, and in the soil with heavier texture, due to lower water permeability, water stagnates in the profile and leads to the formation of Pseudogleyic Dystric Cambisols subtype. In the conifer forest zones on the northern, more humid aspects, on flat terrain, the conditions are suitable for the podzolisation process, and the formation of the Podzolic Dystric Cambisols subtype, with the appearance of the initial E–horizon. Dystric Cambisols in Montenegro are differentiated into two main subtypes:

Fig. 5.33 Dystric Cambisols formed on cherts, Kutska River (*Photo* G. Roganović)

Typical (Haplic) Dystric Cambisols and Humic Dystric Cambisols subtypes are formed on sandstones, schists, chert (Fig. 5.33), flysch sediments, and acid, basic and intermediate igneous rocks.

Distribution. Dystric Cambisols are widespread soils that appear mostly in great continuity in the north-eastern part of Montenegro. They cover an area of 413,580 ha. Dystric Cambisols are the second most distributed soils in Montenegro after Calcomelanosols. Their spatial distribution is presented in Fig. 5.35. Dystric Cambisols varieties formed on schists and sandstones of Paleozoic and Triassic age (werfen and diabase-chert formation) are dominant. Schists and sandstones (Fig. 5.34) are often spatially mixed, which is why it is very difficult to distinguish these varieties of Dystric Cambisols and present them on soil maps. Schists and sandstones spread in the form of certain larger or smaller zones in the Tara basin, from Kolašin and Trešnjevik to Bistrica, from where its river valley is of a canyon–type. Zones of Dystric Cambisols formed on schists and sandstones have an even larger area in the Lim basin from the Komovi and Prokletije mountains to the border with Serbia. In these two basins, schists and sandstones of Paleozoic age predominate, while in the Ibar and Ćehotina basins, in addition to Paleozoic, schists and sandstones of diabase–chert formation are widely distributed.

The chert as a parent material for Dystric Cambisols most often occurs as interlayers, lenses, bumps or intercalations in Triassic and Jurassic limestones and are present in the Piva plateau, the foothills of Durmitor, Ljubišnja, Sinjavina, Kovač and Jelin peak, and then on Bjelasica, Visitor, Zeletin and other mountains in the Lim and Ibar basin up to Hajla and Žljeb mountains. Cherts are more present in the Piva area and on the divide of the Piva and Morača rivers (Malinsko, Javorje, Lola Mountain). In the coastal area, chert appears in a larger area in Devesilje and Vrmac hills, and, to a lesser extent, in other parts of the Boka–Kotorska Bay and Paštrovići.

Dystric Cambisols formed on various eruptive rocks have a larger distribution in the area of Bjelasica, Ljubišnja, Visitor, Sjekirica, and Mokra mountains. The eruptive rocks have predominantly a high SiO_2 content, which means that they are naturally acid, and were formed during the Triassic and Jurassic periods in the area of the mentioned high

Fig. 5.34 Dystric Cambisols formed on sandstones, Slijepač Most (*Photo* G. Roganović)

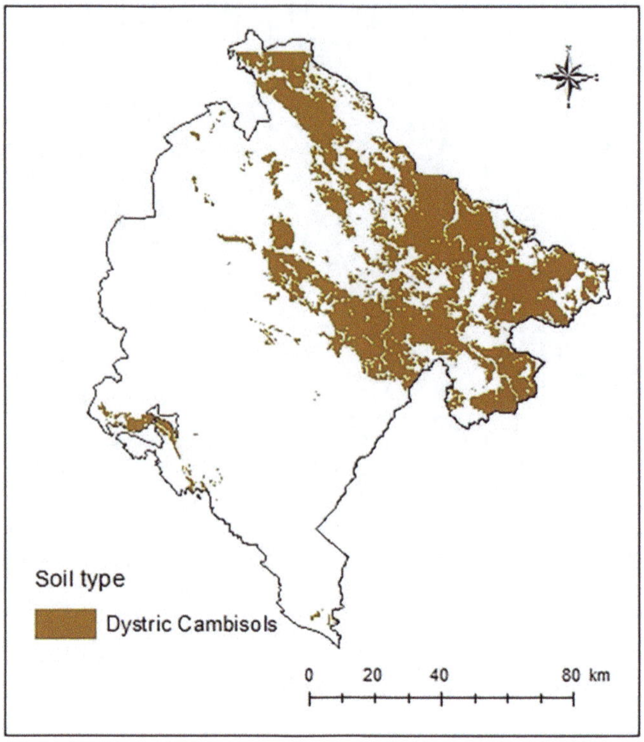

Fig. 5.35 The distribution of Dystric Cambisols in Montenegro

mountains. To a smaller extent, these rocks appear in the central and coastal part of Montenegro (Mratinje, Nikšićka Župa, Krnja Jela, and Mišići).

Dystric Cambisols on flysch (Fig. 5.36) are widespread in the central hilly–mountainous part of Montenegro. Flysch sediments, known as flysch of Durmitor Mountain, of Upper Cretaceous age, extend from Volujak and Maglić mountains, on the border with Bosnia and Herzegovina, and over Piva, Drobnjaci and Morača to Komovi and Prokletije mountains. Predominant members of the Upper Cretaceous flysch series are sandstones, clays, schists, breccia, and conglomerates, whereas marls, clay, and sandy limestones are less present. This kind of parent materials, along with the leaching of $CaCO_3$, caused the formation of Dystric Cambisols in conditions of colder climate and more abundant precipitation.

The siliceous quartz substrates on which the Dystric Cambisols were formed are not widely distributed. They appear in several places in the area of the mountains Starac, Bogićevica, and Prokletije. The quartz substrates are breccias, conglomerates, metamorphosed quartz sandstones, and schists. These types of rocks are primarily of Permotriassic age, but among them are also Paleozoic quartz conglomerates that appear on the surface as compact, large waves of quartz and quartzite (Đokić et al. 1976).

Soil characteristics. The parent materials on which Dystric Cambisols are formed have a crucial impact on soil physical and chemical characteristics, but the influence of climate, relief and vegetation cover is also very pronounced. A common general feature for all varieties of Dystric Cambisols is the uneven morphological appearance of soil horizons. The thickness of individual horizons depends especially on the type of parent material and landscape form, as well as on the type of vegetation cover, and as a result, a heterogeneous soil profile is formed. A further common characteristic of Dystric Cambisols varieties is the presence of gravel, the portion of which varies in a wide range (10–40%, sometimes even more). The gravel content and the predominant content of fine sand with an insignificant portion of colloidal clay determine that the soil has mostly loam, silty–loam or sandy–loam texture. The texture of the A–horizon, and often also the (B)–horizon, causes the soil to be loose and water permeable, with a low to moderate grade of structure development. Dystric Cambisols are not compacted, hard, sticky, or plastic and have medium to low water–holding capacity, except on flat landforms, where the thickness of

Fig. 5.36 Dystric Cambisols formed on Flysch Sediments, Mateševo (*Photo* G. Roganović)

the soil is greater, and conditions are suitable for the process of leaching, pseudogleysation, and podzolisation.

A characteristic of all varieties of Dystric Cambisols is the low base saturation, which is caused by the lack of base ions in the parent material, and then by the leaching and uptake of cations by roots of vegetation. The variable content of humus is determined by the type of vegetation cover and the relief, which together contribute to the stronger or weaker development of erosion processes. The characteristics of humus and the transformation of organic matter depend on climatic conditions, which contribute to additional acidification of the soil and decrease its already acid or highly acid reaction values. The weak mineralisation of humus and other factors, such as soil cultivation and natural acidity, contribute to the low nutrient content of Dystric Cambisols, which generally deficient in nitrogen, phosphorus, and in many cases, potassium.

Characteristics of some Dystric Cambisols are presented in Table 5.10. The first four profiles are formed on schists, the following two on sandstones, and one on igneous rocks and quartz rocks.

Soil use and management. Dystric Cambisols are mainly found in hilly and mountainous areas. This is why this type of soil is less used as arable land, for field crops, orchards, and meadows (Figs. 5.37 and 5.38). These types of land use occur up to about 1000 or 1200 m a.s.l due to climatic and relief conditions. Dystric Cambisols are mostly areas under deciduous forests and are known as forest soils in the Balkans (Fig. 5.39). The type of cultivation is limited by the slope gradient, whereas the choice of agricultural crops is limited by climate. Recently, the use of this soil type for arable farming has been minimal, which is beneficial for its protection. However, given the significance of this soil for agricultural production in the valleys of the Ibar, Lim, Tara, and Ćehotina, it is necessary to take appropriate measures for its better use and valorisation. In this regard, the best way to use, preserve and maintain soil fertility involves the implementation of certain measures of ameliorative and protective character, such as liming and manuring, the establishment of artificial lawns to provide fodder, introduction of protective crop rotation, contour cultivation, sowing and planting, terracing, etc.

5.1.3.3 Calcocambisols

The name of soil type. Calcocambisols' name is derived from the German word *kalk*—lime, which implies that the soil is formed on limestones and dolomites, and the Italian word *cambio*—change, which emphasises the existence of a cambic horizon. In previous classifications, this type of soil was also called Brown or Redish Soil on hard limestones and dolomites. The soil has a characteristic colour that can be yellowish to reddish and a very stable polyhedral (subangular and angular blocky) soil structure in the cambic horizon.

Soil genesis and soil horizon sequence. Calcocambisols are soils formed on limestones and dolomites, like Calcomelanosols and Terra Rossa Soils, and it represents a special stage in the evolutionary series. The chemical composition of Mesozoic limestones has a significant impact on the genesis of Calcocambisols, of which the Triassic limestones have 0.5–1.0% of insoluble residual material, and the Jurassic and Cretaceous have only 0.2–0.5%. The main process that leads to the formation of this soil is the dissolution of limestones, during which the insoluble residual material accumulates, and this is the material for soil formation. In addition to the parent material, research indicates that the ratio SiO_2: R_2O_3 is also important. This ratio is greater than two in limestones with a higher percentage of insoluble residual material. This ratio indicates that the residual material contains less iron and aluminium, so the properties of the formed soil depend on it. The soil properties are affected by the forms of iron that do not contain dehydrated hematite forms but are more hydrated, which is why the colour of the

Table 5.10 Particle size distribution and chemical characteristics of Dystric Cambisols (adopted from Fuštić and Đuretić 2000)

Location	Soil horizon	Soil depth (cm)	Gravel (%)	Sand (%)	Silt (%)	Clay (%)	pH in water	pH in KCl	$CaCO_3$ (%)	Humus (%)	T (meq/100 g)	BS (%)	P_2O_5 (mg/100 g)	K_2O (mg/100 g)
			>2 mm	2–0.02 mm	0.02–0.002 mm	<0.002 mm								
Zminac	A	0–20	15.2	52.1	34.1	13.8	5.25	4.20	0.0	6.29	27.0	29.8	2.2	13
	(B)	20–50	40.2	39.6	38.5	22.0	5.20	4.50	0.0	2.92	14.8	37.3	1.2	7.8
Pasji vrh	A	0–25	51.0	40.5	50.4	9.2	5.48	4.66	0.0	5.15	22.7	46.1	2.0	4.6
	(B)	30–60	48.6	51.9	40.1	8.0	5.37	4.43	0.0	4.32	19.4	31.8	1.6	2.7
Hoti	A	0–12	30.6	51.8	36.7	11.5	5.01	4.19	0.0	5.18	20.5	49.9	2.3	37.6
	(B)	12–25	51.2	48.1	38.1	13.9	4.96	4.01	0.0	3.11	17.0	41.5	2.0	30.3
Mateševo	A	0–18	21.8	60.9	31.1	8.1	4.87	3.86	0.0	6.30	20.8	32.9	1.1	8.0
	A(B)	20–35	6.0	58.2	30.0	11.8	4.61	3.65	0.0	5.59	16.4	21.5	0.5	4.4
	(B)	40–60	36.4	63.5	21.6	14.9	4.96	3.53	0.0	1.54	9.7	25.4	0.2	5.0
Trnovice	A	0–18	24.1	56.8	28.5	14.8	5.00	4.54	0.0	8.28	26.8	36.9	0.2	22.4
	(B)	20–40	25.3	42.0	29.1	28.9	5.55	4.54	0.0	4.85	23.2	35.9	0.3	12.5
Kosanica	A	0–20	17.0	63	31.1	5.9	5.21	4.04	0.0	11.86	49.2	36.7	0.4	10.5
	(B)	20–50	22.4	60.6	31.80	7.6	5.51	4.16	0.0	10.20	41.8	34.0	0.3	7.0
Velika	A	2–14	58.6	46.1	43.6	10.3	4.3	3.45	0.0	9.30	47.8	20.2	4.7	14.4
	(B)	14–31	41.7	52.6	37.1	10.4	4.87	4.01	0.0	6.93	37.5	13.1	2.6	6.3
Bogićevica	A	2–16	16.1	56.5	35.7	7.9	4.81	4.01	0.0	6.82	30.0	34.2	1.6	5.6
	(B)	16–35	22.4	57.3	35.2	7.5	5.11	4.06	0.0	4.64	21.2	41.8	0.5	3.7

Fig. 5.37 Aronia plantation on Dystric Cambisols formed on schists (*Photo* M. Knežević)

Fig. 5.38 Fruit production on Dystric Cambisols formed on cherts, Đurđevića Tara (*Photo* G. Roganović)

Fig. 5.39 Bilberries at Dystric Cambisols formed on Flysch Sediments, Veruša (*Photo* M. Knežević)

soil is not red as in Terra Rossa Soils, but brown or reddish. Calcocambisols are formed from Cambic Calcomelanosols in which the formation of cambic horizon starts at lower depths, where the soil moisture is higher, and fluctuations in temperature are lower. These conditions trigger the activity of microorganisms, which further cause higher mineralisation of humus (de-humification), and lower humus content in this zone. Consequently, fine granular and medium granular aggregates which were enveloped and linked by organic colloids, break down, and mineral clay particles are released and form dense, polyhedral, or subangular soil aggregates. In fact, this is the formation of cambic $(B)_{rz}$–horizon throughout the process of argillo–accumulation.

Calcocambiosols are usually formed on the milder forms of relief and with a variable percentage of rock outcrops (up to 50%) on the surface, and the soils are most often overgrown with transitional woodland shrubs and grasses in the lower zones and forests in the higher ones. The type of vegetation affects the content of humus and its form in the soil. The soil horizon sequence is the following: A_{mo}–$(B)_{rz}$–R or A_{oh}–$(B)_{rz}$–R. The profile is mostly moderately deep, rarely shallow or deep. Depending on the stage of soil development, landscape and topography, and percentage of rock outcrops, the soil horizon sequence might include transitional $A(B)_{rz}$–horizons. The thickness of the A–horizon is usually 15–20 cm. It has dark brown or dark reddish–brown colour, which is a reflection of the presence of humus. (B)–horizon has a brown-reddish or reddish colour and finer soil texture due to a higher portion of clay. Calcocambisols represent a transitional stage of soil formation, preceding Red Soils, and therefore have similar (B)–horizons, which is especially noticeable in the coastal area, the Lake Skadar basin, and the karst area towards Bilećko Lake and the border with Bosnia and Herzegovina. In the continental part of Montenegro, Calcocambisols are rarely found with a reddish (B)–horizon, but with a horizon less intense in colour. This is a consequence of the colder climate and the greater presence of Triassic limestones with 0.5–1% of insoluble residual material, which produces lower amounts of dehydrated iron oxides.

The WRB corresponding RSG—Calcocambisols were reported not to have an adequate name in WRB for soil resources by Balkan scientists (Bašić 2013; Pavlović et al. 2017). Knowing the soil characteristics and morphology, we can deduce that these soils can correlate to RSG's of Phaeozems, Umbrisols, and Cambisols. If they have a mollic horizon, they can be Cambic Leptic Rendzic Phaeozems ([5]Rhodic or Chromic). The supplementary qualifiers Rhodic

[5]The difference between Calcocambisols and Red Soils in the National Classification is in the ratio of SiO_2 to R_2O_3 in residual material, which is higher than 2 in Calcocambisols, which contain more goethite mineral, whereas in Red Soils this ratio is smaller than 2 and Red Soils are characterised by dehydration of Ferrous hydroxides and formation of hematite which is known by its red colour. Goethite on the other side could have yelowish to reddish, reddish-brown to dark brown and even black colour. Therefore, reddish colour found in Calcocambisols could sometimes be confused with red colour of Red Soils. Tentatively we decided to add Chromic and Rhodic qualifiers for Calcocambisols, and Rhodic qualifer for Red Soils.

or Chromic might be used in the case of appropriate soil colour. Also, Vertic and Luvic can be used in some soil varieties. If Calcocambisols have an umbric horizon, then these soils can correlate to Cambic Leptic Umbrisols with the addition of Chromic or Rhodic as supplementary qualifiers. Umbrisols cannot be Rendzic, and therefore, the parent material cannot be explained in the name of this soil. Also, they can correlate to Leptic Cambisols, and Chromic or Rhodic can be added as principal qualifiers to explain soil colour, whereas the Rendzic qualifier cannot be used to explain the characteristics of calcareous parent material, as soils do not have mollic horizon in this case. We avoid using the textural differentiation in national Calcocambisols to refer them to Luvisols, as the pedogenic alteration in Calcocambisols appears in situ, whereas in Luvisols, it is mostly obtained by clay accumulation after migration. Classification of Calcocambisols in WRB requires more research from local soil scientists.

Position in an evolutionary series. Calcocambisols are formed from Cambic (Brownised) Calcomelanosols. Increased soil moisture content and microbiological activity trigger faster mineralisation of humus and decomposition of humus–mineral complexes, and these processes increase the content of clay and foster the creation of a polyhedral soil structure. The increase in clay content also occurs throughout the process of argillo–accumulation, which means that the clay accumulated in (B)–horizon takes a part of the insoluble residual material of the parent rock. The formation of clay after the decomposition of humus–mineral complexes is a characteristic of Calcocambisols, whereas in Eutric Cambisols, the content of clay increases exclusively throughout the process of argillo–genesis (argillo-synthesis), which means through the processes of weathering of primary minerals and the synthesis of secondary minerals from weathered products.

Typical (Haplic) Calcocambisols can further evolve into the next stage of soil development, i.e., into Luvisols, and this occurs mainly in depressions, karst sinkholes, and dolines, or on flat terrain, where the depth of the soil is greater (>60 cm), and topographic conditions promote descendent migration. The evolution of Calcocambisol is as follows:

Cambic (Brownised) Calcomelanosols (increased mineralisation of humus, release of clay minerals, and accumulation of insoluble residual material)—Typical (Haplic) Calcocambisols (thicker (B)–horizon than A–horizon, the mineral component is derived from insoluble residual material)—Illimerised (Luvic) Calcocambisols (decreased humus content, differentiation of clay in the soil profile, migration of clay throughout the thickness of soil profile, initial formation of E–horizon)—Luvisol.

The subtypes of Calcocambisols distributed in Montenegro are Typical (Haplic) Calcocambisols and Illimerised (Luvic) Calcocambisols, where the E–horizon begins to form. The soil varieties are differentiated according to the depth of the soil, which can be shallow, moderately deep, and deep (>50 cm), whereas the forms are differentiated based on soil texture to loamy and clayey Calcocambisols.

Distribution. Calcocambisols cover an area of 43,236 ha. They are distributed all over the Montenegrin territory where hard limestones and dolomites are present on variable elevation ranges, from 150 to 1700 m a.s.l. They are distributed in associations with Calcomelanosols and Terra Rossa's, but are less encountered compared with these two soil types. The larger areas of Calcocambisols are located around Sutorman and Rumija mountains, on sideslopes above Lake Skadar, in Lješanska Nahija (area), and Komani. Also, they are distributed in the Kuči and Malesija areas. In the Banjani area, they are found within the 500–1000 m a.s.l elevation range. In the northern part of Montenegro, Calcocambisols are distributed in the basins of Tara, Lim, and Ćehotina rivers, on sideslopes of Sinjavina Mountain in Lipovo and Štitarica, as well as on the eastern sideslopes of Mountain Komovi and Vujeva, and northern aspects of Visitor, Zeletin and Pogana Glava (Fig. 5.40). They are also distributed in Šekular, from Planinica to Mokra Mountain, around Rožaje, in the area around Pljevlja, like Otilovići, Krće, and Krupice. Nevertheless, Calcocambisols are distributed in a fragmented manner, like lenses in the bodies of Calcomelanosols and Red Soils, and are not encountered as big soil complexes. The spatial distribution of Calcocambisols is presented in Fig. 5.41.

Soil characteristics. The thickness of Calcocambisols varies depending on the presence of rock outcrops and the irregular appearance of rocks in the subsurface, but the soil is mostly thin to moderately deep (25–50 cm). If there are rock outcrops, then the earthy material is located between exposed rocks that emerge on the surface of the terrain (Fig. 5.42). The total depth of the soil can be from the shallowest to the deepest, at very short distances, because the boundary between the soil and the limestone rock is broken. There are noticeable differences between the A and (B)–horizons in terms of soil texture. A–horizon is mostly loamy, whereas (B)–horizon is clayey, which means that the soil has a heavy texture, but despite this, it does not have bad soil water characteristics. Calcocambisols have a well-developed soil structure and stability, enhancing water permeability and aeration, similar to Terra Rossa Soils (Red Soils).

The content of humus varies from 4 to 8% at lower altitudes and on arable land, whereas under natural vegetation, it reaches 12–15%. The humus content decreases with depth, but the (B)–horizon also contains 3–5% of humus. In

Fig. 5.40 A zone of Brown Soils formed on hard limestones (Calcocambisols), Kamena Luka, (*Photo* G. Roganović)

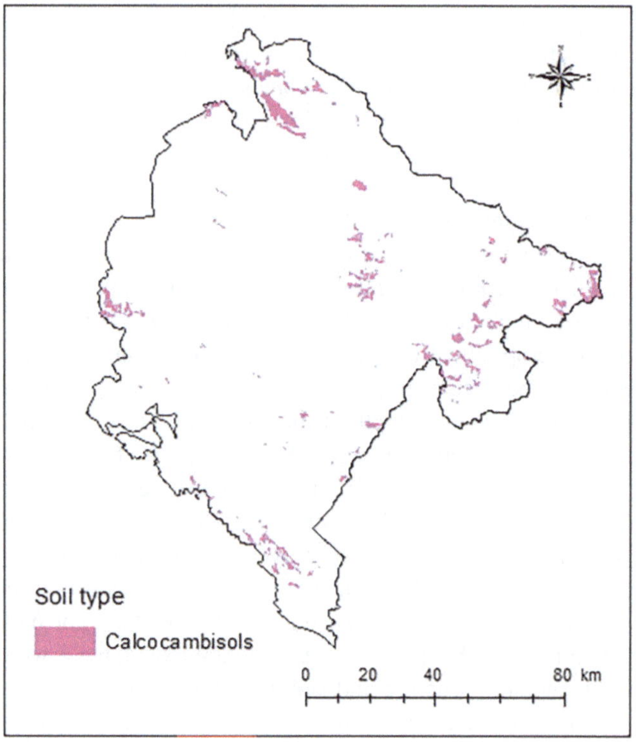

Fig. 5.41 The distribution of Calcocambisols in Montenegro

terms of humus content, Calcocambisols are between Terra Rossa and Calcomelanosols but are closer to Black Soils. These ratios are similar to the position in an evolutionary series. This is understandable because these soils are genetically closer, and the climatic conditions and vegetation favour the creation of more organic matter in the higher and colder areas. Although the soil is formed without the presence of $CaCO_3$, it sometimes contains calcareous gravel that affects the soil reaction. Therefore, soil reaction varies from acid to neutral. The reaction is closely related to the soil development stage. In the initial formation stages, there are enough Ca^{2+} and Mg^{2+}—ions that neutralise humic acids. In the later stages, $CaCO_3$ is leached, and the type of plant cover, altitude, slope, aspect, and higher amounts of precipitation contribute to soil acidification. Base saturation is high (> 60%), and so is the cation exchange capacity. This type of soil is poor in phosphorus, as all soils on limestones. The A–horizon is moderately or well supplied with potassium, while its content in the (B)–horizon is somewhat lower. Nitrogen supply depends on the degree of mineralisation of humus.

Characteristics of some Calcocambisols are presented in Table 5.11.

Fig. 5.42 Brown soils formed on hard limestones (Calcocambisols) (*Photo* Lj. Životić)

Soil management and utilisation. Calcocambisols are encountered at different altitudes and in different landscape and climate conditions (Fig. 5.43). The thin solum and rock outcrops on top of the terrain influence its utilisation mainly as woodland shrubs, degraded forests, and pastures, which is especially pronounced in the southern drier area of Montenegro under the influence of the Mediterranean climate. Only the deep soils of karst sinkholes, dolines and flat terrain are used for cultivation, but yields are low and unstable, especially in dry years. In the high–altitude and continental areas, there are good pastures and forest communities on Calcocambisols, as well as meadows, where the bare rocks are less present on the top of the terrain. These soils are not quite suitable for arable use, except in varieties thicker than 50 cm, and also not quite adequate for intensive commercial orchards that require soil cultivation.

5.1.3.4 Terra Rossa

The name of the soil type. The name Red Soils or Terra Rossa Soils comes from the red colour of this soil, which is related to the presence of the mineral hematite. This term, originating from the Italian word *terra,* which means land or soil, and *rossa,* which means red, entered

Table 5.11 Particle size distribution and chemical characteristics of Calcocambisols (adopted from Fuštić and Đuretić 2000)

Location	Soil hori-zon	Soil depth (cm)	Gravel (%)	Sand (%)	Silt (%)	Clay (%)	pH in water	pH in KCl	$CaCO_3$ (%)	Humus (%)	T (meq/100 g)	BS (%)	P_2O_5 (mg/100 g)	K_2O (mg/100 g)
			>2 mm	2–0.02 mm	0.02–0.002 mm	<0.002 mm								
Gornje Drume	A	0–16	0.0	41.9	18.3	39.9	5.87	5.10	0.0	7.46	24.9	71.6	2.8	15.7
	(B)	16–25	0.0	29.0	10.1	60.9	6.55	5.59	0.0	3.17	25.3	88.1	3.3	14.2
	(B)	25–53	0.5	25.7	12.3	62.0	6.38	5.44	0.0	2.46	24.9	79.8	5.8	13.4
Poljane	A	0–18	0.0	37.1	34.1	28.8	5.36	4.19	0.0	5.00	27.5	57.9	0.8	20.7
	A(B)	20–40	0.0	31.2	29.00	39.8	5.81	4.49	0.0	4.66	24.4	54.5	0.3	9.4
	(B)	60–80	0.0	28.8	20.8	50.4	5.20	3.99	0.0	0.78	26.0	46.8	0.2	12.7

Fig. 5.43 Soil sampling in the winter period–sampling at a customer's request (*Photo* M. Knežević)

the international soil literature. Accordingly, the local term Crvenica is also derived from the word—*crveno*, which means red. Terra Rossa Soils are also described as Mediterranean Red Soils. When referring to Red Soils, we are referring to and considering only the soils formed on compact limestones and dolomites, mostly of Mesozoic age, and not to red soils formed on siliceous parent materials.

Soil genesis and soil horizon sequence. There are many theories about the genesis of Red Soils, but two are the most interesting. According to one theory, the red earthy material is formed from the insoluble residual material that remained after the chemical dissolution of limestones that was then subjected to another process—rubification or reddening. The process of rubification consists, according to Gračanin (1951), of several phases, namely: (1) decarbonisation—leaching of $CaCO_3$ and the accumulation of residual material, (2) destruction of the alumino-silicates of the residual material and formation of clay minerals, (3) desilication—removal of SiO_2 by intense weathering and leaching, and (4) accumulation of iron and aluminium sesquioxides.

These processes took place in the conditions of the tropical climate that prevailed during the Tertiary period, which is why it was considered that Red Soils are relict soil, and according to some other interpretations, Red Soils are older than bauxite, i.e., bauxite was formed from Terra Rossa Soils by its further weathering and desilicification (Marić 1965).

According to the second theory, Red Soils are formed by the chemical weathering of limestones, which also occurs in current conditions under the influence of a warm Mediterranean climate but does not go through all specific processes of rubification. After the dissolution of limestones, $CaCO_3$ is lost in the form of bicarbonates, and from the insoluble residue, which was formed before it was incorporated into the limestone rock, Red Soils are formed (Resulović et al. 2008). According to Pavićević (1958), rubification is a recent process in Montenegrin

Mediterranean environments. Therefore, the Red Soils genesis is similar to that of Calcocambisols, with the difference being that the characteristics of the insoluble residual material as products of further chemical weathering are slightly different.

Regarding the modern pedo conditions in which the Red Soils were formed in Montenegro, the climate, parent material, and relief certainly had the greatest importance and impact, whereas the vegetation and human activities had a somewhat lesser influence. These conditions and factors were described in detail by Fuštić (2000), who stated that Red Soils are a type of soil that, under the conditions of the Mediterranean climate, is formed exclusively on limestones and dolomites of Cretaceous, Jurassic, and Triassic age. These carbonate rocks are pure (> 99% $CaCO_3$) and with minimal content of insoluble residual material, i.e., 0.2–0.4% in Cretaceous and Jurassic, and < 1% in Triassic limestones and dolomites. The influence of the Mediterranean climate on the formation of Red Soils is manifested in the coastal area and the Lake Skadar basin. The basic characteristics of this climate are hot and dry summers and a very humid and relatively warm autumn–winter period.

Karst relief, as a dry habitat with a high percentage of rocks and stones, pronounced aridity during the growing season, and other factors, does not create suitable conditions for higher biomass production. That is why the vegetation is dominated by maquis and transitional woodland shrubs, degraded low forests, i.e., xerothermic communities of deciduous trees and grasses with bushes and weeds (Fig. 5.44).

The karst relief has specific landforms such as dolines, sinkholes, small fields, and plains above which craggy sides, ridges, and peaks arise. These forms of relief, without or with very sparse vegetation, are exposed and subjected to erosion by water and wind (Fig. 5.45). The material transported by intense water erosion is deposited in depressions but also lost underground through numerous cracks, fractures, sinkholes, caves, and other forms. Depending on the aspect, Red Soils occur at an elevation of up to about 500–600 m. The relief with its forms has the greatest influence on the thickness of the soil layer, which has an A–$(B)_{rz}$–R soil horizon sequence. In natural conditions, the thickness of the A–horizon is up to 15 cm; it usually has a dark brown colour or with a reddish nuance and most often crumbly or

Fig. 5.44 Shallow Terra Rossa, Plavi Horizonti (*Photo* M. Knežević)

Fig. 5.45 Shallow Red Soils at Spuška Glavica, Spuž (*Photo* G. Roganović)

fine granular structure. The (B)–horizon has various thicknesses, red colour, polyhedral structure, and it is much more compacted than the A–horizon. When observing the soil profile along the roads, on cut sideslopes, it is noticeable that the Red Soils depth is very uneven, and that at very close distances shallow, moderately deep and deep soils are interspersed with wedge-shaped or pocket-like limestone rocks. The greatest depth of Red Soils is found in plains, dolines, sinkholes and smaller fields, where the colluvial material is deposited, and also on terraces where humans played the main role.

The WRB corresponding RSG—Red Soils and Calcocambisols were reported, according to some authors, to not have an adequate name in WRB for soil resources (Pavlović et al. 2017; Bašić 2013).

Knowing the soil characteristics and morphology, we can deduce that these soils can correlate to RSG's of Phaeozems, Umbrisols and Cambisols, whereas in the older version of WRB (IUSS Working Group WRB 2006), Bautista et al. (2011) stated that Red Soils are classified as Cambic Leptosols, Cambisols, Luvisols, Nitisols and Haplic Vertisols (Chromic). If Red Soils have a mollic horizon, they can be Cambic Leptic Rendzic Phaeozems ([6]Rhodic). The supplementary qualifiers Rhodic might be used in the case of appropriate soil colour. Also, Luvic can be used in some soil varieties. If Red Soils have an umbric horizon, then these soils can correlate to Cambic Leptic Umbrisols with the addition of Rhodic. Umbrisols cannot be Rendzic and therefore the parent material cannot be explained in the name. Also, they can correlate to Leptic Cambisols, and Rhodic can be added as principal qualifiers to explain soil colour. Also, Eutric or Dystric can be used to refer to base saturation, and the supplementary qualifier Colluvic can be used for Colluvial Red Soils (Fig. 5.46). Some authors are relating Red Soils to Chromic and Rodhic Cambisols (Miloš and Bensa 2019; Durn et al. 2019), Leptic Chromic Rhodic Luvisols (Clayic, Colluvic), or Chromic Luvisols (Agadzhanova et al. 2021). We avoid classifying Red Soils as Luvisols since the term Luvisol in the national classification is strictly related to clay illuviation.

[6]The difference between Calcocambisols and Red Soils is in the ratio of SiO_2 to R_2O_3 in residual material, which is higher than 2 in Calcocambisols, which contain more goethite mineral, whereas in Red Soils this ratio is smaller than 2 and Red Soils are characterised by dehydration of Ferrous hydroxides and formation of hematite which is known by its red colour. Goethite on the other side could have yelowish to reddish, reddish–brown to dark–brown colour. Therefore, reddish colour found in Calcocambisols could sometimes be confused with red colour of Red Soils. Tentatively we decided to add Chromic and Rhodic qualifiers for Calcocambisols, and Rhodic qualifer for Red Soils.

Fig. 5.46 Buried Red Soils, Ulcinj, Briška Gora (*Photo* J. Vasin)

Position in an evolutionary series. There are no unified theories about the genesis and evolution of Red Soils. One point of view, which is acceptable to us, is that the Red Soils is a member of the evolutionary series of soils formed in the karst landscape. This can be presented as:

Lithosols—Calcomelanosols—(Organic—Organo–Mineral—Cambic or Brownised, Chromic or Rhodic according to colour)—Terra Rossa Soils (Typical—Luvic)—Luvisols.

According to some opinions, Red Soils were formed in an earlier development cycle, in Paleo-climate, and incorporated into limestones as an insoluble residue, with Red Soils being the parent materials for further soil development, following this pathway:

Terra Rossa Soils (Cambic—Luvic)—Luvisols (Typical—Acric)—Pseudogleys.

The formation of Pseudogley Soils from Red Soils is not typical for Montenegro.

In addition to these theories, there are opinions and some evidence suggesting that Red Soils are formed on siliceous rocks, Eocene sediments and that insoluble residual material was formed of the remaining aeolian materials and tertiary tuffs found in the Dinaric karst, i.e., tuff layers in "Crvena Stijena" (Red Rock) deposits near Bileća and aeolian materials in the Red Cave near Trebinje (Basler et al. 1966). These locations are in the border zone of Montenegro and Bosnia and Herzegovina.

Some pedologists, both in the world and from the territory of the former Yugoslavia, do not dispute the possibility that Red Soils are inherited or relict soil, but they believe and provide concrete evidence that these soils are formed under modern conditions. Some other authors position the Brown Soils on limestones before Red Soils in the developmental series of soils formed on limestone, whilst others, again, consider that Red Soils can represent the first stage of soil formed on limestones.[7]

Distribution. Terra Rossa Soils cover an area of 84,149 ha on the territory of Montenegro and, according to the surface area they cover, are ranked third among the soils distributed in the country. On the Montenegrin coast, Red Soils occur in a larger area in the Bay of Kotor. From the border with Croatia and Debeli Brijeg, these soils cover the Vitaljina hill up to the sea coast and then to the Luštica Peninsula and Grbalj up to Tivatsko and Mrčevo fields on one side, and the sea coast on the other side. Upper Cretaceous limestones with interlayers of dolomite are represented here, and typical Red Soils were formed on them.

In a much smaller area, Red Soils are distributed in the area between Budva and Bar, where the Triassic limestones are interspersed with the flysch zone of Paštrovići, above Petrovac, Buljarica and Sutomore. From Bar to Bojana River, Typical Red Soils appear on a larger area, also on the Cretaceous limestones of Volujica, Dobra Voda, Pećurice, Bijela Gora, Mavrijen, Šasko hill, Briška hill, and Fraskanjel.

In the Lake Skadar basin, Red Soils are distributed on the islands and along its southern coast through Krajina and Šestane up to Virpazar, where the parent material is Jurassic limestones, while from Virpazar to Rijeka Crnojevića, the parent material are Triassic limestone. Jurassic limestones occur from the valley of Rijeka Crnojevića up to Malo Blato, while the Cretaceous limestones are found in Lješanska Nahija, Komane, Zagarač, and other hills, spreading up to Bjelopavlićka Plain.

[7] In the coastal area of Montenegro Shallow Red Soils are found without the noticable presence of A–horizons, and evidence of $A\text{–}(B)_{rz}\text{–}R$ soil horizon sequence. It seems more as A–R soil horizon sequence. Surface horizon has red colour, thickness from few to 20 cm, and it is moderately rich to rich in humus, has often slightly acid soil reaction and moderately high to high cation exchange capacity. These Red Soils are considered eroded but in fact they may be formed as a first stage of development of soils on limestones, where accumulation of humus is higher but the colour of the soil is masked due to prounanced formation of hematite in the soils.

Along the edge of the Zetska Plain, Red Soils extend from the border of Albania to Mala Rijeka and Morača, covering the area of Malesia, Kuči, Bratonožići, and Piperi. The continuous appearance of Red Soils is somewhere split with flysch sediments, as on the edges of Bjelopavlićka Plain, from Martinići to Glava Zete and Bogetići. Spatial distribution of Red Soils is presented in Fig. 5.47.

Soil characteristics. The depth of the Red Soils varies spatially depending on the rock outcrops and relief. Differences in soil texture between A and (B)–horizons are noticeable, especially under natural vegetation. That is a consequence of the higher humus content of the surface horizon, which has a slightly lighter texture than the (B)–horizon, which has more clay. Thanks to its stable polyhedral soil structure, these soils are well water–permeable, with a uniform ratio of micro and macropores, and also well aerated, although sometimes with very heavy subsoil. The stable structure of this soil is the basis for the most important physical characteristics—water capacity, water infiltration, and thermal properties. The stable structure is explained by the saturation of the adsorptive complex with Ca^{2+} and Mg^{2+}–ions, which are very effective coagulators of microaggregates. The joining of microaggregates and the formation of macroaggregates occurs through irreversible coagulation of iron and aluminium hydroxides. Red Soils rarely contain gravel, which is a consequence of weak physical weathering of limestones. There is also a low content of coarse sand, usually 1–3%, and it is slightly higher in Red Soils formed on dolomites and dolomitic limestones. Fragments of limestones are found in the soils on terraces, where they were left behind after the clearing of stones and building of the dry subwalls of the terraces. The porosity of Red Soils is more favourable in A than in (B)–horizon, and this enables good infiltration in the surface and weaker filtration throughout the subsurface horizon. Nevertheless, when moistened during a rainy period or after irrigation, Red Soils aggregates do not break down to the extent they do in other clay soils.

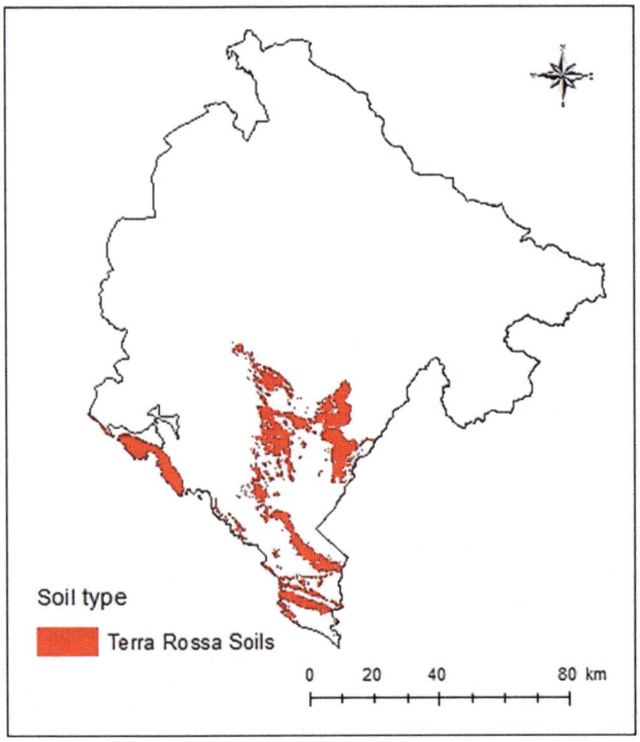

Fig. 5.47 The distribution of Terra Rossa Soils

Although formed on limestones, Red Soils do not contain $CaCO_3$, even in the initial stages of formation, which is the case with other soils formed on compact limestones. Soil reaction is acid, slightly acid to neutral (5.5–7), but it can also vary, more in the direction of higher acidity (pH in water < 5.5) than on the side of neutral or alkaline. Fluctuation of pH, not only spatially but also in the same soil profile, indicates that the soil reaction and other chemical properties are affected by the stage of development, altitude and aspect, type of vegetation cover and humus content, as well as with land use. Red Soils with shallower sola have lower pH values. This applies also to soils in northern aspects, which are more humid and colder, and soils at higher altitudes. The pH is also lower under dense natural vegetation and with higher humus content unless erosion is more pronounced. Red Soils are rich in bases, 20–45 meq/100 g of soil, whereas the base saturation is often between 70 and 95%. The ability to adsorb basic cations is explained by the fact that Red Soils are rich in clay, have a higher content of humus and clay, and amphoteric colloidal particles are present, which in Red Soils behave as basoids, attracting certain anions, including the phosphoric acid anion. The content of humus in Red Soils varies. Higher content of 4–11% has a surface horizon of the shallow varieties, which are dominant in Montenegro, and the thickness of this layer is usually up to 15 or 20 cm, under natural vegetation and without significant erosion. Moderately deep and deep Red Soils of dolines, sinkholes, plains, and even terraces are rich in humus. When they are used as arable land in the surface layer (up to 30 cm), they contain 3.5–6% of humus and the humus content gradually decreases with depth, but the layer up to 60 cm depth also contains 2.5–5% of humus. Shallow Red Soils varieties are poor in phosphorus, because they contain less than 10 mg/100 g of soil of this element in an easily available form. Moderately deep and deep Red Soils have moderate content of phosphors in the surface layer (10–20 mg/100 g), which was increased through cultivation and regular fertilisation, but the deeper layers are poor, which is unfavourable for plants with a deeper root system, especially for orchards and vineyards. They are most often moderately and well supplied with potassium, not only in surface layers but throughout the entire soil profile thickness.

Characteristics of some Terra Rossa Soils are presented in Table 5.12. Terra Rossa in Briska Gora is eroded.

Soil management and utilisation. Red Soils are of great importance as agricultural soils in the Montenegrin coastal narrow strip and the Lake Skadar basin. Both zones are exposed to the influence of the Mediterranean climate, which enables the cultivation of subtropical plants, i.e., citrus and olives in the coastal area, and other types of fruits in both zones — figs, pomegranates, kiwi, Japanese apples, peaches and others, as well as all kinds of vegetables, vines and tobacco.

All the plants grown on the Red Soils have high-quality yields, the height of which depends on the soil and application of irrigation, especially during the dry growing season (Fig. 5.48). The olives grown on Red Soils produce olive oil of better quality, and the Vranac and Kratošija grape varieties provide exceptional quality for the famous Montenegrin wine "Vranac" and "Lozova Rakija" (grape brandy). The increasingly intensive tourism development also requires better quality, not only of products—wine, rakia, brandy, olive oil with geographical indication but also of fruits and vegetables that can be grown very successfully on the Red Soils. Very shallow and shallow Red Soils in both areas are overgrown with garrigue (scrubs) and maquis, low forests and pastures, with several types of forest fruits like wild pomegranate, plantain, chestnut, or are rich in honey and medicinal plants, which is important for beekeeping.

Moderately deep and deep Red Soils, also those with colluvial material, are good agricultural land despite the lack of phosphorus (Fig. 5.49). However, agricultural crops do not have stable yields due to high temperatures, lack of moisture due to insufficient rainfall, and high evaporation. Therefore, irrigation is the most important measure for more intensive production. The aridity of the karst region limits the water supply for irrigation, which requires greater investments for the construction of small reservoirs and accumulations and the construction of village water supply systems that will utilise water from karst springs, often found in the area of Red Soils. In case irrigation is not feasible, it is necessary to utilise manure and other organic fertilisers, including peat from the shores of Lake Skadar, which in its processed state would be very suitable for nursery production, greenhouses and perennial fruit and vine plantings, fertilisation, deep tillage, ploughing and other agricultural measures, including mulching, allow the soil to be enriched with organic matter and contribute to a better accumulation and conservation of moisture in the soil. By taking all these measures, it is possible to achieve two

Table 5.12 Particle size distribution and chemical characteristics of Red Soils (adopted from Fuštić and Đuretić 2000)

Location	Soil horizon	Soil depth (cm)	Gravel (%)	Sand (%)	Silt (%)	Clay (%)	pH in water	pH in KCl	$CaCO_3$ (%)	Humus (%)	T (meq/100 g)	BS (%)	P_2O_5 (mg/100 g)	K_2O (mg/100 g)
			>2 mm	2–0.02 mm	0.02–0.002 mm	<0.002 mm								
Briska Gora	A	0–15	0.0	51.2	29.9	18.9	6.80	5.60	0.0	6.12	25.7	90.6	0.8	30.5
Gornji Kokoti	A	2–13	2.0	43.9	33.9	22.2	6.79	5.56	0.0	7.49	64.6	95.9	0.7	49.2
	(B)	20–30	0.0	24.9	11.8	63.3	6.69	5.45	0.0	3.2	54.2	90.0	0.2	16.3
Fraskanjel	A	0–20	0.0	30.9	18.8	50.3	6.93	5.91	0.0	2.54	29.6	63.4	5.0	14.0
	(B)	40–70	0.0	32.3	14.7	53.0	6.78	5.95	0.0	0.95	25.1	90.6	4.0	9.6
Progonovići	A	0–17	0.2	32.5	35.5	32.0	5.62	4.68	0.0	4.48	16.5	58.2	2.0	48.0
	(B)	17–40	0.0	34.3	26.9	38.8	6.10	5.42	0.0	3.73	13.4	71.8	1.7	19.4

Fig. 5.48 New olive groves at Terra Rossa Soils, Grbalj (*Photo* G. Roganović)

Fig. 5.49 Buried Terra Rossa at the toeslopes of Hum hill (*Photo* G. Roganović)

harvests per year on Red Soils, because the thermal and light conditions are very favourable during the very long vegetation period.

Red Soils are prone to soil erosion by water and wind. Without vegetation cover or where it is scarce and thinned—devastated, rainfall, which are abundant in the autumn–winter period, causes strong surface erosion. Also, the strong and powerful northerly wind named Bura (Italian Bora) quickly dries up soil surface layers, removes and carries away particles of detached soil. Effective protection is possible only by maintaining the vegetation cover, which is achieved by using it in such a way that the time without plant cover is shortened. Where possible, afforestation should be carried out.

5.1.4 Class of Anthropogenic Soils [P–C or P–(B)–C Sequence of Horizons]

The name of this soil class comes from the Greek word *anthropos,* which means man, and the Greek word *genesis*, which refers to creation or origin. This implies that these soils are formed and modified by human activities, and consequently acquired new characteristics.

In historical papers and soil classifications in Yugoslavia, soils on terraces were described and presented on soil maps as Anthropogenic Soils. Terraces were created by humans, most often by raising walls transversely to the slope of the terrain and then filled with earthy material, so the terrace plateau was formed through soil erosion and ploughing. In this way, humans created arable soil and, at the same time, protected it from the destructive effects of erosion. A new Croatian soil classification system distinguishes Soils of terraces as a soil type in the class of Anthropogenic Soils (Bašić 2013). The area under terraces in Croatia is significant, although not reported. The exact area under terraces in Montenegro is not known, but it is not negligible in the coastal zone and around Lake Skadar. Therefore, these soils might be included in some new revised version of soil classification in Montenegro, especially the terraces formed on flysch and limestones.

According to the current national classification, this class of Autogenic Soils includes soils that exhibit changes in their natural properties due to human impact, like long-term cultivation and other agrotechnical measures. These changes affected the entire thickness of the soil profile or part of the solum, and sometimes also part of the parent material, and based on the modified soil properties, the soils formed in this way are classified as Anthropogenic Soils. These soils differ from Anthropogenised Soils, in which the cultivation has changed only the properties of the surface arable layer. According to national classification, at the level of soil type, this soil class comprises only those soils that represent new soil formations and which cannot be classified into any known soil type due to their soil characteristics.

Anthropogenic Soils are distinguished from other classes and types by soil horizon sequence. They are characterised by the P–horizon, formed by cultivation, mixing, and homogenisation of one, two, or more natural soil horizons or layers, including the loose parent materials in some cases.

The class of Anthropogenic Soils in Montenegro distinguishes Deeply Ploughed Soils—Rigosols and Vegetable Garden Soils—Hortisols.

5.1.4.1 Rigosols

The name of the soil type. The name Rigosols is derived from the French word *rigole* which means furrow or drainage ditch, and denotes a process that is carried out to improve the productive capacity of soil, and the Latin word *solum,* which means soil. "Rigolovanje" is an internationalised word and it means deep ploughing (from 60 to 120 cm) with special ploughs called "Rigolers". Deep ploughing at this depth mixes soil layers, and the soil becomes homogenised over time. That is how a new type of soil with a new soil horizon sequence and its own characteristics is formed (Fig. 5.50).

Soil genesis and soil horizon sequence. The most important process of the genesis of this soil type is the formation of the anthropogenic P–horizon (P comes from the English—to plough) by deep cultivation and mixing of two or more horizons of the original soil in combination with other measures, fertilisation, removal of gravels and stones, etc.

The soil horizon sequence, therefore, has a P–horizon, which has been completely altered in comparison to the previously naturally formed horizon or the horizon above which it lies. When the P–horizon lies above a shallow layer with abundant stones or gravel, this layer should be removed in order to deepen the physiologically active layer for root growth. Rigosols can be formed from different types of soils, and in Montenegro, the following are mostly prone/suitable to anthropogenisation: Brown Eutric Skeletal Soils, Skeletal Rendzina Soils, Alluvial–Colluvial Soils and Luvic Eutric Cambisols.

The WRB reference RSG—Terric Anthrosols.

Position in an evolutionary series. The formation and evolution of Rigosols depend primarily on the original type of soil from which it was formed and the stage of its development. Thus, in the zone with pronounced leaching and vertical descendent fluxes, the evolution goes in the direction

Fig. 5.50 Rigosols near Savina, Herceg Novi (*Photo* J. Vasin)

of leaching, acidification, eluvial–illuvial migration, and textural differentiation. In Rigosols formed on shallow skeletal or skeletal soils, as well as Alluvial–Colluvial Soils, migration of individual components and textural differentiation of the soil profile is possible due to excess precipitation in the autumn–winter period and due to irrigation. The criteria for classifying Rigosols into subtypes according to the national classification is conducted by land use, i.e., agricultural systems into: Vineyard Soily—Vitisols, Orchard Soils—Frutisols, and Arable Deeply–Ploughed Ameliorated Soils.

Distribution. In Montenegro, Rigosols have small distribution, but these soils appear in several areas on smaller complexes where the soils have been improved by the application of agrotechnical and reclamation measures. The largest complex, with an area of about 2500 ha, is the Ćemovsko Field near Podgorica (Fig. 5.51), where grapevine and peaches are grown in modern plantations, and olive orchards have been planted in recent years. Smaller complexes under orchards and vineyards are also distributed around Podgorica (Sadine, Pavlovine, Lješkopolje and Donja Gorica), then Crmničko Field and Bjelopavlićka Plain (Jelenak, Velje Polje). There were several complexes on the Coast (Sutorina, Tivatsko, and Mrčevo fields) and the surroundings of Ulcinj (Ulcinjsko Field, Zoganje), where there were mainly citrus and olive groves and vineyards.

Soil characteristics. Deep ploughing and gradual soil homogenisation enabled the formation of the P–horizon, and many properties of the original soils were changed. Most of the Rigosols of Ćemovsko (Fig. 5.52) and other fields around Podgorica, Bjelopavlićka Plain, Crmnica, and in the coastal area have altered soil characteristics, starting from soil morphology and soil colour, over the content of humus and $CaCO_3$ to particle size distribution, soil structure, compaction, aeration and other soil properties. Deep ploughing ("rigolovanje") and other agrotechnical and reclamation measures were mandatory in the design of intensive orchards and vineyards. The nutritive fund in soil was enriched through ameliorative fertilisation, which was an obligatory measure, and regular annual fertilisation. Many macro- and microelements got into the Rigosols through the use of chemicals for plant nutrition and protection. Their utilisation was regular and more intensive in plantation orchards and vineyards compared to field crops.

Soil management and utilisation. Rigosols are mostly used for orchards and vineyards. Since tree plants have a deep root system, they need a deep physiologically active layer in the soil in which the root system will penetrate and strengthen. Without deep ploughing, the parent material of glaciofluvial gravel with conglomerates, as in Ćemovsko Field (Fig. 5.53), cannot provide favourable conditions. The situation is similar with the diluvial loams of the Bjelopavlićka Plain, which have a lot of clay and are water impermeable, as well as the soils that are formed on them.

Recommendations for better use of Rigosols are quite similar to those for the original soil types they were formed of. In addition, the intensity of changes in the characteristics of Rigosols should be monitored in order to gain new knowledge and determine their characteristics, and thus valorise this type of soil in the best way possible. Rigosols are formed anthropogenically in order to be characterised by good fertility, and after soil reclamation, their productive capacity is higher, and Ćemovsko Field is a good example of this—a case study.

5.1.4.2 Hortisols

The name of the soil type. The name Hortisols comes from the Latin words *hortus,* which means garden, and *solum,* which means soil, therefore referring to soils in vegetable gardens and gardens.

Soil genesis and soil horizon sequence. Hortisols or Vegetable Garden Soils are formed through a combination

Fig. 5.51 Vineyards of Plantaže Company, Ćemovsko Field (*Photo* G. Roganović)

of different agro-technical measures, such as intensive fertilisation and manuring, irrigation, and (deep) tillage, i.e., through long-term anthropogenic impact, in order to enhance the production of vegetables, flowers, fruits and vines around utilised kitchen gardens and/or gardens. Soil cultivation and incorporation of larger doses of organic and mineral fertilisers have caused changes in the soil properties up to a depth of 50 cm. In this way, the P–horizon is formed, enriched with nutrients, and biologically very active since it is also regularly fertilised with organic fertilisers (Fig. 5.54). Therefore, the conditions for the growth and development of microorganisms and fauna, primarily earthworms, are favourable. The organisms contribute to the fact that the P–horizon is very loose, with a very favourable crumbly structure and homogeneous. The deeper layer of Hortisols that lies below the P–horizon represents a part of the original soil, and it can be a part of a horizon or, most often, the parent material, which is why the soil horizon sequence is P–C, but also P–(B)–C.

The WRB corresponding RSG—Hortic Anthrosols.

Position in an evolutionary series. The formation and evolution of Hortisols depends on the primary soil they were formed of. As these soils are characterised by high and sometimes excessive inputs of water and organic and mineral fertilises, such measures jointly affect the process taking place with vertical descendent fluxes, typical of Montenegro, such as leaching of calcium carbonate, decarbonisation, acidification, and clay migration, etc. Excess amounts of water and added nutrients can jeopardise soil development and lead to soil acidification or even alkalisation, and trigger soil degradation. These processes can be followed by soil structure deterioration and an increase in soil compaction. However, such soils remain Hortisols in terms of classification, but with specific characteristics. There are no criteria for classifying Hortisols into subtypes and lower classification units.

Distribution. In the beginning, Hortisols appeared most often in utilised kitchen gardens and gardens where vegetables and flowers were grown. Recently, vegetable gardens have also been expanding into purely agricultural areas. Vegetable gardens have been recorded in the cadastre for a long period as a separate "cadastral culture" and classified into a separate class, like arable soils, meadows, pastures, orchards, and vineyards, respectively. In the past decades, Hortisols have also been formed in greenhouses.

Soil characteristics. Hortisols are usually very loose soils with a good structure and loamy texture. They have good soil water and air regimes, so the overall physical properties of this type of soil are significantly improved compared to the original soil they were formed from. Hortisols are characterised by a high level of nutrients, resulting from regular

Fig. 5.52 Rigosols of Ćemovsko Field (*Photo* B. Vrščaj)

fertilisation with high doses of organic and mineral fertilisers. For this reason, this soil is highly productive and certainly surpasses the soil it was formed from.

Soil management and utilisation. Regular application of agrotechnical measures should be continued on this soil in order to obtain high, stable, and healthy yields. It is important to monitor the nutrient level in the soil and soil's chemical and physical characteristics in order to avoid decreasing the productivity of Hortisols, their degradation, or exclusion from agricultural production due to bad soil management.

5.1.5 Class of Technogenic Soils [I–II–III Sequence of Horizons]

Technosols are soils mainly formed as a result of the exploitation of various mineral raw materials, primarily coal, ores, and others. Exploitation is mainly carried out in open pits, where overburden or mine tailings are disposed of—deposited outside or on a part of the open pit. During ore processing, mine tailings are also deposited in special landfills, where Deposols are formed. In addition to these types of materials, industrial and communal waste which may contain toxic and polluting substances dangerous and harmful to the environment, are also disposed of and present parent materials for Deposols formation.

Technosols, as new soil formations, have specific properties. They were formed in very different ways and are significantly different from naturally formed soils. Investigations of their properties are more recent, and the findings indicate that all man-made materials differ from each other from a pedological point of view: by colour, chemical reaction, the content of organic matter and $CaCO_3$, particle size distribution and content of gravel, the content of nutritive elements, presence or absence of heavy metals, pollutants, etc. In order to form soil from man-made materials, that is, to start the processes of pedogenesis, complex measures are needed to bring it to a specific purpose. Until the necessary measures are taken to remediate the individual components of man-made materials, these deposits can pose a danger to groundwater and surface water, the surrounding soil, and thus to the food chain and the environment.

In the class of Technogenic Soils, the national classification distinguishes the following types: Landfill Soils—Deposols (Soils of Disposal Sites), Technogenic Recultivated Soils—Recultisols,[8] Flotation Deposits—Flotisols, and Atmospheric Depositions—Aeroprecipitate Soils (Aeroprecipitates).

5.1.5.1 Deposols

The name of the soil type. The name Deposols is derived from the Latin words *deponere,* which means to deposit, and the word *solum,* which means soil. It indicates a deposit of overburden, waste material from coal, lead, zinc, and bauxite mines. The waste is usually a mixture of earthy material and parent material, which have very different properties and are deposited as layers on disposal sites.

Soil genesis and soil horizon sequence. Deposols are unsystematised mine tailing or overburden material that is practically a pile material of different dimensions in terms of shape, size, composition, and other characteristics. On such deposited materials, soil genesis begins after the stabilisation and the appearance of weedy (pioneer) vegetation, i.e., biotisation processes. The weathering and decomposition of waste material is a normal process followed by the accumulation of humus and the formation of the initial (A)–horizon, whereas the unchanged waste deposit represents the C–horizon. The genesis of Deposols is gradual,

[8] Recultisols do not exist in former Yugoslav classification (Škorić et al., 1985).

Fig. 5.53 Vineyards of Plantaže Company (*Photo* G. Roganović)

Fig. 5.54 Hortisol formed by agricultural management of heavy Eutric Cambisol, Jelenak (*Photo* G. Roganović)

but it depends on the type of waste that has been deposited. Marls, clays and sands mixed with earthy material are suitable for faster soil genesis. Unfavourable substrates are limestones and dolomites, which present an overburden in bauxite mines. The accumulation of organic matter, thus the formation of humus, is continuous only when the disposal site is stabilised, that is, when it is not covered with new waste above it.

The WRB corresponding RSG—Technosols.

Position in an evolutionary series. Deposols are actually the first stage of soil formation, and their genesis and evolution are closely linked to deposited materials. If formed on dense and compacted materials, they can remain at the initial stage of development, with very difficult formation of the (A)–horizon. If they are formed on some loose calcareous materials with favourable water holding characteristics, then the formation of these soils can go into the direction of the A–C soil horizon sequence, or A–(B)–C in the second stage of development. The process of humus accumulation is the most important one, with continuous vegetation growth and weathering of primary minerals and formation of secondary minerals. Therefore, Deposols have various pathways in their development, which depend on the constellation of pedogenic factors, and above all, on the characteristics of the deposited material.

Distribution. Deposols are mostly distributed near mines and factories where ore processing takes place. For example, the majority of landfills are formed in the Pljevlja area, where coal is exploited in surface mines (Potrlica, Borovica and Šumani). Coal has been exploited at the "Potrlica" mine since 1950. The coal layers are at a great depth, and huge amounts of mine tailings or overburden material pose a major problem, and disposal is carried out at several landfills in the Pljevljsko Field. Due to the lack of space and the displacement of the Ćehotina River bed, disposal is carried out by a transport system to the "Jagnjilo" landfill, which is at a significantly higher height than the open pit (Fig. 5.55).

The "Borovica" and "Šumani" open pits were completed in a relatively short time and in such a way that the "Borovica" mine was recultivated, while the "Šumani" open pit recultivation covered an insignificant area. Coal from this mine meets the needs of the "Pljevlja" Thermo–Power Plant (TPP), from which ash and slag are deposited in dedicated landfills.

The "Šuplja Stijena" mine is an open pit, and ore flotation is carried out in the location Gradac, whereas the tailings are disposed of in a landfill built along the Ćehotina River.

The "Brskovo" mine near Mojkovac carried out ore extraction by underground mining, but in one area, it is an open pit. The tailings flotation and deposition are located in the town of Mojkovac, right next to the banks of the Tara River, whose clear and clean water has been called the "tear

Fig. 5.55 Deposols, Jagnjilo—Pljevlja (*Photo* G. Roganović)

of Europe". The tailings dump was restored some time ago with colmation (silting), i.e., by applying soil deposits of a certain thickness above the tailing. During the construction works, the embankment on the Tara River and the landfill were covered with impermeable film cover, and the measures applied transformed this ecological "hotspot" into an area with standard conditions.

Bauxite is exploited in several quarries in the area of Nikšićka Župa, and the overburden consists mainly of limestones and dolomites with some earthy material, deposited in the immediate vicinity of the quarries. Separation of bauxite takes place in the city of Nikšić, while processing is carried out at the Aluminium Plant in Podgorica. The red mud dump is located next to the factory and right next to the bank of the Morača River, but it was not completely safe, which was the reason why, in accordance with the decision of the Government of Montenegro, the settlement of Velji Brijeg was evicted due to the pollution from the landfill ash and other harmful substances.

Landfills formed by industrial and communal waste occur in several places, and most of them are not constructed in accordance with the regulations and pose a potential danger to the environment.

Soil characteristics. Deposols do not have homogeneous physical and chemical properties due to the heterogeneous composition of the deposited and/or tailings material. The overburden of marls, marly clays, and clay lacustrine sediments is strongly calcareous, whereas those from siliceous materials have an acid reaction. Marls are the raw material for the cement industry. There is limited information on other properties of Deposols, as there are no adequate studies to determine their characteristics in detail, which might be done through case studies introducing regular monitoring of Deposols characteristics.

5.1.5.2 Recultisols

The name of the soil type. The name Recultisols comes from the English word *cultivate*, which means to prepare soils for growing crops, the prefix *re*—which means again, and the Latin word *solum*, which means soil and refers to the reutilisation of some soil after the application of reclamation measures. This soil type is an addition to the national soil classification in Montenegro, and it was not a part of the Yugoslav soil classification system.

Soil genesis and soil horizon sequence. Recultisols are formed when reclamation or recultivation measures are carried out on deposited materials. Therefore, they can be formed by recultivation or reclamation of Deposols and Flotisols. In Montenegro, Recultisols are distributed in a small area. Successive reclamation of a small part of the open pit Potrlica, right next to the urban settlement of Pljevlja, and of the open pit Borovica resulted in a new pedological formation—Recultisols (Fig. 5.56). This is a zone of open pits that have been exploited since 1952.

Fig. 5.56 Recultisols in Plevlja area—an area right from the road (*Photo* G. Roganović)

Reclamation measures should be carried out as a mandatory measure prescribed by the Law on mining activities, but in practice, they are conducted with a delay. During recultivation, the most important procedure should be the separate removal (cut) of soil and its deposition as a layer to the formed landfill. The next important procedure is the sowing of grasses. The process of pedogenesis, in this case, takes place faster due to the accumulation of organic matter that contributes to the formation of the initial (A)–horizon. Without the deposition of soil layer over deposited material, it takes a much longer time to cover the landfill with vegetation, except when forest trees are planted (Dožić 1978). The soil horizon sequence of Recultisols is (A)–C, and they are undeveloped soils.

The WRB corresponding RSG—Technosols. Since the process of recultivation can alter soil characteristics in Recultisols, then supplementary qualifiers like Cambic, Dystric, Eutric, Humic, Ochric, Mollic, Umbric, Skeletic, Dolomitic or Calcaric, or those that refer to soil texture, can be used.

Position in an evolutionary series. Recultisols are soils developed from Deposols and Flotisols, and they are the next stage of soil development. Their formation and development should go in the direction of humus accumulation and formation of an A–horizon, which should have a good and strongly developed structure. The development of the A–horizon with appropriate characteristics and restoration of vegetation will push also the development of the deeper parts of the soil profile. This means that the soil can undergo various soil processes, transformation of mineral and organic components and migration of matter throughout the soil profile. Hence, as the human impact is important at the first stages of Recultisols development, the anthropogenic impact is also important further on. The addition of manure and mineral fertilisers, as well as irrigation practices, can lead to the formation of soil with quite good characteristics. However, this is strongly related to previous characteristics of Deposols or Flotisols. If the soil is polluted, then all the processes will be interrupted. There are no data on the genesis of Recultisols in Montenegro.

Distribution. As already said, Recultisols are distributed in a small area of open pits around the town of Pljevlja.

Soil characteristics. The physical properties of Recultisols depend on the type of tailings and the material applied to the landfill. The disposal of marls, marly and clayey materials from coal mines to landfills is done in layers and those layers are compacted by excavators. Because of this, the deposited material is extremely dense and has unfavourable soil water and air characteristics. This favours the soil erosion processes on uneven terrain, and the processes are more intense if the vegetation cover is damaged or absent. Recultisols on calcareous parent materials, i.e., with marly substrates, often contain a high percentage of $CaCO_3$, and can potentially contain a high proportion of active lime, which causes the chlorosis of plant leaves. Recultisols are poor in nutrients, notably nitrogen, and phosphorus, especially in the initial stage of their formation. When the parent material contains some heavy metals and other types of pollutants, which can reach the food chain, it is necessary to apply remediation measures. That particularly applies to Recultisols formed at flotation materials (Flotisols) during the processing of lead and zinc ores in Mojkovac (Fig. 5.57) and Gradac, as well as at ash and slag landfills of TPP "Pljevlja".

Recultisols are anthropogenic formations created by the recultivation—improvement—of Deposols and Flotisols, and due to a better production capacity, they can be reused in forestry and agriculture in certain cases.

5.2 Order of Hydrogenic Soils

This soil order encompasses all soils that show signs of excessive wetting from rainfall, surface runoff water, subsurface water flow, flood water and groundwater, at a certain depth in the soil solum. The water that remains in the soil profile stagnates in it for a shorter or longer time and causes certain hydrogenic processes, which are the most important characteristic of these soils.

In addition to water, the genesis of these soils is also significantly influenced by topography. The flat relief and depressions favour additional moistening of soils by surface or groundwater, in addition to rainwater. In the river valleys and on the shores of lakes Skadar, Plavsko, Šasko and other lakes, the relief and hydrographic conditions have a combined effect on the redox processes, flooding, the addition of new alluvial material and the presence of hydrophilic plant communities. Excessive wetting causes the processes of pseudogleysation and the appearance of the horizons over impermeable layers, in alternating oxidation–reduction conditions, and the processes of gleysation and the formation of a G–horizon under the impact of groundwater. Gleysation causes the formation of reduced compounds of iron and manganese, and in special conditions, Peat Soils are also formed due to a large accumulation of plant residues.

5.2.1 Class of Undeveloped Hydrogenic Soils [(A)–C, (A)–I–II–III–G or (A)–G Sequence of Horizons]

This class of hydrogenic soils comprises young river, lacustrine and sea sediments without developed genetic horizons but with evidence of stratification. In the case of older stabilised sediments, the appearance of the initial (A)–horizon or A_p–horizon, as well as G–horizon, is possible.

Fig. 5.57 Recultisols in Mojkovac made by recultivation of tailings flotation of mine Brskovo (*Photo* G. Roganović)

5.2.1.1 Fluvisols

The name of the soil type. The name Fluvisols is derived from the Latin word *fluvius,* which means river, indicating the way the soil was formed, originally from river sediments, but more recently, the name Fluvisols includes, as a rule, multi-layered river, lacustrine and maritime sediments that are recent formations. In pedology, this distinguishes Fluvisols from the term "alluvium", which in geology has a chrono-stratigraphic meaning, and is a broader term because it also includes older river and lacustrine sediments, on which Cambisols and other types of soils are formed in Montenegro.

Soil genesis and soil horizon sequence. Fluvisol is a young (recent) soil formed by the deposition of alluvial material along the alluvial plain of rivers, on the shores of lakes and seas (Fig. 5.58). Fluvisols soil profile usually do not encounter developed genetic horizons, but rather layers with different texture which depend on the type of deposited material and power of water transfer. The initial humus-accumulative (A)–horizon is formed after the stabilisation of sediments and poor accumulation of organic matter. When the soil is cultivated, the initial horizon is designated as A_p, and it could have some characteristics of an A–horizon, while other layers are designated with Roman numerals, which actually represent the sedimentation layers (I–II–III…). Therefore, Fluvisols could have soil profiles with evidence of stratification and weak

Fig. 5.58 Fluvisol—Alluvial Soil, Gornji Štoj (*Photo* M. Knežević)

horizon differentiation, but a distinct topsoil horizon may be present.

Fluvisols are often renewed with a newly deposited material, and this material contains certain characteristics of the soil it originates from. That is why sediments contain organic matter, clay and other minerals, colloidal particles, biogenic elements, and are often characterised by fertility. Alluvial sediments depend on the type and composition of the material and the transmission power of the watercourse, which vary over time and along the length of the watercourse. Upstream, water speed is higher, and finer material is easily moved and carried away, while the coarser material remains. Coarse material is also deposited in the bed and on the banks of the watercourses, whereas finer material is deposited further from the banks. In the lower parts of the alluvial plains, depressions, under the influence of flood or groundwater, hydrophilic grasses or marshy vegetation might prevail, and in such conditions, Fluvisols might evolve into Fluviatile–Meadow Soils, Humogley Soils, and Marsh–Gley Soils, and sometimes even into Peat Soils. In the coastal zone of the sea, processes of gleysation are accompanied by the processes of salinisation.

The WRB corresponding RSG—Fluvisols. Some of the Fluvisols in Montenegro could have Calcaric, Eutric, Dystric, Tidalic, Gleyic, or Skeletic principal qualifiers.

Position in an evolutionary series. The position of Fluvisols in the evolutionary series is specific. The dominant process in soil formation is sedimentation, and sediment stabilisation is followed by the beginning of pedogenesis with the formation of the initial (A)–horizon. However, with the introduction of new fresh alluvial material, this process is interrupted, and the soil genesis, in this case, returns to the beginning. Floods, which occur annually, are not a desirable process because they prevent the normal development of vegetation or soil cultivation during the period affected by floods. Therefore, floods affect the speed of soil formation and pedogenic processes. The construction of embankments changes the direction of Fluvisols genesis. Fluvisols waterlogged by floodwater or wetted by groundwater can evolve into Marsh–Gley Soils or Peat Soils, or some other soils from the Hydrogenic soil order.

The division of Fluvisols into subtypes was made on the basis of the chemical composition of the deposited material and the processes of gleysation caused by the dynamics of water movement. Based on this, the following subtypes of Fluvisols are distinguished: (1) Calcareous, (2) Calcareous Gleyic, (3) Calcareous, Gleyic and Salic, (4) Non-Calcareous Fluvisols, (5) Non-Calcareous Gleyic Fluvisols. Fluvisols soil varieties for each subtype are distinguished on the basis of the depth of the active soil layer until the appearance of gravel layers, into shallow (up to 40 cm deep), moderately deep (40–80 cm), and deep (80–120 cm). Fuvisols soil forms are distinguished based on soil texture and gravel content into sandy (arenic), loamy (loamic) and clayey (clayic), and weakly, moderately and strongly skeletal, respectively.

Distribution. The distribution of Fluvisols is not wide due to the prevalence of the hilly–mountainous relief and the large expansion of karst over the territory of the country. Nevertheless, Fluvisols cover an area of around 28,607 ha. They are found in the valleys of the bigger rivers, like Lim, Tara, Bojana and Morača (Fig. 5.59), and in the smaller areas in the valleys of Ibar, Ćehotina, Ljuča and other watercourses. Fluvisols, in association with Colluvial Soils (previously described as Alluvial–Deluvial Soils), occupy significant areas of coastal bays and fields like in Sutorina, Kuti, Tivatsko, Mrčevo and Barsko fields, fields around Ulcinj and others, then in the Plav–Gusinje basin, Maoče, Ćehotina and Ljuboviđa valleys. Fluvisols are also present in the flat coastal terrain around lakes Skadar, Plavsko and Šasko. They are mostly associated with Marsh–Gley Soils and sometimes Peat Soils. These soils are exposed to long–term flooding or are constantly waterlogged throughout the year. In the Bojana River delta, Fluvisols are salinised during the period of tides, when salty seawater penetrates along the Bojana watercourse. The spatial distribution of Fluvisols and other soils from Hydrogenic Soil Order is presented in Fig. 5.60.

Soil characteristics. The most important morphological characteristic of Fluvisols is evidence of stratification. Alluvial deposits in river valleys have various thicknesses and usually a heterogeneous composition. Soil texture can be gravelly, sandy, loamy and clayey, and large differences can be observed on smaller distances, but also in the same soil profile. This is conditioned primarily by the power of the watercourse carrying the sediments and a location where the deposition of a smaller or larger amount of material is possible. This is why Fluvisols' texture can be from skeletal and gravely to clayey. The physical characteristics and soil water–physical characteristics of Fluvisols depend on the particle size distribution. The most favourable are loamy and sandy loam Fluvisols. Very skeletal and some clayey Fluvisols have poor physical properties. Most of the medium-textured Fluvisols, which are used for agricultural production, have good porosity, water permeability, and favourable aeration, which is reflected in their fertility.

The chemical characteristics of Fluvisols are quite variable. They can be calcareous and non-calcareous, and the chemical reaction varies from weakly acid and neutral to weakly alkaline. The content of humus is low in skeletal and sandy Fluvisols, but varies in the range of 1–4% in loamy and clayey Fluvisols. Sometimes, the deeper layers

Fig. 5.59 A zone of Alluvial Soils—Fluvisols, Ponari (*Photo* G. Roganović)

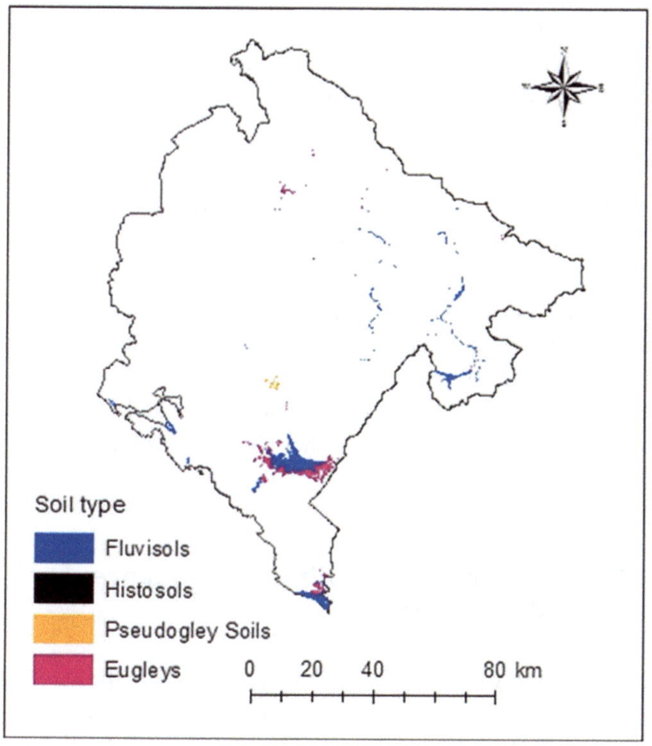

Fig. 5.60 The distribution of Hydrogenic Soils

contain more humus than the surface, which is a consequence of the manner, time, and origin of the deposited material. The content of other nutrients is related to the chemical and mineralogical composition of the deposited material.

Characteristics of some Fluvisols are presented in Table 5.13. The first profile in Gusinje is Sandy Calcareous Fluvisol. The next profile in the table is Loamy Calcareous Fluvisol, followed by two profiles of Calcareous Fluvisols (Gleyic), one profile of Calcareous Fluvisols (Gleyic, Salic), and Fluvisols (Loamic).

Soil management and utilisation. The productive capacity of Fluvisols is variable and depends on physical and chemical characteristics, primarily on the depth of the physiologically active soil and soil texture. Loamy Fluvisols with a thickness above 50 cm and a groundwater level of 1–2 m below the surface can be considered very good soil. Very intensive agricultural production takes place on Fluvisols in the coastal zone and the Lake Skadar basin. In these and other areas, their importance is reflected in the fact that they are located on a flat relief that enables the application of intensive agrotechnical measures and the small distance of water source for irrigation. High yields of cultivated plants,

Table 5.13 Particle size distribution and chemical characteristics of Fluvisols (adopted from Fuštić and Đuretić 2000)

Location	Soil horizon	Soil depth (cm)	Gravel (%)	Sand (%)	Silt (%)	Clay (%)	pH in water	pH in KCl	$CaCO_3$ (%)	Humus (%)	P_2O_5 (mg/100 g)	K_2O (mg/100 g)
			>2 mm	2–0.02	0.02–0.002	<0.002						
Gusinje	Ap1	0–20	0.0	80.9	13.8	5.3	8.01	6.76	8.4	3.61	-	-
	Ap2	20–40	0.0	77.6	18.3	4.1	8.21	6.72	10.1	2.58	-	-
	I	40–60	0.0	78.4	17.1	4.5	8.20	7.00	10.1	1.50	-	-
Crmničko Field	(A)	0–30	7.0	64.1	33.2	2.7	7.92	6.80	12.5	1.87	1.0	11.3
	I	34–54	62.8	91.7	6.3	2.0	7.92	6.50	8.6	0.90	-	-
	II	80–100	1.3	61.2	32.6	6.2	8.09	6.80	11.2	0.03	-	-
Žabljak Crnojevića	(A)	0–20	0.0	52.2	36.8	11.0	7.68	6.52	24.1	1.55	5.3	50.0
	I	20–45	0.0	72.7	19.7	7.6	7.70	6.27	27.1	1.08	5.7	36.3
	II	50–80	0.0	86.7	9.0	4.3	7.96	6.72	34.4	0.66	5.7	35.2
	III	100–130	0.0	52.4	37.4	10.2	7.80	6.52	22.4	1.13	4.0	40.0
Dodoši	(A)	0–15	0.0	65.0	27.2	7.8	7.99	7.02	25.4	1.60	5.0	30.0
	I	20–50	0.0	70.2	21.8	8.0	7.80	6.64	26.7	1.04	4.3	20.5
	II	70–110	0.0	79.6	15.20	5.2	8.01	6.87	27.5	0.78	3.7	30.0
	III	110–140	0.0	59.7	20.3	10.0	8.17	6.94	24.1	1.09	4.0	18.5
Sveti Đorđe	(A)	0–25	0.0	47.8	44.1	8.1	8.40	8.33	8.7	1.45	0.7	5.6
	I	25–50	0.0	43.8	46.4	9.8	8.54	8.45	7.4	1.03	0.4	3.8
	II	55–75	0.0	71.3	23.3	5.4	8.36	8.19	7.8	0.64	0.8	3.2
	III	80–100	0.0	50.6	42.6	6.8	8.50	8.05	3.7	0.83	0.6	3.9
	IV	110–130	0.0	61.9	25.0	13.1	8.71	8.26	2.9	0.80	1.1	4.7
	V	140–160	0.0	61.0	13.5	25.5	8.26	8.08	9.5	0.45	0.5	2.6
Gusinje	Ap	0–15	0.0	80.3	15.8	3.9	6.70	5.8	0.0	6.39	2.1	4.6
	I	15–25	26.7	82.0	14.6	3.4	6.70	5.8	0.0	2.77	2.8	2.8

especially vegetables and fruits (Fig. 5.61), indicate that loamy Fluvisols can be considered the most fertile soils in Montenegro. Limiting factors for the use of Fluvisols for agricultural purposes are the floods that occur every year in the coastal zone of the Skadar, Plavsko and Šasko lakes and the Bojana River, and occasionally in the valley of Lim (Berane, Zaton), Tara and Zeta rivers. Also, the deposition of new, fresh and rough alluvial material and the removal of fertile soil by erosion of river banks due to a sudden rise in water level, and linear erosion, if there are no embankments, are constraints. Restrictions in the use of Fluvisols should be reduced to a minimum by flood protection, watercourse regulation, and soil reclamation. Complex conservative measures should be applied not only in an inundation area where Fluvisols are distributed, but also more broadly, in the area of whole watersheds of a certain watercourse. This will prevent or at least mitigate the sudden surface water runoff and soil erosion. The most important measures to improve Fluvisols should be focussed on the regulation of soil moisture regime, and improvement of soil physical and chemical properties. This implies regulation of the level of groundwater, provision of irrigation, and fertilisation with larger doses of phosphorus and nitrogen fertilisers. A necessary agrotechnical measure is the introduction of organic fertilisers, i.e., manure and siderites to increase the content of humus, which contributes to the biological activity of the soil and improves soil structure. The content of humus is particularly low in skeletal and sandy Fluvisols, whereas the soil structure is either weakly developed or presented by single grains.

5.2.2 Class of Gleyic Soils [A–G Sequence of Horizons]

Class of Gleyic Soils is characterised by the process of gleysation and sometimes pseudogleysation, which occur in the first 100 cm of topsoil. This class consists of three soil types: Pseudogley–Gleyic Soils, Humogley Soils, and Eugley Soils. Pseudogley–Gleyic Soils are characterised by both processes, pseudogleysation, and gleysation in the soil profile. Humogley Soils are affected by the gleysation

Fig. 5.61 Protection of mandarins from frost—Fluvisols, Gornji Štoj (*Photo* M. Knežević)

process at depths of 50–100 cm from the surface of the terrain, and groundwater fluctuates during the year. Eugley Soils are affected by groundwater in the first 50 cm of the thickness of the soil profile and are, in some situations, affected by flood waters. Fluctuations of groundwater in Eugley Soils are not as large as in Humogley Soils and are formed in more humid conditions.

5.2.2.1 Eugleys

The name of the soil type. The name comes from the Russian word *glej* (gley), which means mucky mass, and the Greek prefix *eu,* which means good or well, and it denotes muddy fertile soil formation. The sub-term gleysation as the main pedogenic process occurring in these soils also arose from the word gley. As these soils are found around marshes and have a higher portion of mineral soil component than organic component, they are also called Mineral-Marsh Soils, Marsh-Gley Soils, or simply Wetland Soils.[9]

Soil genesis and soil horizon sequence. Large amounts of water and topographic position play the most important role in the formation of Eugley Soils (Fig. 5.62). Eugley Soils are most often formed in the coastal zone of lakes, slow watercourses and ponds, where the soil subsurface is saturated with water throughout the year. In depressions, a high

[9] Definition of wetlands says that wetlands are areas where water covers the soil, or is present either at or near the surface of the soil all year or for varying periods of time during the year, including during the growing season. Therefore, Eugley Soils can be callled Wetland Soils like in Pavlović et al. (2017). However, the same authors call Humogley Soils as Wetland Black Soils, which is debatable, because in this soil type groundwater does not reach the depth of 50 cm below the surface.

Fig. 5.62 Swampy-Gley Soils—Eugleys (*Photo-archive* CZM)

level of groundwater or flood waters causes very intensive redox processes. In these processes, the genesis of the soil is mostly reflected and influenced by the change of the dry and wet phases in the soils. The withdrawal of flood waters or fluctuations in the groundwater level during the summer enables the change of oxidation–reduction processes in the upper part of the profile, while the deeper part of the profile is constantly saturated with water. Hydrophilic vegetation of cattail (*Typha latifolia*), rushes (*Schoenoplectus lacustris*), reeds (*Phragmites australis*), sedges (*Carex*) and willows (*Salix alba* and *Salix fragilis*) grows and develops in these humid conditions. This vegetation produces a lot of organic residues that are transformed and accumulated in the surface layer of the soil. In this way, the pedogenesis of Eugleys is characterised by a series of specific processes. However, the most important ones are the weathering of primary minerals, the slow decomposition of organic matter, and the development of reduction processes that are dominant when there is a high level of water in the soil layer, whereas oxidation processes are limited during the year to the upper part of the profile, and occasionally in the lower part.

As a result of specific processes, an A_h–horizon is formed in Eugley Soils, in which organic matter accumulates, and water is found in capillary pores. Below the A_h–horizon lies the G_o–horizon, oxidised part of the gleyic horizon, in which oxidation and reduction processes alternate during the year. When the soil layer is not completely saturated, air enters the soil pores, Fe^{3+} and Mn^{4+} oxides form in oxidation processes, which are recognisable by the rusty colour and as dark brown spots and mottles that give this horizon a mosaic pattern. Below the G_o–horizon lies the G_r–horizon, always in reducing conditions, which is saturated with water throughout the year, so anaerobic conditions prevail, which leads to the formation of reduced compounds of Fe^{2+} and Mn^{2+}, and the horizon acquires a blue–greyish or green–greyish colour. The gleyic pattern is essentially made up of reddish, brownish or yellowish colours at aggregate surfaces and/or in the upper parts of solum, in combination with greyish/bluish colours inside the aggregates and/or deeper in the soil profile.

The origin of water and the source of wetting is the basis for the division of Eugley Soils into subtypes, namely: (1) Hypogley Soils—gleysation occurs under the impact of groundwater, (2) Epigley Soils—the process of gleysation occurs due to surface runoff water and flood waters, and (3) Amphigley Soils—the process of gleysation occurs under the impact of all water sources (rainfall, surface runoff, flood and groundwater). Varieties of Eugleys for all subtypes are distinguished on the basis of humus content: (1) Mineral—up to 10% of humus, Humus—10–30% of humus, and Peat–Gley variety with more than 30% humus, but the thickness of the soil horizon is lower than 30 cm. The forms of Eugley Soils are distinguished on the basis of their chemical composition and the occurrence of vertic properties.

Position in an evolutionary series. Eugley Soils can be formed from Fluvisols or Humogleys and Humofluvisols. The most important factor for their formation is the level of groundwater and humid conditions in the upper 50 cm of the soil profile. If vegetation growth is especially triggered in these conditions, a lot of dead biomass residues could be formed and accumulated, in which case Eugley Soils can evolve into Peat Soils. The other evolutionary pathway in Eugley Soils might appear in the case of the decrease of groundwater levels by drainage when Humogley Soils or Humofluvisols can be formed, and then these soils can be used for intensive agricultural production.

Distribution. The total area of Eugley Soils is about 10,170 ha. The largest area is located in the coastal zone of Lake Skadar (Zeta and Ceklin cassettes and Crmničko Field, Fig. 5.63). In the coastal area, these soils are found in depressions (local—knete) next to the Bojana River, Šasko Lake and pools of saltern. A smaller area of Eugley Soils is found in Buljarica and Soličko Field near Tivat. In these

Fig. 5.63 The area of Mineral–Marsh Soils (Eugleys)—Northern border (rim) of Lake Skadar (*Photo* G. Roganović)

locations, in the marshes, some of the fresh and saltwater stagnates, so the Eugley Soils are saline. In the interior of Montenegro, Eugley Soils occur on the shores of Plavsko Lake on a slightly larger surface and in other places occasionally in depressions on the plateaus, like Jezera near Žabljak, Kosanica and Glibaći near Pljevlja, and others.

The WRB corresponding RSG—Gleysols. The possible principal qualifiers for Eugleys in Montenegro are Tidalic, Folic and Histic, Mollic, Stagnic, Calcaric, Eutric or Dystric.

Soil characteristics. Differences in morphology, i.e., gleyic pattern, are noticeable in Eugley Soils. Sometimes there is no clear differentiation between the A_h and G–horizons, but the soils undergo gleysation from the top of the surface and a mixed A/G–horizon appears. This is the case with Eugleys formed from alluvial sediments that contain several layers of different thicknesses and textural differentiation, and are deposited in different periods and under different power of water courses. The texture of Eugley Soils varies spatially in certain areas but also within the same soil profile as a consequence of layering, if they are developed from Fluvisols. The physical properties of clayey Eugley Soils are unfavourable.

The chemical properties of Eugley Soils are diverse because these soils can be calcareous or non–calcareous, and this affects the soil reaction, which can be in a wide range from acid to alkaline. Also, Eugley Soils could have a wide range of base saturation and cation exchange capacity, which depend on the content of humus and clay, and the type of clay minerals. The organic matter content amounts from several to 6% in the area of Ulcinj and up to 10–30% in other places. Eugley Soils are poor in phosphorus, and moderately or well supplied with potassium. Gleyic horizons can also contain toxic compounds.

Characteristics of some Calcareous and Non–Calcareous Eugley Soils are presented in Table 5.14.

Soil management and utilisation. Eugley Soils that are not ameliorated are unsuitable for cultivation, so they are mainly used for pastures and sometimes also for meadows, but their hay is of poor quality. Swampy vegetation dominated by marsh reed and willow grows on the shores of Skadar, Plavsko, and Šasko lakes, in the knetes next to the Bojana River and Buljarica near Petrovac. It is an outstanding waterfowl habitat, so it is used exclusively for that purpose as a reserve. Lake Skadar is the largest freshwater bird reserve in Southeast Europe. Due to the unregulated water regime, the Eugley Soils are left to natural vegetation, and with hydrotechnical and agrotechnical melioration, these potentially fertile soils can become cultivated soils and also used as highly productive soils, especially in the coastal zone and in the Lake Skadar basin. The most important reclamation measure on Eugley Soils is drainage, in the

Table 5.14 Particle size distribution and chemical characteristics of Eugleys (adopted from Fuštić and Đuretić 2000)

Location	Soil horizon	Soil depth (cm)	Gravel (%)	Sand (%)	Silt (%)	Clay (%)	pH in water	pH in KCl	$CaCO_3$ (%)	Humus (%)	T (meq/100 g)	BS (%)	P_2O_5 (mg/100 g)	K_2O (mg/100 g)
			>2 mm	2–0.02	0.02–0.002	<0.002								
Žabljak Crnojevića	A	0–7	0.0	23.7	65.8	10.5	7.70	6.45	25.4	4.89	-	-	5.0	39.2
	G	20–50	0.0	11.7	67.6	20.7	7.92	6.84	27.1	2.30	-	-	3.7	34.2
Gropa Smajlova	A	0–30	0.0	34.1	57.3	8.6	7.55	7.40	14.6	6.63	-	-	5.1	8.6
	G	30–60	0.0	25.0	59.9	15.1	7.94	7.60	13.8	3.68	-	-	1.4	5.7
Bare	A	0–10	0.0	65.9	29.8	4.3	6.19	5.13	0.0	21.6	45.4	78.8	13.0	1.3
	A/G	10–23	0.0	63.8	28.9	7.3	6.11	5.23	0.0	9.86	38.5	83.7	11.7	0.7
	G	23–45	0.0	79.9	10.0	10.1	6.32	5.68	0.0	1.79	32.5	86.7	13.0	0.6

form of channels and horizontal pipe drainage in case of integrated lowlands management. In these situations, new Anthropogenic Hydrogenic Soils can be formed after reclamation measures.

5.2.3 Class of Peat Soils [T–G Sequence of Horizons]

The class of Peat Soils includes soils that are characterised by the presence of a T–horizon, which consists of undecomposed or slightly decomposed organic matter accumulated in anaerobic conditions, under the water. Groundwater usually fluctuates slightly during the year and contributes to saturation of the soil with water. The decomposition of organic matter is slowed down due to reducing conditions and, in some situations, because of the low temperatures during the season. In these conditions, high amounts of organic matter, residues of plants, are accumulated year-on-year, and this is how Histosols and the T–horizon are formed. This process is also called paludisation, from the Latin word *paludio*, which means marsh, indicating that this process occurs in marshes and around swamps. Soils are classified as Peat Soils in national classification if the T–horizon has more than 30% of organic matter and a thickness above 30 cm. Otherwise, the soils belong to the Peat-Gley variety of Eugley Soils. Histosols are typical of wetlands and they have high ecological importance. Three types of soils are distinguished within the Peat Soils class: Plano–Histosols, Acro–Histosols and Transitional Histosols. Plano–Histosols are usually groundwater (fen) peats, located on lower flat terrain. Acro–Histosols are called raised peats (bogs) as they have the shape of a hillock (mound) and, in the conditions of Montenegro, are usually located on higher elevations. These Histosols are usually rainwater peats. Transitional Histosols are both rainwater and groundwater peats.

5.2.3.1 Histosols

The name of the soil type. The name Histosols is derived from the Greek word *histos,* which means tissue, and the Latin word *solum,* which means soil. This term indicates that Histosols have a peat horizon—T,–which is recognised by the large undecomposed amounts of plant residues that are the main contributors in their formation.

Soil genesis and soil horizon sequence. Peat Soils are formed in conditions of excessive wetting, which occurs in depressions near watercourses, in marshes, and on lake shores, as well as in closed drainage basins in the hilly-mountainous terrain. These are wet habitats—wetlands—in which hydrophilic vegetation develops and accumulates in anaerobic conditions. Most often, the peat horizon is found in the upper part of the profile, but sometimes, for example, on the Lake Skadar shores, peat deposits are found both on the surface of the terrain and at greater depths, with mineral soil layers between them. The main pedogenic processes affecting the genesis of Histosols are the formation of an organic and often mineral layer/horizon in anaerobic conditions and gleysation. Therefore, the Histosols soil profile has a T–horizon containing weakly decomposed organic matter and a G–horizon with a higher content of mineral matter. The thickness of the peat horizon exceeds 30 cm, but in the Lake Skadar zone, the thicknesses range from 0.4 to 7 m (Bogdanović et al. 1988; Fuštić 2000). In this case, the formation of peat and semi-peat deposits of different thickness and properties, occurred over a long period of time, and these are eutrophic peats of the postglacial period. The processes of peat formation are more or less intense depending on the ground level, which affects the thickness of the T–horizon.

WRB corresponding RSG—Histosols. Histosols in Montenegro can be Tidalic, Fibric, Hemic or Sapric, Dystric or Eutric.

Position in an evolutionary series. Histosols are formed in the process of paludisation. Nevertheless, the paludisation process might also affect Eugley Soils, and therefore, an evolution of the A–horizon of Eugleys into a T–horizon can and does occur. Histosols evolution is slow under the current constellation of pedogenic factors, and Histosols can change their characteristics only if the wetting conditions change, which may occur rapidly because they contain a very high content of organic matter, which can oxidise quickly. Therefore, the conditions of decreased groundwater level can trigger high oxidation of organic matter in Histosols, and these soils can change their characteristics, which is also followed by subsidence of the land. So, the groundwater depth influences the Histosols evolution.

Distribution. Peat Soils are weakly distributed in Montenegro. The locations where Peat Soils are formed are very specific, such as the area surrounding Lake Skadar, where the conditions for the formation of Histosols appear in an extended area (Fig. 5.64). The rare appearance of Histosols in Montenegro is caused by the fact that the rivers of Montenegro do not make meanders, they do not cause permanent annual waterlogging after floods, and there are no depressions, meander cutoffs, meander scars or oxbow lakes that would be filled with water following floods. Exceptions are the lowest parts of the alluvial plain of the Bojana river, called "knete" and the parts of Šasko Lake adjacent to the Bojana River, as well as parts of the former Lake Zoganje, where the Ulcinj salt pan (1500 ha) is located. Nevertheless, at these locations, at Buljarica near Petrovac, near Plavsko and Šasko lakes, a small area of Histosols is associated with Eugley Soils, which were formed by the action of flood and groundwater. In the Lake

Fig. 5.64 The area of Peat Soils, Podhum (*Photo* G. Roganović)

Skadar area, Histosols are found in Podhum Bay, near Malo Blato and Razovac. At higher elevations, Acrohistosols are formed only sporadically on flat siliceous terrain of high mountains, which is not often in the north of Montenegro, and in the conditions of close drainage basins.

As stated above, national classification recognises Lowland Peat Soils—Plano–Histosol, Transitional Peat Soils and Mounded Peat Soils (Acro–Histosols). These three types are divided into subtypes based on the thickness of the T–horizons, which can be thin (30–50 cm), moderately thick (50–100 cm), and thick (>100 cm).

Soil characteristics. Histosols of Podhum Bay have peat layers containing 50–70% of organic matter, but also layers with more than 80% of organic matter. They contain very decomposed and highly humified organic matter. These peats and semi–peats have a neutral and weakly alkaline reaction, and are usually non–calcareous. They are rich in humic acids and content of macronutrients and micronutrients. Microbiological activity is very high, and they have low radioactivity, which is why they are suitable for the creation of organic fertilisers (Fuštić and Đuretić 2000). Histosols of Malo Blato and Razovac dominantly contain between 30 and 50% of organic matter, but there are also locations with more than 70%. These Histosols are calcareous in one part of this area. Soil reactions vary widely from 4.89 to 8.59, thus from acid to moderately alkaline, but most often, soil reaction is neutral or slightly acid. There is a great interest in the exploitation of peat material and its further utilisation in horticulture.

Soil management and utilisation. Peaty areas of Montenegro are left to nature (nature–free), and they are overgrown with marsh vegetation and present habitat for birds and other species. Lowland Histosols represent an oligotrophic environment and, therefore, have an important ecological role and importance. Histosols are known as a rich source of organic matter, which is especially true for the Lowland Peat Soils of Lake Skadar, which is, after processing, used as a substrate or fertiliser in horticulture, nursery production, and for growing vegetables (Fig. 5.65). However, this utilisation is nowadays neglectable since Lake Skadar is a protected area. In order to be used in agricultural production or to change the habitat, natural peats need to be drained, and usually also limed and fertilised in order to permit cultivation of crops. However, the experience of reclamation projects in the temperate zone has shown that in many instances, a gradual degradation of Histosols was provoked, and ultimately, the loss of the precious peat, as well as the formation of sulphuric acid, reducing the overall fertility of these soils. Regarding the conditions of Montenegro, it is desirable to protect and conserve fragile peat lands because of their intrinsic value presented through their common functions as water flow regulators and wetland supporters, and because their sustained agricultural use is under question.

Fig. 5.65 Cleaning of trenches and removal of peat (*Photo* G. Roganović)

5.2.4 Class of Pseudogleyic Soils [A–Eg–Bg–C or A–g_1–g_2 Sequence of Horizons]

This soil class includes soils with stagnant ("upper ") groundwater that is occasionally retained in the soil above the impermeable soil horizon or layer. Longer water retention in the part of the profile is caused by poor water permeability or heavy rainfall in the autumn–winter period. In Montenegro, Primary Pseudogley Soils formed on diluvial loams are found in the Bjelopavlićka Plain.

5.2.4.1 Pseudogleys

The name of the soil type. The Pseudogley's name comes from the Greek word *pseudos,* which means false, wrong, and the Russian word *gley*, which means mucky mass. The name was introduced in the literature by Kubiena (1953), because the soil has a number of properties that are similar to Gley Soils. Pseudogleysation is the process that takes place under the influence of surface waters (precipitation, snowmelt) stagnating in the soils in the upper part of the soil profile. For pseudogleysation process, stagnation up to a depth of 100 cm is important from the point of view of soil classification.

Soil genesis and soil horizon sequence. The conditions under which Pseudogleys occur in Montenegro are different. Diluvial loams as parent material and the flat relief play an important role, because they enable the profile to be texturally differentiated into a water-permeable surface horizon and an impermeable deeper subsurface horizon which can become very compact and dense. Abundant precipitation in the autumn–winter period in Danilovgrad[10] conditions a long wet period in the soil, and a particularly dry summer period. In these conditions, under favourable soil reaction, clay particles are leached from the surface horizons into the deeper B–horizon. An increase in the clay content increases the subsurface soil compaction and decreases the rate of hydraulic conductivity, which finally causes water stagnation in the upper part of the soil profile above this horizon/layer. Therefore, the main stamp of the genesis of the Pseudogley Soils is given by the alternation of wet and dry phases in the soils during the year-span, and this is why Pseudogley Soils differ from Gley Soils, which are formed due to stagnation of groundwater in the lower part of the profile during most of the time of the year. A longer wet phase, when the soil is saturated with water, without the presence of air, triggers reduction processes and the formation of Fe^{2+} and Mn^{2+} compounds. These compounds are reduced and more soluble in the water, with which they start to move by diffusion and appear (accumulate) as greenish to bluish colours. Their transport causes the formation of whitish, black pale, and grey spots in reduced zones. During the season, the wet soil phase transposes into the moist phase, and reaches the dry phase in the summer period. In the dry phase, there is no plant available water in the soil. Therefore, oxidation processes take place gradually during the year, and complete loss of moisture occurs during the summer. In the dry phase, previously reduced iron and manganese compounds are oxidised, and transformed into Fe^{3+} and Mn^{4+}, and they are deposited in the form of rusty-red and yellowish-orange spots or blotches, called mottles, or as Fe and Mn-concretions and grains of dark brown or black colour—ornstein. In this manner, the pseudogley horizon (g–horizon) generally gets a mosaic appearance due to the occurrence of greyish-white and rusty-brown colours, looking like marble, i.e., marmoreal appearance. Apart from iron and manganese ions, organic and other substances can be bound in concretions and grains. The alteration of oxidation–reduction processes in Pseudogley Soils produce specific morphological signs—hydromorphisms, which are a typical characteristic of this soil—the alternate appearance of rusty-brown and pale colours when the soil is dry (see marmoreal appearance above), and bluish–grey and greenish–grey colours resembling snakes skin when the soil is saturated with water. The main colour of the soil matrix is light brown or light grey. Pseudogleysation is a characteristic of the Bg or g–horizons, and it occurs at the top of the impermeable layer affecting a part of its thickness and upper parts of the soil, which also face the alternation of wetting–drying cycles. Therefore, soil horizons Eg or Ag are also identified in the field.

The duration of particular phases (wet, moist and dry) depends on the depth and position of the impermeable horizon or layers. The textural differentiation (abrupt textural difference) of the profile in the climatic, relief, and geological conditions of the Bjelopavlićka Plain has caused the following soil horizon sequence in Pseudogley Soils: A–Eg–Bg–C.

The WRB corresponding RSG—Stagnosols and Planosols. Among others, less often, Dystric and Eutric principal qualifiers can be used to describe Pseudogleys in Montenegro.

Position in an evolutionary series. Pseudogley Soils in the Bjelopavlićka Plain (Fig. 5.66) are formed from Leached (Luvic) Brown Eutric Cambisols, which appear not only in this plain but also in the neighbouring Lješkopoljski Lug near Podgorica, Strganica and Drezga in Piperi, where smaller brackish lakes existed during the diluvium. In fact, Pseudogleys appear in association with Eutric Cambisols,

[10] An average amount of precipitation in Danilovgrad (1961–1990) is 2212.5 mm per year: June–August: 209.4 mm; September–November: 718.5 mm; December–February: 776.7 mm; March–May: 507.9 mm (Knežević, 2008).

Fig. 5.66 Pseudogley Soils of the Bjelopavlićka Plain (*Photo-archive* CZM)

with the fact that it is found in the lowest parts of this plain, where the wet phase certainly lasts longer, regardless of the depth of the impermeable layer or horizon. The position of the impermeable layer or horizon affects the duration of the wet and dry phases. This means that it is important at what depth the impermeable horizon or layer is, i.e., at 20, 50 or 70 cm, which was observed by Muckenhauzen (1959) and Škorić, et al., (1973). Resulović, et al., (2008), soil scientists from Bosnia and Herzegovina, classified Pseudogley Soils into Autogenic soil order.

The combined effects of impermeable diluvial loams, clayey (B)–horizons of Eutric Cambisols, flat relief, and high amount of precipitation affected the formation of Pseudogley Soils in the Bjelopavlićka Plain. Pseudogleys of the Bjelopavlićka Plain belong to the subtype of Pseudogley formed on flat terrain, whereas the subtype formed on undulating terrain is less common in Montenegro. The varieties of Pseudogley Soils in Montenegro are differentiated by the appearance of an impermeable layer, with shallow (up to 25 cm) and moderately deep (25–50 cm) varieties prevailing in Montenegro, whereas deep (50–70 cm) varieties are less frequently identified. Pseudogley Soils of the Bjelopavlićka Plain belong to the eutric form of Pseudogley Soils, with a pH>5.5 in water. The eutric forms are dominant since the parent material is very calcareous loams (> 50% CaCO3), so the soil is also rich in bases (V>70%). Pseudogley is the final, last type of soil in the pedo-evolutionary series on lacustrine sediments deposited in the Bjelopavlićka Plain during the diluvium.

Soil characteristics. Pseudogleys are soils with poor physical characteristics and especially poor soil water characteristics. The surface horizon is texturally a bit lighter; it usually has a loamy or loamy–clay texture, is 10–30 cm thick, under natural vegetation, and has unstable crumbly or granular soil structure. The clay content ranges from 10 to 20%, and the silt content is 2–3 times higher, so the total physical clay content ranges from 56 to 72%. This particle size distribution causes poor soil water–physical characteristics, especially low air capacity (2–6%) but moderate to high water retention capacity (25–40%). The Bg–horizon is 15–30 cm thick, nonstructured (massive structure) due to an even greater portion of the clay fraction. It is very sticky and swells, and soil aggregates break down into an amorphous mass in wet conditions, decreasing water permeability and practically reducing it to a minimum (10^{-4}–10^{-5} cm/s). The air capacity drops even below 1%, and the water retention capacity ranges from 28 to 35%. These properties and the properties of the Bg–horizon cause the stagnation of surface water during its vertical movement through the thickness of the soil profile. Poor water permeability is also affected by compacted diluvial loams with a higher clay content, which are layered and very hard—compact. In the dry phase, the entire soil profile, due to moisture loss by evapotranspiration, turns into a very hard and compacted mass with the occurrence of cracks. Cultivation of Pseudogley Soils in both wet and dry stages is very difficult. The moist stage with lower soil moisture lasts a short time, and then ploughing is possible. Although Pseudogley Soils in Bjelopavlićka Plain are formed on calcareous loams with 50–60% $CaCO_3$, soils under natural vegetation are not calcareous. However, in several complexes where land reclamation measures were applied, like clearing, drainage, and deep ploughing, the horizons were mixed, which changed the soil horizon sequence and the chemical properties, including the calcium carbonate content and soil reaction, which ranges from weakly acid to slightly alkaline in these situations. The content of humus in the surface horizons of Pseudogley Soils is low to medium, which is also the case with potassium, while the soil is poor in phosphorus.

Distribution. Pseudogleys appear only in the Bjelopavlićka Plain (Fig. 5.67) on lacustrine sediments, i.e., diluvial loams. Their formation took place in conditions of humid, semi-humid and arid climate and on completely flat terrain in the central widest part of this plain (Kosovi Lug, Kosić, Duboke lazine, Begovina, Ćurilac and Drenovica). It covers an area of 569 ha.

Fig. 5.67 A zone of Pseudogley Soils in the central part of the Bjelopavlićka plain (*Photo* G. Roganović)

Soil management and utilisation. Pseudogley Soils have lower productive capacity due to the physiologically shallow depth of the solum and poor soil water, and soil physical characteristics, which affect soil water and thermal regime(Pavićević 1962). Pseudogley Soils of the Bjelopavlićka Plain have been converted into arable land by land reclamation on several complexes of former social property, but they are still characterised by lower production capacity compared with Eutric Cambisols. Reclamation of Pseudogley Soils is a very important measure because they cannot be intensively used in agriculture without the drainage of excess water in the soil through the application of horizontal pipe drainage. In order to prepare Pseudogley Soils for high and stable yields in given climatic conditions, the water regime should be regulated by drainage and irrigation followed by deep tillage and other agrotechnical measures that should improve soil structure, water permeability, and fertility.

References

Agadzhanova NV, Izosimova YuG, Kostenko IV, Krasilnikov PV (2021) Indicators of pedogenic processes in red clayey soils of the Cape Martyan Reserve, South Crimea. Eurasian Soil Sci 54(1):1–12

Basler Dj, Malez M, Brunnacker K (1966) Die rote Hohle (Crvena stijena) beo Bileća (Jugoslavien) Eiszeitaler und Gegenwart, Band 17

Bašić F (2013) The soils of Croatia, World Soils Book Series. Springer, Berlin

Bautista F, Palacio-Aponte G, Quintana P, Zinck JA (2011) Spatial distribution and development of soils in tropical karst areas from the Peninsula of Yucatan, Mexico. Geomorphology 135:308–321

Blečić V, Lakušić R (1976) Prodromos biljnih zajednica Crne Gore. Glasnik Republičkog Zavoda za zaštitu prirode i prirodnog muzeja u Titogradu, 9, Titograd

Bogdanović M, Tešić Ž, Gigov A, Stojanović S, Tančić N, Todorović M, Jakovljević M (1988) Geneza i svojstva treseta Skadarskog jezera. Plenarni referati i izvodi VIII kongresa JDPZ. Cetinje

Buol SW, Hole FD, McCracken RJ (1973) Soil genesis and classification. The Iowa State University Press, Ames, p 360

Burić M, Fuštić B, Bulajić P, (2017): Atlas zemljišta Crne Gore, Crnogorska akademija nauka i umjetnosti (CANU), Podgorica

Dožić S (1978) Ogled na jalovištu rudnika uglja u Pljevljima. Zemljište i Biljka, Beograd 27(1–2):157–164

Durn G, Škapin SD, Vdović N, Rennert T, Ottner F, Ružičić S, Cukrov N, Sondić I (2019) Impact of iron oxides and soil organic matter on the surface physicochemical properties and aggregation of Terra Rossa and Calcocambisol subsoil horizons from Istria (Croatia). CATENA 183:104184

Duschafour Ph (1982) Pedology, George Allen and Unwin, London

Đokić V, Živaljević M, Perović Z (1976) Osnovna geološka karta 1:100000. Tumač za list Gusinje. Zavod za geološka israživanja Crne Gore, Titograd, 1976

Filatov M (1945) Geografija počv SSSR, Moskva

Fuštić B (2000) Atlas pedoloških karata 1:50:000. Univerzitet Crne Gore, Biotehnički institut, Podgorica

Fuštić B, Đuretić G (2000) Zemljišta Crne Gore. Univerzitet Crne Gore, Biotehnički institut, Podgorica

Gerasimov IP, Ivanova EN (1959) Comparison of three scientific trends in resolving general questions of soil classification. Soviet

Soil Sci 11:1190–1205 (Transl. 1960 from Pochvovedenie 11 (1959):1–18)

Gračanin M (1945) Mali pedološki praktikum. Poljoprivredna naklada Zagreb

Gračanin M (1951) Pedologija, III deo, Školska knjiga, Zagreb

IUSS Working Group WRB (2006) World Reference Base for Soil Resources 2006. World Soil Resources Report No. 103, FAO, Rome.

Knežević M (2008) Planiranje i projektovanje sistema za odvodnjavanje i navodnjavanje u Bjelopavlićkoj ravnici. Univerzitet u Beogradu, Poljoprivredni fakultet Beograd, Doktorska disertacija

Kubiena WL (1953) Bestimmungsbuch und Systematik der Boden Europas. Stutgart, Germany

Marić L (1965) Terra Rossa u karstu Jugoslavije. Aauta. Geol. JAZU, Zagreb

Miloš B, Bensa A (2019) Background variation and threshold values for cadmium concentration in Terra Rossa Soil from Dalmatia, Croatia. Eurasian Soil Sci 52(12):1622–1631

Mirković M, Kalezić M, Pajović M, Živaljević M, Škuletić D (1968) Osnovna geološka karta 1:100000. Tumač za list Bar i Ulcinj. Zavod za geološka israživanja Crne Gore, Titograd

Muckenhauzen E (1959) Die wichtigsten Boden der Bundesrepublik Deutschland. Frankfurt an Main

Pavićević N (1956) Buavice na Crnogorskom kršu. Beograd

Pavićević N (1958) Crvenice Crne gore i Hercegovine. Arhiv za poljoprivrerdne nauke, god. XI, sv. 32

Pavićević N. (1959) Pedologija-skripta II i III, Beograd

Pavićević N. (1962) Tipovi zemljišta Jugoslavije i problemi njihovih melioracija, Beograd

Pavlović P, Kostić N, Karadžić B, Mitrović M (2017) The soils of Serbia. World Book Series. Springer, Dordrecht, p 225

Radmanović S, Životić Lj, Nikolić N, Đorđević A (2017) Classification of Rendzina soils in Serbia according to the WRB system. In: Congress Proceedings, 2nd International and 14th National Congress of Soil Science Society of Serbia "Solutions and Projections for Sustainable Soil Management", 25–28th Sept 2017, Novi Sad, Serbia, pp 1–9

Resulović H, Čustović H, Čengić I (2008) Sistematika tla/zemljišta - Nastanak, svojstva i plodnost, Univerzitetski udžbenik, Univerzitet u Sarajevu, Poljoprivredno-prehrambeni fakultet

Škorić A, Filipovski G, Ćirić M (1973) Klasifikacija tala Jugoslavije, Zagreb

Škorić A, Filipovski G, Ćirić M (1985) Klasifikacija zemljišta Jugoslavije. Akademija nauka i umjetnosti Bosne i Hecegovine. Posebna izdanja, urednik: Tihomir Vuković, Sarajevo, knjiga LXXVIII. Odeljenje prirodnih i matematičkih nauka. Međuakademijski odbor za proučavanje zemljišta

Životić Lj, Đorđević A, Mohlala DB, Bogosavljević J, Kaluđerović L (2021) Correlation between Ranker Soil Type of National Classification System and Leptosols Reference Soil Group of World Reference Base For Soil Resources—theoretical approach. In: 3rd International and 15th National Congress of Serbian Society of Soil Science: Soils for Future under Global Challenges, Sokobanja, Serbia, 21–24 Sept 2021, Book of Proceedings, pp 53–62

Životić Lj, Gajić B, Žarković B, Radovanović V, Nešić Lj, Đorđević A (2019) Variation of soil structure in the foot and toe slopes of Mt. Vukan, East-central Serbia. In: Proceedings of the X International scientific agricultural symposium "Agrosym 2019". Jahorina, 03–06 Oct 2019, pp 1318–1324

Životić Lj, Radmanović S, Gajić B, Mrvić V, Đorđević A (2017) Classification and spatial distribution of soils in the foot and toe slopes of mountain Vukan, East-Central Serbia. CATENA 159:70–83

6 Soil Issues and Perspective of Land Management

Abstract

This chapter gives a review of soil conservation, land degradation, and sustainable agriculture in Montenegro, emphasising the crucial role of soil as an essential, non-renewable resource pivotal for food production and ecosystem health. It outlines the extensive studies by the Biotechnical Faculty on Montenegro's diverse soil profiles, predominantly nutrient-poor, spotlighting the urgent need for soil fertility preservation. Various institutions, including the Ministry of Ecology and the University of Montenegro, are pivotal in implementing policies, conducting research, and enforcing land protection and management measures in alignment with international conventions like the UNCCD. Also, the chapter provides an in-depth analysis of the legal and strategic frameworks aimed at countering land degradation and achieving land degradation neutrality (LDN), highlighting the significance of synergised efforts, restoration of degraded land, and sustainable land management practices. It also addresses Montenegro's specific vulnerabilities (such as forest fires) and proposed nature-based transformative solutions, emphasising the integration of LDN into national planning. Additionally, the chapter delves into Montenegro's comprehensive strategy to combat drought and promote sustainable land use through the Montenegro National Drought Plan. The significance of legal frameworks, strategic planning, institutional capacity building, and eco-friendly farming practices is underscored.

Soil is a valuable natural resource that is easily destroyed and difficult to restore, making it important to preserve for future generations. Producing enough food for a growing global population requires healthy soil resources.

As cited in the first chapter, the Biotechnical Faculty (formerly Agriculture Institute) conducted a soil survey in Montenegro, resulting in a detailed soil map at a scale of 1:50,000. The survey included the description and analysis of more than 2,000 soil profiles, with all other relevant information on the presence and distribution of different soil types. The soils of Montenegro are ranked in five categories of effective fertility. The majority of soils are shallow and poor in nutrients. Namely, Montenegro has a lack of fertile arable land, and over 85% of agricultural areas are natural meadows and pastures of low fertility. The importance of sustainable land and soil management is a cornerstone of future development of Montenegro.

6.1 Institutional Framework

University of Montenegro—Biotechnical Faculty, Centre for Soil and Amelioration, is a primary institution that integrates teaching, professional, and scientific research work in the field of soil science (Fig. 6.1). It is focused on studying, and solving problems of land use, land arrangement, and conservation and proposes recommendations for agricultural practices.

Besides Biotechnical Faculty, there are several institutions dealing with soil management from different perspectives.

The Ministry of Ecology, Spatial Planning, and Urbanism is responsible for spatial planning and implementation of the UN Convention to Combat Desertification (UNCCD), while the Ministry of Agriculture, Forestry and Water Management formulates programmes for development policy, protection, exploitation, and improvement of agricultural and forest land. The Directorate for Property and Legal Affairs of the Ministry of Finance is responsible for the use, supervision, and management of state property, including state-owned land.

The Cadastre and State Property Administration assess the market value of state-owned land and are in charge of real estate cadastre, registration of real estate rights, cadastral classification and land evaluation, and geodetic and

M. Knežević et al., *The Soils of Montenegro*, World Soils Book Series,
https://doi.org/10.1007/978-3-031-82919-2_6

Fig. 6.1 Field work with students of the Biotechnical Faculty (*Photo* G. Roganović)

agronomic works on land consolidation. Other institutions, such as the Directorate for Food Safety, Veterinary and Phytosanitary Affairs, Forest and Hunting Management Administration, Water Administration, and Environment Protection Agency of Montenegro, perform tasks related to plant nutrition and protection, forest management, water regulation and protection, and environmental monitoring, respectively. The Directorate for Inspection Affairs performs supervision in the areas of spatial protection, urbanism, construction, ecology, agriculture, water management, and forestry.

The Geological Survey of Montenegro deals with the examination of the formation, composition, and structure of the earth's crust in Montenegro, research on mineral reserves, hydrogeological and engineering-geological research and testing, and projects in applied geology. The Institute of Hydrometeorology and Seismology of Montenegro (IHSM) monitors the impact of meteorological conditions on the quality and quantity of yields, conducts phenological observations and analyses, and prepares agrometeorological studies.

Six laws tackle a majority of soil and land issues. The Law on Environment in Montenegro protects soil, i.e., its physical, chemical, and biological properties, and ensures that land use suits its natural characteristics. Land protection measures are required before and after the implementation of projects, and remediation measures are prescribed in the Law on Environmental Liability. The Law on Agricultural Land regulates land use and land conversion, while the Law on Spatial Planning and Construction centralises spatial planning at the state level. The Law on Agriculture and Rural Development establishes and maintains records of entities, agricultural holdings, parcels, and requests for support and payments. Legislative acts address state-owned agricultural land, land tenure, and sale procedure and conditions, and agricultural land is leased exclusively for agricultural activity. The Law on Ionising Radiation Protection and Radiation Safety mandates the systematic testing of radionuclide content in arable and non-arable land every six months.

6.2 Land Degradation

Land degradation is a worldwide problem. The main human drivers of direct land degradation are soil sealing (urbanisation and infrastructure development), soil mismanagement in agriculture, land conversion, deforestation, industrial activities, waste disposal, and mining activities, while natural drivers are droughts, floods, and temperature variations.

Climate change has increased the frequency of forest fires, floods, droughts, torrents, and landslides. Forest fires indirectly cause severe soil erosion, leading to the loss of earthy material and the formation of barren landscapes. Floods and erosions have also caused damage and losses to agriculture and natural resources. Droughts have reduced crop yields, depleted water resources, and indirectly triggered soil erosion and wildfires. Inappropriate use of fertilisers, pesticides, and chemicals causes chemical soil degradation and environmental pollution.

Montenegro ratified the UNCCD in 2007 and established a strategic framework for soil protection by adoption of the National Action Plan (NAP) to Combat Desertification in 2015. The NAP includes an analysis of soil degradation in the country and outlines activities to combat land degradation, with strategic and operational goals.

Montenegro supported the UNCCD initiative Land Degradation Neutrality (LDN) Target Setting Process in 2016 (Fig. 6.2). LDN prioritises the reduction of land degradation caused by human activities and climate change. This project consists of three main components: improving the institutional, legislative and policy framework for land degradation neutrality (LDN), capacity building and raising awareness for climate-responsive planning and development, and implementation of agro-environmental measures.

The goal of NAP and LDN is to achieve a balance between anticipated and on-going land degradation and future efforts to improve degraded land, leading to environmental, social, and economic benefits. The key activities leading to LDN achievement are integrated landscape management, sustainable land management, and restoration and reclamation of degraded land, which require the involvement of various level stakeholders (Fig. 6.3). The twelfth session of the Conference of Parties of the United Nations Convention to Combat Desertification (UNCCD), in October 2015, endorsed sustainable development goal target 15.3 and the concept of land degradation neutrality (LDN) as a strong vehicle for the implementation of the Convention (Životić et al. 2017). Target 15.3 aims to combat desertification, restore degraded land and soil, including land affected by desertification, drought, and floods, and strive to achieve a land degradation-neutral world by 2030 (UNCCD 2016). The achievement of this target is measured through indicator: "Proportion of land that is degraded over total land area". LDN aims to maintain or enhance land-based natural capital compared to the reference state. Therefore, it is important to correctly define the reference state, also called the baseline. All the parties were

Fig. 6.2 UNCCD/CRIC: LDN inception workshop, Nairobi (Kenya) 2016

Fig. 6.3 LDN TSP Workshop—field trip, Sinjavina 2017 (*Photo* Archive CZM)

invited to formulate national voluntary targets to achieve LDN, and that work was done after the analysis of changes in sub-indicators. The COP.12 decided that as a means to understanding the status of land degradation, it would be necessary to monitor three UNCCD progress indicators: trends in land cover, trends in land productivity, and trends in carbon stocks above and below ground. These indicators are used for the calculation of the baseline conditions and again in the next monitoring process for the detection of changes. These indicators are important also because they are recommended as sub-indicators to compute the proportion of land that is degraded over the total land area.

Global datasets were used for the analysis of land degradation in Montenegro and the calculation of LDN baseline due to the lack of national data (Fig. 6.4). The analysis of three LDN sub-indicators, applying the one-out all-out principle, showed that there has been a degradation of 6.39% of the country's territory or 85,840 ha (LDN report 2018). Namely, there was the loss of 800 ha of forests and their conversion to shrubs, and conversion of 1,700 ha of forests to croplands. Also, land productivity dynamics data indicated the decline in land productivity on 4,500 ha of croplands and 3888 ha of shrubs, grasslands, and sparsely vegetated areas. In total, 74,331 ha of Montenegro territory were found to be under the three potentially degradable LPD classes: declining, early signs of decline, and stable but stressed (Figs. 6.5 and 6.6). The level of confidence in input datasets was assessed as moderate. The analysis also showed a change in SOC stocks at 2457 ha across the country territory. The global SOC stocks data are just a step in the determination of this indicator, which requires measurements. Especially, the SOC stocks vary a lot in the conditions of dissected topography on small distances (Životić et al. 2020), as is the case on Colluvial Soils and Colluvial Calcomelanosols, but also on other terrains affected by water erosion, which is pronounced in Montenegro. The overall analysis of three sub-indicators identified the existence of 15 hotspots in the country, and the most common land degradation type is biological degradation caused by wildfires (Figs. 6.7 and 6.8). The watersheds level analysis

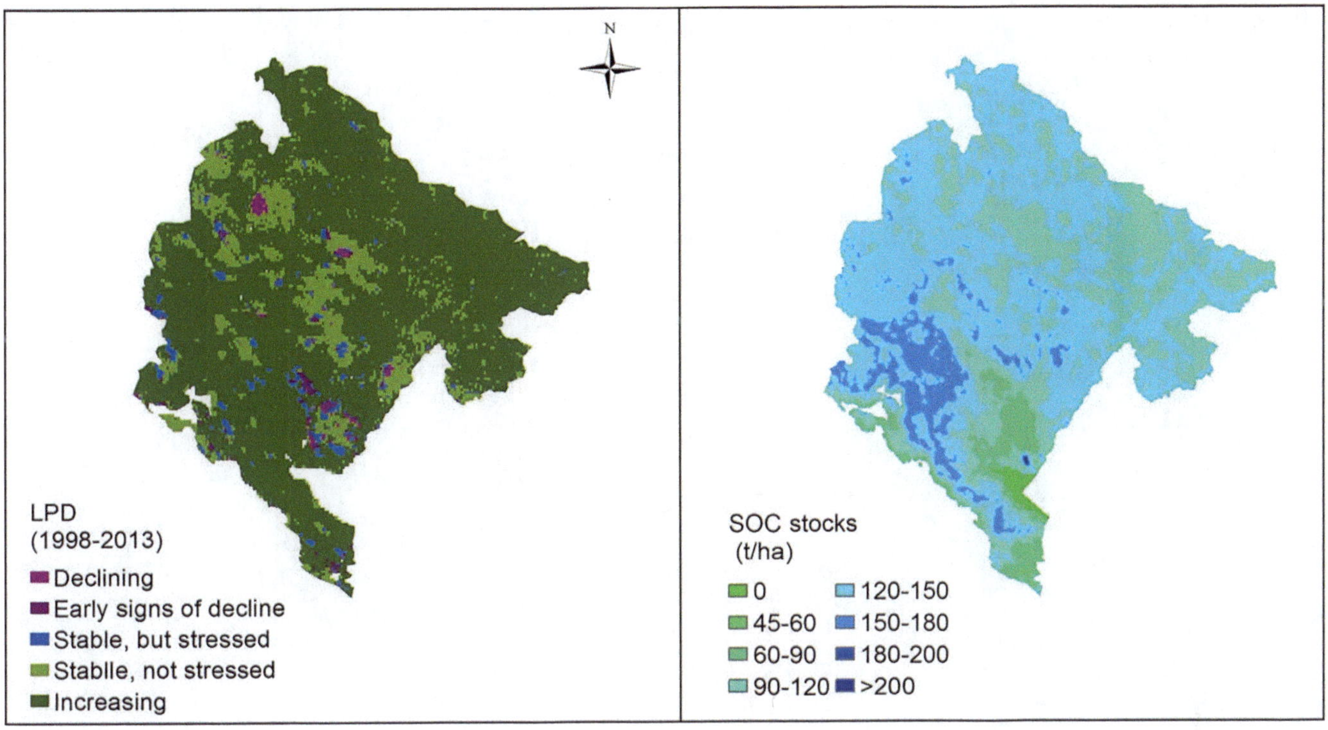

Fig. 6.4 Land productivity dynamics and soil organic carbon stocks map in Montenegro obtained via global datasets

showed that in the Morača and Bojana rivers, around 12% of the territory was degraded. The voluntary target of the country was to achieve LDN as compared to the baseline up to 2030, with four specific voluntary targets and twenty-five associated measures. They are related to the enhancement of LDN baseline in Montenegro, environmental legislations, direct measures to prevent, minimise land degradation, and restore degraded land, sustainable agriculture and forestry, land use changes, and social awareness (Knežević et al. 2018a, b). Three defined measures are recognised as drivers of potential transformative projects/programmes: (1) designing micro-accumulations to cope with wildfires and water scarcity, (2) production and use of biochar in sustainable land management, and (3) improving the agro-forestry sector through raising new orchards of hazel, pomegranate, or other perennial species in areas subjected to fires. The responsible institutions for the measures are the Ministry of Ecology, Spatial Planning and Urbanism, the Ministry of Agriculture, Forestry and Water Management, the University of Montenegro—Biotechnical Faculty, and the Institute for Hydrometeorology and Seismology.

LDN should be embedded into the basic national planning document towards UNCCD implementation, which can be a trigger to increase investments in land degradation or establish land degradation neutrality-related partnerships.

Considering that over 60% of Montenegrin territory is covered by forests and an additional 9.7% by forest land, the protection of forests through afforestation, arrangement, and seedling production is of crucial importance. Montenegrin forests are significant carbon sinks—annually accumulating about 4.6 million tonnes of CO_2 from the atmosphere, which is roughly equivalent to the country's total annual greenhouse gas emissions of 5.3 million tonnes of CO_2 in 2003 (MARD 2014). However, climate change increases the risks of droughts, fires, and pests. The fires are a serious threat to forests and forest land in Montenegro, especially in the coastal and central regions.

The struggle against **drought** is a fundamental challenge that requires the implementation of several measures. In the drought plan (MSDT 2020), inter alia, there are five measures to address water management issues in Montenegro. The first measure, Integrated Water Management, aims to achieve a harmonised water regime throughout the country, ensuring sufficient quality drinking water and meeting economic needs while protecting the aquatic and water-dependent ecosystems. The second measure, Development of Supporting Systems, focuses on the establishment of the Decision Support System, which includes a Water Information System to facilitate decision-making based on predictive models. The third measure, connecting land use planning and drought management, emphasises the importance of linking land use planning and drought management to avoid the expansion of drought-prone areas and encourage holistic management practices on farms. The fourth measure, irrigation of agricultural land, highlights the transition from extensive to intensive agricultural production

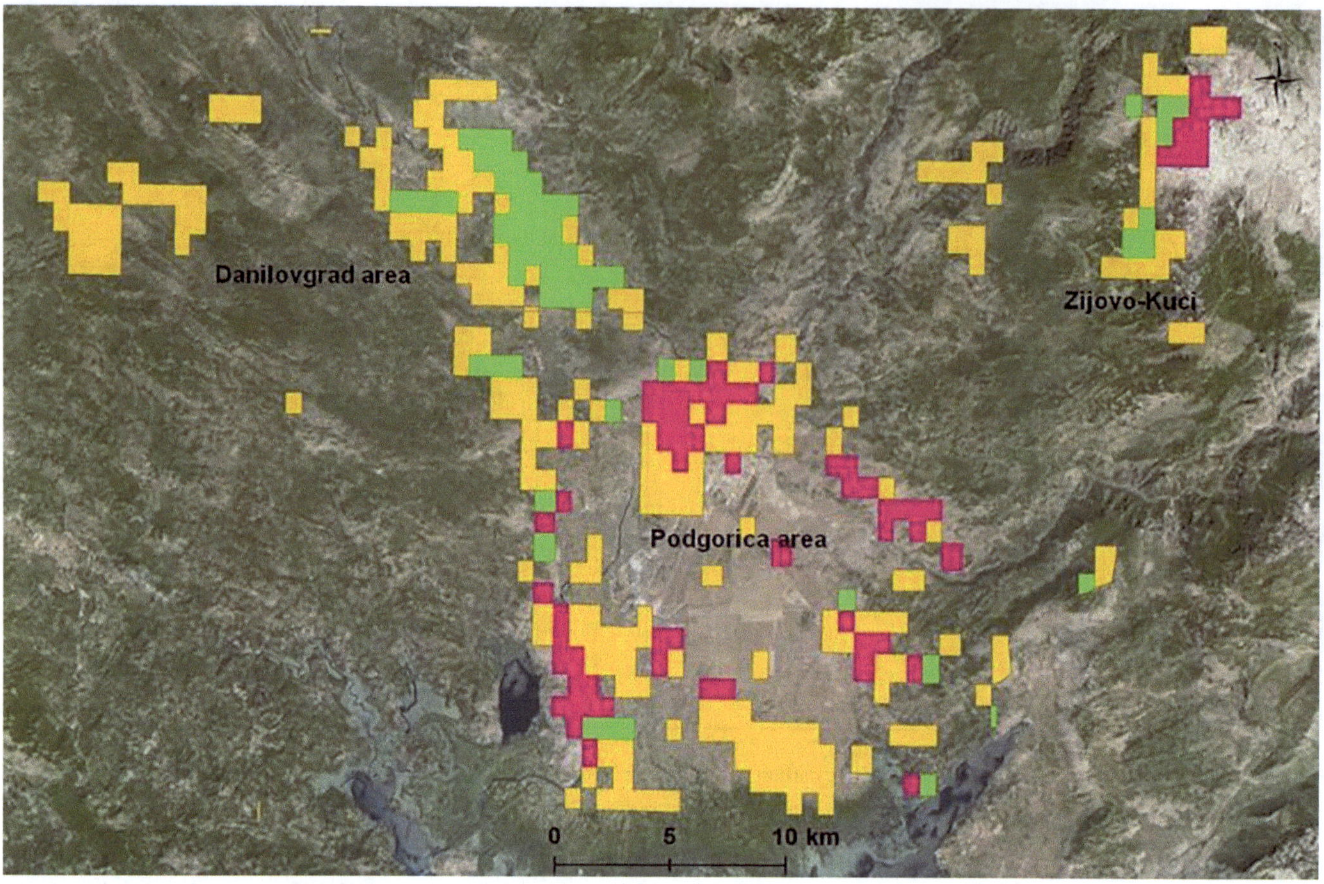

Fig. 6.5 Podgorica area, Danilovgrad area, and Žijovo–Kuči hotspots[1]

and the need for irrigation and drainage systems. The fifth measure, implementation of agri-environmental measures, includes establishing and protecting forests and implementing agri-environmental measures to encourage farmers to adopt sustainable farming practices, including the environmentally favourable intensification of farming, preservation of high-value habitats, and the conservation of landscapes and historical features.

The Montenegro National Drought Plan aims to improve drought management in the country by establishing relevant institutions, developing a monitoring and early warning system, assessing vulnerability and risk, and implementing measures to limit the impact of droughts. The plan recommends the creation of a National Drought Authority, Intersectoral Drought Advisory Board, and Department for Drought in IHMS to manage drought-related issues.

The plan also suggests improving communication channels and expanding the drought monitoring and forecasting system to include vulnerable economic sectors, pressure on communities, and potential damage or loss. It proposes connecting land use planning and drought management, coordinating communication between government and the private sector, and improving legislation on droughts and the environment.

The Centre for Soil and Amelioration of Biotechnical Faculty in Podgorica stands out as the most important organisation in the country dealing with the influence of drought on agricultural production. During the last decade, the research team of this Centre was engaged in the application of soil water balance models (Knežević 2008; Knežević et al. 2012, 2013) and crop growth models, but also in real-time soil moisture monitoring for the optimisation in irrigation with the purposes of increasing yield and quality of agricultural products (Topalović et al. 2018; Ivanović et al. 2018; Topalović et al. 2020; Ivanović et al. 2021; Topalović et al. 2021; Topalović et al. 2024). As regards the automation in irrigation and through cooperation with business entities, significant progress was made within the BIO-ICT project—Centre of Excellence in Bioinformatics (the implementation period 2016–2019) (Salković et al. 2018; Topalović et al. 2018). Further work is directed towards finding simpler and less demanding solutions in the determination of

[1] More than 11,700 ha is found under degraded LPD classes in Podgorica area and more than 4600 ha in Danilovgrad area; the main drivers of degradation are urbanisation, abandonment of agriculture, and forest fires. In Žijovo–Kuči area more than 2350 ha of forests and pastures are degraded, and the main driver of degradation is wildfires.

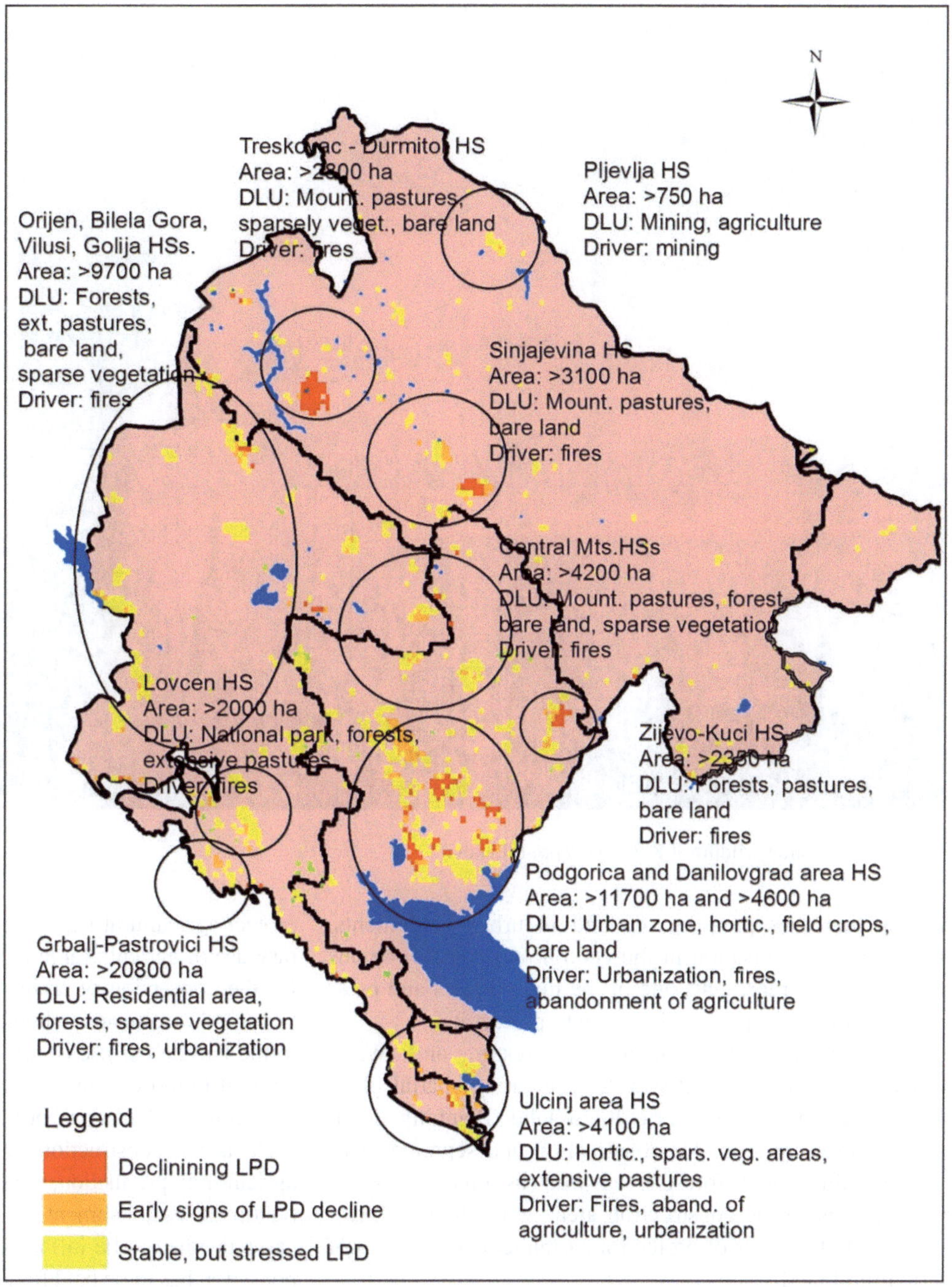

Fig. 6.6 National map of selected hotspots

irrigation time and amounts. Besides, the same group of authors dealt with predicting the effects of drought due to climate change on olive (Knežević et al. 2017) and potato production (Knežević et al. 2018a, b). According to their studies, the territory for olive groves would expand in future towards the north, which might bring to new investments. Given the climate change, a similar prediction in fruit growing was conducted in the region (Ćosić et al. 2021), which is of great importance for the intensification of agricultural land use (Eulenstein et al. 2022). Also, new ideas will be incorporated into soil investigations by simplifying soil description (Knežević et al. 2021). The efforts should be oriented towards soil protection because the land use conversion without appropriate measures leads to the degradation of soil properties (Tolimir et al. 2020; Gajić et al. 2023).

The Montenegro National Drought Plan, along with the Danube Drought Strategy, serves as a guideline for governmental institutions to lead this process. The plan provides various recommendations and implementation actions to continuously mitigate and prepare for drought in Montenegro.

Urbanisation leads to the loss of high-quality agricultural land, while migration from less developed

Fig. 6.7 Forest after wildfire at Kostica (*Photo* I. Laković)

regions increases pressure on land in urban settlements. Accelerated urbanisation in the central and coastal regions of Montenegro represents one of the main threats and pressures on soil, so steps must be taken to protect it.

The Ministry of Agriculture, Forestry, and Water Management plays an active role in creating spatial plans and protecting the best agricultural land. Planning documents demarcate areas for construction and separate areas for agriculture and forestry, as well as other activities planned outside the construction area, and define the principles and objectives of protection, arrangement, and use of agricultural land.

According to the Law on Spatial Planning and Construction of Structures, certain buildings can be built outside the designated construction areas. These include infrastructure and engineering facilities, agricultural production facilities, facilities for processing mineral raw materials, energy production facilities, camps and outdoor sports and recreational playgrounds, and buildings for housing and auxiliary purposes on agricultural farms and rural tourism. The exploration and exploitation of mineral resources can also be conducted outside the construction area. However, arable agricultural land cannot be subdivided into plots with areas less than half a hectare, and land arranged by consolidation cannot be divided into plots less than one hectare. In order to protect agricultural land, compensation is paid for the temporary use of agricultural land for non-agricultural purposes and for the conversion of arable agricultural land. After exploitation, land recultivation should be conducted, and an exception to the ban on the use of arable agricultural land for non-agricultural purposes can only be made with prior consent from the Ministry. This consent may be given for various purposes such as the construction of economic facilities for primary agricultural production, reconstruction of family residential buildings, establishment of forests and nurseries, and other cases deemed to be in the general interest. In all cases, compensation has to be paid for conversion, in accordance with the law. Agricultural land that is temporarily used for non-agricultural purposes must be recultivated to restore it for appropriate agricultural production within the specified time limit. Anti-erosion measures are taken to protect agricultural land from the harmful effects of erosion and floods in the eroded areas.

6.3 Recommendations for Appropriate Land Management

Recommendations for spatial planning present a series of instructions that should be followed by planners when preparing planning documents. It is necessary to define

Fig. 6.8 Soil and forest degradation caused by mismanagement (*Photo* M. Knežević)

appropriate procedures and necessary contents in the "Rulebook on detailed contents and format of the planning document, as well as criteria for land use, elements of urban regulation, and the standardised graphical symbols" that would enable successful spatial planning.

The analysis of agricultural production, the measures for land protection taking into account the principles of LDN, the type of buildings that can be constructed on agricultural land, and the corresponding exceptions, as well as the sequence of actions, are necessary when adopting appropriate spatial plans.

Agricultural experts should conduct an analysis of agricultural production that includes current development, basic development tendencies, and problems. The analysis should cover agricultural land, population engaged in agriculture, primary agricultural production, and characteristics of production. Future development potentials should also be defined along with agrarian policy measures and incentives for sustainable development. It is important to have a separate chapter for agricultural production and land, which should include a detailed description of the possible directions of agricultural production development. The analysis should consider existing natural and socio-economic conditions and available resources. The characteristics of production should cover plant production, animal husbandry, technical equipment and processing capacities, and business entities and associations important for agriculture. The purpose of the analysis is to build sustainable development of agricultural production.

The Ministry of Agriculture, Forestry, and Water Management also determines and keeps records for permanently protected lands and provides guidelines for land protection. The protection of agricultural land includes measures to maintain the natural functions of land, preserve its intended use, and prevent permanent changes in land use. Conservation involves legal, technical, and organisational measures aimed at consolidating and improving the natural and ecological conditions of the land. Land reclamation includes measures to improve the quality of arable agricultural land, and irrigation and drainage are used to improve soil water regime and, in this manner, foster agricultural production.

Montenegro should prohibit the construction of buildings on the best agricultural land and take into consideration land conservation in all development planning. Soil quality monitoring should be introduced to support economic and socio-economic planning. An information system about land should be created. NAP and LDN objectives should be used to balance land degradation and restoration, which means that in the case of a decrease in the quality or conversion of the land of a certain area, it is important to improve the properties of the land on another area in similar bio-geographical region. Funds collected on the basis of land conversion fees could be used for these purposes, i.e., these funds are used for the acquisition of new or intensification of existing agricultural lands. The code of good agricultural practice should be respected, and necessary measures should be implemented to eliminate or mitigate soil degradation. Financial resources and incentives for soil conservation should be provided through fees and taxes prescribed by law, budget-specific funds, grants, and loans. The state, form, and degree of soil and environmental pollution should be monitored, and successive recultivation and complete rehabilitation/restoration of the land after exploitation should be mandatory. Funds should be allocated for the acquisition of new land or the improvement of existing agricultural land.

For all the above issues related to sustainable agriculture and environment and development, the creation of a multipurpose Land Information and Monitoring Service for Montenegro (LIMS-ME) was suggested by GEF Support to UNCCD 2018 national reporting process (Montenegro). The aim was to provide land cover information to state institutions for their needs by using machine learning for mapping environmental variables such as soil properties in 4D (time as the fourth component), vegetation distribution, and similar variables.

Recognising the opportunities for growth in Montenegro's soil domain and understanding the need for a robust legislative and strategic framework, awareness about the importance of addressing land degradation is rising. Embracing agro-environmental principles and investing in human, technological, and financial resources in order to improve soil quality and support soil conservation and sustainable soil management, the implementation of the following activities should be considered:

I. Establishment of a Regional Centre for scientists from the region

- Application of geostatistics and modelling—numerical and process modelling approaches which would allow increasing human and technical capacities for monitoring soil degradation processes (SOC dynamics, land use, and land use change, agro-ecological zoning, soil erosion—intensity and erosion risk management, soil sealing—smart urban planning protection of highly productive soils, soil contamination);
- Establishing and harmonisation of methodologies for soil monitoring schemes as a long-term commitment;
- Establishing regional soil platform.

II. Improvement of the legal and strategic framework related to land management and protection

- Revision of National Action Plan to Combat Desertification, according to LDN and the new UNCCD Strategy by 2030.
- Adoption of a national Strategy for Agriculture and Rural Development, focusing on the vision and objectives of the EU Soil Strategy for 2030 regarding climate neutrality, including Land Use, Land Use Change, and Forestry (LULUCF) and preconditions for good soil health, and to integrate the "land take hierarchy" to avoid converting agricultural or natural land into a built environment.
- Adoption of Law on Soil Protection and revision of Law on Agricultural Land in order to transpose the best European practices and to protect high-quality land against urbanisation.
- Integration of soil and land use management in the river basins and flood risk management plans by deploying nature-based solutions.
- Enhancing coordination mechanisms and clarifying stakeholder roles and responsibilities for effective and sustainable land and drought management.

III. Capacity building and raising awareness of government officials, local communities, and beneficiary communities for climate-responsive planning and development

- Providing education and training to government officials, local communities, and farmers to improve the technical

capacities of relevant institutions and align national climate policies and measures with the LDN targets.
- Promoting sustainable land and forest management.
- Creation of land information and monitoring service for Montenegro.

IV. Support for implementation of agro-environmental measures recognised within the project proposal "Application of agro-environmental measures for the purpose of sustainable land use, forest management, mitigation and achievement of the goals of land degradation neutrality (LDN)"

- Establishment of new orchards on previously degraded land.
- Enhancement of organic agricultural production and change in manure management.
- Reduction of GHG emissions through the adoption of environmentally-friendly farming techniques by farmers; the use of organic fertilisers instead of synthetic nitrogen fertilisers, which contribute to N_2O emissions.
- Investments in water supply (wells, reservoirs) in order to combat forest fires.
- Support for the improvement of the livestock fund (small ruminants—sheep and goats) is key to ensuring appropriate management of grasses for the purpose of fire prevention.
- Production and use of biochar and study of its effects on soils.

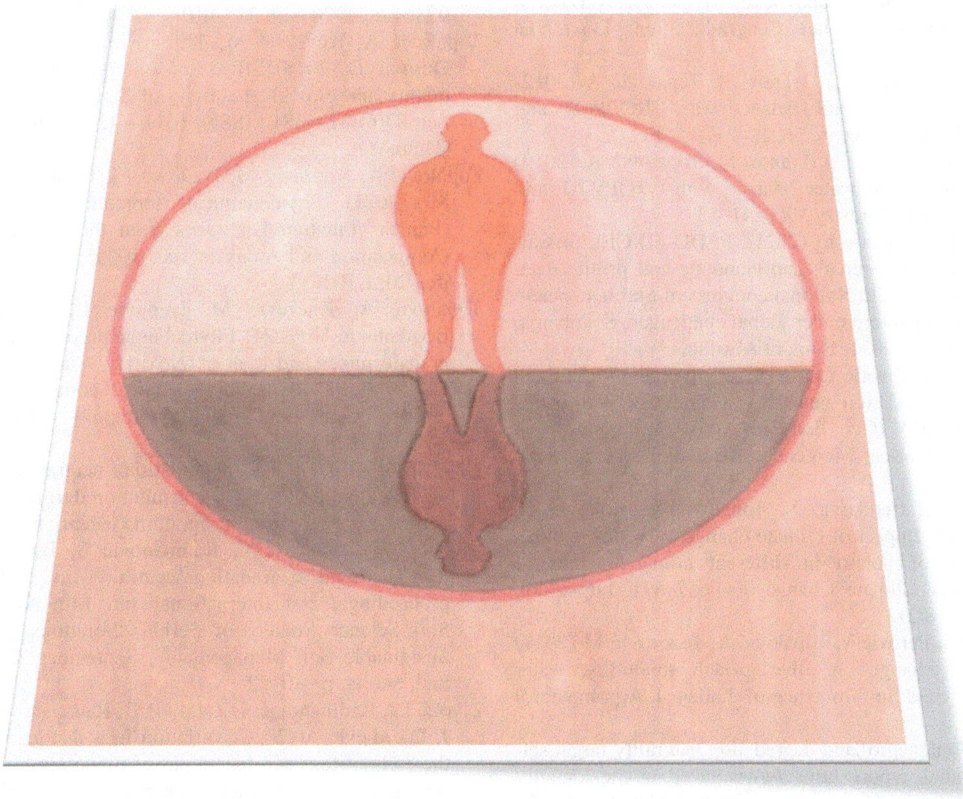

Human and Soil *With hope for a better tomorrow, Ana Topalović's illustration vividly reminds us of the words of a wise man, which the soil also sends us more and more every day:* ***"People have betrayed me the most, yet again, people I trust the most".***

References

Ćosić M, Lipovac A, Vujadinović Mandić M, Stričević R, Đurović N, Ranković Vasić Z, Životić LJ (2021) Water requirements of fruit and vine plantations in the area of the Kolubara District in present and Future Conditions.In: 3rd International and 15th National Congress of Serbian Society of Soil Science: Soils for Future under Global Challenges, Sokobanja, Serbia, 21–24 September 2021, Book of Proceedings, pp 240–247

Eulenstein F, Saljnikov E, Lukin S, Sheudshen AK, Rukhovich O, Schindler U, Saparov G, Pachikin K, Thielicke M, Behrendt A, Armin W, Životić Lj, Müller L (2022) Climate change as the driving force behind the intensification of agricultural land use. Zemljiste i Biljka 71(1):24–39

Gajić K, Kresović B, Tolimir M, Zivotić Lj, Lipovac A, Gajić B (2023) Hydraulic properties of fine-textured soils in lowland ecosystems of Western Serbia vary depending on land use. Geoderma Regional 32, e00603. https://doi.org/10.1016/j.geodrs.2022.e00603

Ivanović Lj, Topalović A, Bogdanović V, Đurović D, Mugoša B, Jadranin M, Tešević V, Beškoski V (2021) Antiproliferative activity and antioxidative potential of Swiss chard from Montenegro, grown under different irrigation and fertilization regimes. British Food Journal 123, 7, 2335-2348. DOI 10.1108/BFJ-11-2020-1062

Knežević M (2008) Planiranje i projektovanje sistema za odvodnjavanje i navodnjavanje u Bjelopavlićkoj ravnici. Univerzitet u Beogradu, Poljoprivredni fakultet Beograd-Zemun, Doktorska disertacija

Knežević M, Perović N, Životić Lj, Ivanov M, Topalović A (2012) Simulation of silage-maize water balance with CROPWAT and ISAREG models. Agricul Forestry, 56, 10(1–4), 5–17

Knežević M, Perović N, Životić Lj, Ivanov M, Topalović A (2013) Simulation of winter wheat water balance with CROPWAT and ISAREG models. Agricul Forestry 59(1):41–53

Knežević M, Topalović A, Životić Lj (2021). PEDO–EXCEL: a simple excel tool/database to prepare and elaborate soil profile data. In: 3rd International and 15th National congress of Serbian society of soil science: soils for future under global challenges, Sokobanja, Serbia, 21–24 September 2021, Book of Abstracts, p 14

Knežević M, Topalović A, Životić Lj (2018) Environmental trends in Montenegro: land degradation neutrality. Green Room Sessions 2018 International GEA (Geo Eco-Eco Agro) Conference, 1–3 November 2018, Podgorica, Montenegro, Book of Abstracts, pp 188

Knežević M, Životić Lj, Čereković N, Topalović A, Koković N, Todorović M (2018b) Impact of climate change on water requirements and growth of potato in different climatic zones of Montenegro. J Water Climate Change 9(4):657–671. https://doi.org/10.2166/wcc.2018.211

Knežević M, Životić Lj, Perović V, Topalović A, Todorović M (2017) Impact of climate change on olive growth suitability, water requirements and yield in Montenegro. Italian J Agrometeorol 2(2017):39–52

LDN report (2018) Montenegro land degradation neutrality target setting process national report https://www.unccd.int/sites/default/files/ldn_targets/2019-05/Montenegro%20LDN%20TSP%20Country%20Report.pdf

Lj I, Milašević I, Topalović A, Djurović D, Mugoša B, Knežević M, Vrvić M (2018) Nutritional and phytochemical content of Swiss chard from Montenegro, under different fertilization and irrigation treatments. British Food J 121(2):411–425

Ministarstvo poljoprivrede ruralnog razvoja - MARD (2014) Strategija sa planom razvoja šuma i šumarstva 2014. – 2023. godina - Nacionalna šumarska strategija (Predlog)

Ministry of Sustainable Development and Tourism (MSDT), Montenegro National Drought Plan, November 2020

Salković E, Djurović I, Knežević M, Popović-Bugarin V, Topalović A (2018) Digitization and mapping of national legacy soil data of Montenegro. Soil and Water Research 13 (2), 83–89. DOI: 10.17221/81/2017-SWR

Tolimir M, Kresović, B, Životić Lj, Dragović S, Dragović R, Sredojević Z, Gajić B (2020) The conversion of forestland into agricultural land without appropriate measures to conserve SOM leads to the degradation of physical and rheological soil properties. Scientific Reports (2020), 10:13668. https://doi.org/10.1038/s41598-020-70464-6

Topalović A, Knežević M, Bajagić B, Ivanović LJ, Milašević I, Đurović D, Mugoša B, Podolski-Renić A, Pešić M (2020) Grape (*Vitis vinifera* L.): Health benefits and effects of growing conditions on quality parameters. Biodiversity and Biomedicine: Our Future. Editedby Munir Ozturk, Dilfuza Egamberdieva, Milica Pešić, Academic Press (an imprint of Elsevier), eBook ISBN: 9780128223703, pp 385–401

Topalović A, Knežević M, Trifunović S, Novaković M, Pešić M, Đurović D (2018) Effects of soil properties and fertilization on quality and biological activity of Swiss chard. Eur J Horticultural Sci 83(6):374–381. ISSN 1611–4426 (Print), ISSN 1611–4434 (Online)

Topalović A, Knežević M, Ivanović Lj, Ganik S, Mikulič-Petkovšek M (2021) Phytochemical screening of wild pomegranate (Punica granatum L.) juices from the market. Journal of Food Composition and Analysis 100, 103933. https://doi.org/10.1016/j.jfca.2021.103933

Topalović A, Knežević M, Ivanović Lj, Mikulic-Petkovsek M, Bogdanović V (2024) Phytochemical composition of wild pomegranate juices and their cytotoxicity. Nutrition and Food Science 54 (8), 1498–1514. Phytochemical composition of wild pomegranate juices and their cytotoxicity. https://doi.org/10.1108/NFS-02-2024-0073

UNCCD (2016) Land degradation neutrality target setting—a Technical guide. draft for consultation during the land degradation neutrality target setting programme inception phase. May 2016

Životić LJ, Golubović S, Radmanović S, Belić M, Djordjević A (2017): Land degradation neutrality in Serbia. In: Congress proceedings, 2nd International and 14th National Congress of Soil Science Society of Serbia "Solutions and Projections for Sustainable Soil Management", September 25–28th, 2017, Novi Sad, Serbia, pp 80–92

Životić LJ, Radmanović S, Žarković B, Radovanović V, Bogosavljević J, Djordjević A (2020) Variation in soil organic carbon stocks in the soils of foot and toe slopes of mountain Vukan, East—Central Serbia. Book of Abstracts: Symposium—Irrigation and drainage in the light of climate change. Vršac, 9–11 September 2020., p 57

Appendix A

See Figs. A.1, A.2, A.3 and A.4.

M. Knežević et al., *The Soils of Montenegro,* World Soils Book Series, https://doi.org/10.1007/978-3-031-82919-2

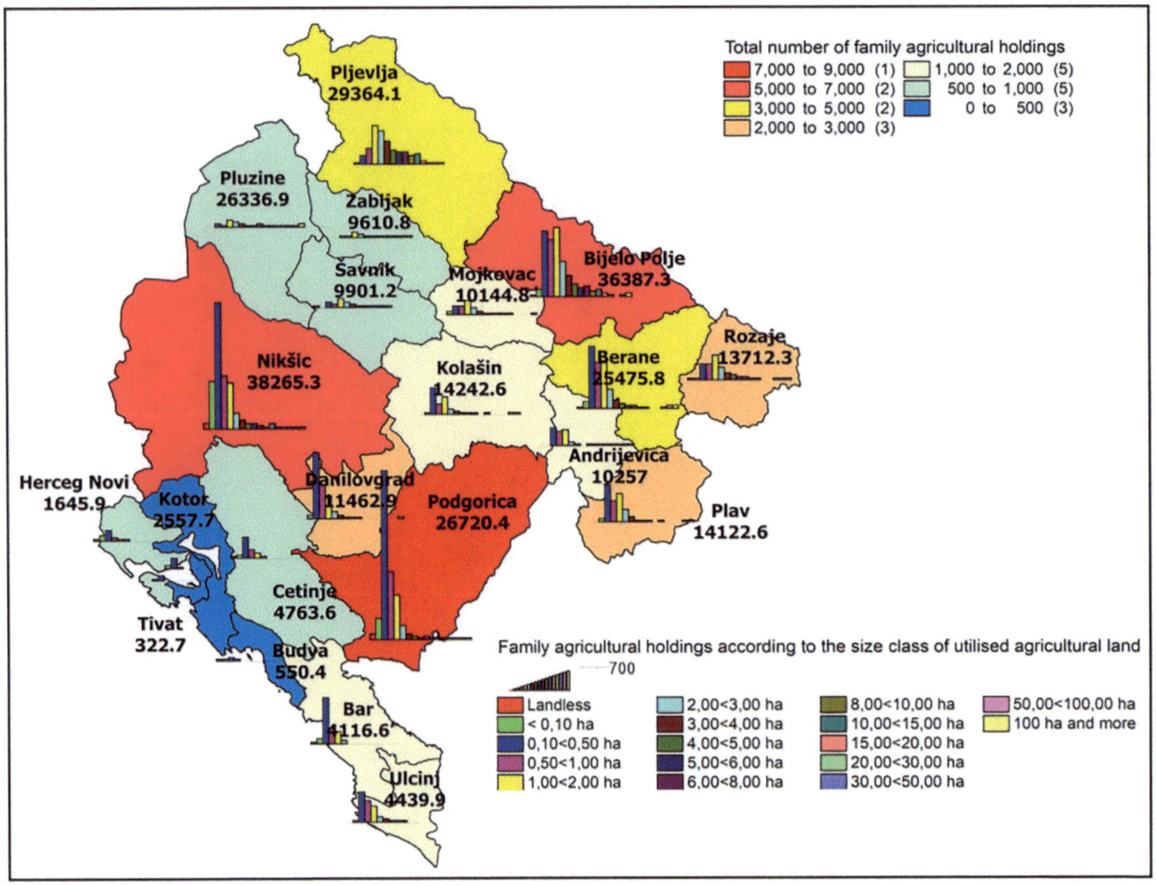

Fig. A.1 Family agricultural holdings according to the size class of utilised agricultural land, the total number of family agricultural holdings, and areas (ha) of total available land by municipality—LD5

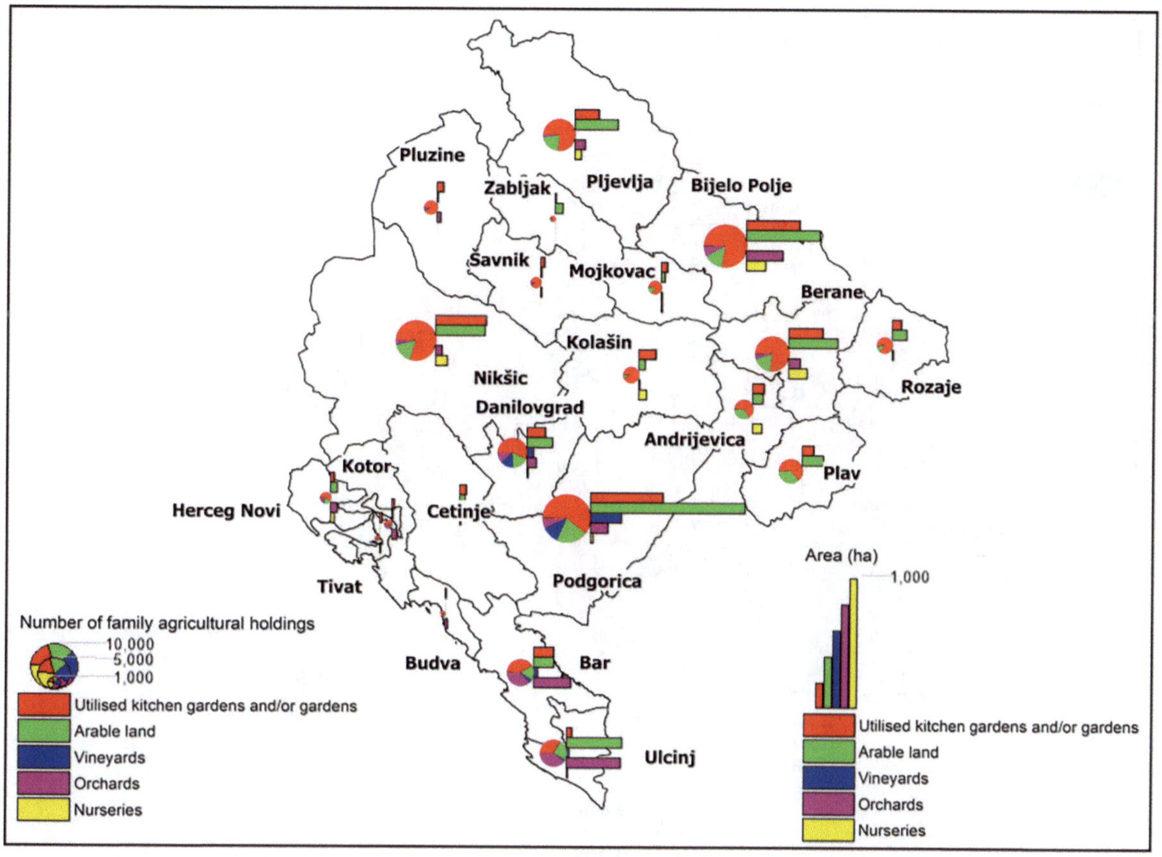

Fig. A.2 Family agricultural holdings by the place of establishment and manner of use of agricultural land by municipality—LD4

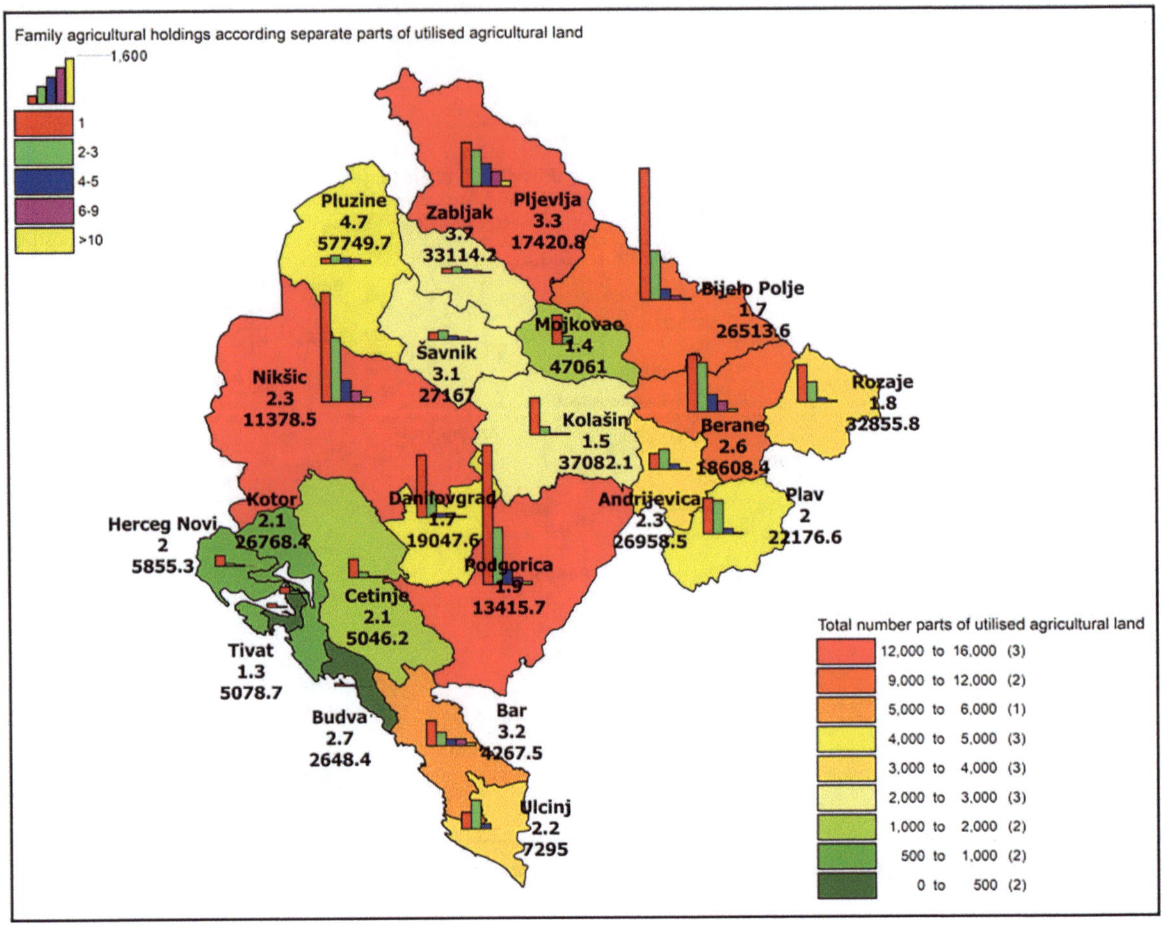

Fig. A.3 Family agricultural holdings according to the total number and class of separate parts of utilised agricultural land, average number of parcels, and average area (in m^2) by municipality—LD7

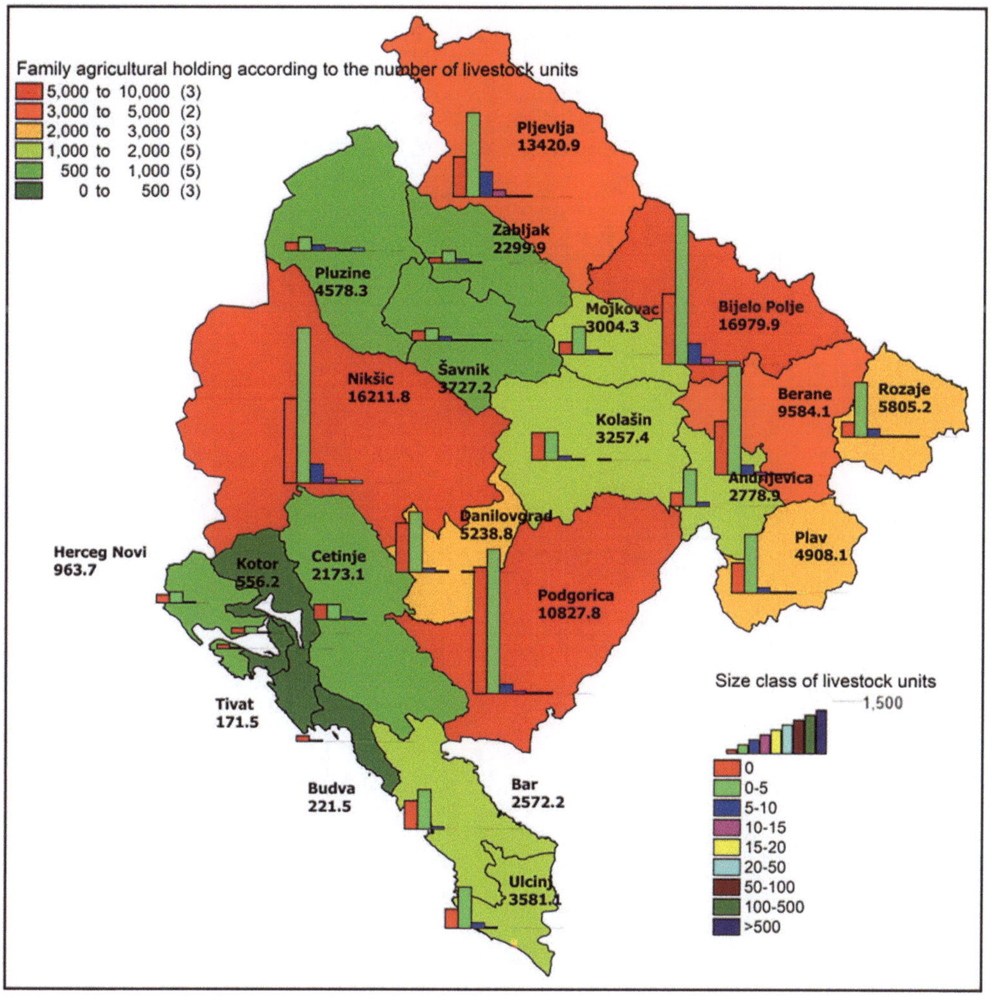

Fig. A.4 Family agricultural holding according to the number and the size class of livestock units, and the number of livestock units by municipality—LD1

[illegible]

Appendix B

See Tables B.1 and B.2.

M. Knežević et al., *The Soils of Montenegro,* World Soils Book Series, https://doi.org/10.1007/978-3-031-82919-2

Table B.1 Discharges of the main rivers in Montenegro

n	HS	Watercourse	Basin	Start of operations[a]	Minimum annual discharge Q (m^3/s)	Maximum annual discharge Q (m^3/s)	Mean annual discharge Q (m^3/s)
Adriatic basin							
1	Podgorica	Morača	Skadar L	1948–	7.93	2073	157.16
2	Zlatica	Morača	Skadar L	1983–	0	1369	58.33
3	Pernica	Morača	Skadar L	1956–	1.14	812	28.69
4	Međuriječje	Mrtvica	Morača	1948–	0.48	603	14.48
5	Duklov Bridge	Zeta	Morača	1955–	0.045	286	18.63
6	Danilovgrad	Zeta	Morača	1948–	4.68	577	76.58
7	Gornje Polje	Sušica	Zeta	1948–	0.05	94.5	5.46
8	Brodska Njiva	Rijeka Crnojevića	Skadar L	1987–	0.390	228	6.29
Black Sea basin							
1	Dobrakovo	Lim	Drina	1961–	6.46	1518	74.19
2	Bijelo Polje	Lim	Drina	1948–	8.20	1077	62.7
3	Ravna Rijeka	Ljubovića	Lim	1948–	0.240	310	9.03
4	Gubavač	Bjelopoljska Bistrica	Lim	1949–	0.360	65.6	6.05
5	Andrijevica	Zlorečica	Lim	1985–	0.310	198	6.29
6	Plav	Lim	Drina	1948–	1.30	324	19.21
7	Vusanje	Grlja	Vruja (Plavsko Lake)	1959–	0.058	81.8	3.68
8	Rožaje	Ibar	West Morava	1968–	0.040	102 1972g	2.50
9	Bać	Ibar	West Morava	1978–1993	0.160	98.3	5.22
10	Crna Poljana	Tara	Drina	1957–	0.720	468	11.72
11	Trebaljevo	Tara	Drina	1959–	1.55	701	24.39
12	Pljevlja	Ćehotina	Drina	1948–	0.32	145	6.27
13	Gradac	Ćehotina	Drina	1963–	1.35	414	12.79
14	Gornja Bijela	Bijela	Pridvorica (Komarnica)	2007–	0.06	113	1694
15	Timar	Bukovica	Pridvorica (Komarnica)	2007–	0.056	89.8	1.95

[a] Up to 01.01.2022

Table B.2 Water levels of Lake Skadar and the Bojana River

n	HS	Basin	Start of operations[a]	Spot level "0" a.s.l	Minimum annual water level (cm/m a.s.l.)	Maximum annual water level (cm/m a.s.l.)	Mean annual water level (cm/m a.s.l.)
1	Fraskanjel Bojana	Adriatic Sea	1960–	– 0.07 cm	7 cm	636 cm	179 cm
2	Plavnica	Skadar L	1948–	4.56 asl	4.54 asl	10.4 asl	6.41 asl
3	Ckla	Skadar L	1950–	4.86 asl	4.66 asl	10.55 asl	6.47 asl

[a] Up to 01.01.2022

Appendix C

See Table C.1.

M. Knežević et al., *The Soils of Montenegro,* World Soils Book Series, https://doi.org/10.1007/978-3-031-82919-2

Table C.1 Extended climate indicators

ULCINJ	Jan	Feb	Mar	Apr	May	June	July	Aug	Sep	Oct	Nov	Dec	Annual
Hor.visibility (km)	19.5	19.5	19.2	19.2	19.8	19.9	19.9	19.7	20.3	19.8	19.6	19.3	19.6
Cloudiness	5.3	5.2	5.1	5.3	4.3	3.1	1.8	2.0	3.3	4.4	5.4	5.8	4.2
Sunshine duration (h)	138.0	144.8	181.3	207.6	265.7	294.2	336.5	319.6	246.4	198.2	141.3	117.1	2590.6
Number of days with precipitation	12	12	11	12	8	6	4	4	8	11	12	14	113
Number of summer days	0	0	0	1	9	20	30	30	21	5	0	0	116
Number of tropical days	0	0	0	0	1	5	13	14	2	0	0	0	35
Number of tropical nights	0	0	0	0	1	10	22	22	6	1	0	0	60
Number of frost days	3	3	1	0	0	0	0	0	0	0	0	1	7
Number of icing days	0	0	0	0	0	0	0	0	0	0	0	0	0
KF m/a	15.6	15.4	10.0	7.4	3.2	2.4	1.0	1.8	4.6	7.9	13.1	17.5	74.4
Humidity	ph	ph	h	h	a	a	a	a	sa	h	h	ph	SH
BAR	Jan	Feb	Mar	Apr	May	June	July	Aug	Sep	Oct	Nov	Dec	Annual
Hor.visibility (km)	18.6	18.6	18.1	18.2	19.1	19.5	19.7	19.3	19.5	18.8	18.4	18.1	18.8
Cloudiness	5.3	5.3	5.3	5.2	4.3	3.1	1.9	2.0	3.4	4.5	5.5	5.8	4.3
Sunshine duration (h)	129.8	137.0	177.2	201.6	269.2	306.3	351.8	322.0	246.2	188.7	129.0	111.2	2570.0
Number of days with precipitation	12	11	11	11	8	7	4	5	8	11	12	14	113
Number of summer days	0	0	0	0	8	20	30	30	21	5	0	0	114
Number of tropical days	0	0	0	0	0	4	11	10	1	0	0	0	27
Number of tropical nights	0	0	0	0	0	5	14	13	2	0	0	0	35
Number of frost days	2	2	1	0	0	0	0	0	0	0	0	1	5
Number of icing days	0	0	0	0	0	0	0	0	0	0	0	0	0
KF m/a	14.6	15.2	11.0	8.2	3.7	2.6	1.2	2.0	6.1	8.9	12.5	17.1	81.7
Humidity	ph	ph	h	h	sa	a	a	a	sh	h	h	ph	H
HERCEG NOVI	Jan	Feb	Mar	Apr	May	June	July	Aug	Sep	Oct	Nov	Dec	Annual
Hor.visibility (km)	18.4	18.6	18.2	18.3	19.1	19.5	19.7	19.5	19.6	18.7	18.4	18.1	18.8
Cloudiness	5.3	5.2	5.4	5.4	4.5	3.4	2.1	2.2	3.4	4.6	5.6	5.7	4.4
Sunshine duration (h)	116.2	130.9	164.9	191.0	249.4	289.4	326.7	312.8	234.5	175.0	118.0	100.1	2409.0
Number of days with precipitation	13	12	12	13	10	8	5	6	9	12	13	15	128
Number of summer days	0	0	0	0	8	20	30	30	20	4	0	0	113
Number of tropical days	0	0	0	0	0	5	14	14	2	0	0	0	36
Number of tropical nights	0	0	0	0	1	7	18	19	4	0	0	0	49
Number of frost days	1	1	0	0	0	0	0	0	0	0	0	1	3
Number of icing days	0	0	0	0	0	0	0	0	0	0	0	0	0
KF m/a	19.8	20.6	15.2	11.0	5.2	3.3	1.4	2.8	6.3	11.2	19.1	25.9	109.6
Humidity	ph	ph	ph	h	sh	sa	a	a	sh	h	ph	ph	H
PODGORICA	Jan	Feb	Mar	Apr	May	June	July	Aug	Sep	Oct	Nov	Dec	Annual
Hor.visibility (km)	17.3	17.7	18.3	18.6	19.4	19.6	19.8	19.7	19.3	18.7	17.7	16.9	18.6
Cloudiness	5.1	5.2	5.3	5.6	4.8	3.8	2.6	2.6	3.7	4.5	5.4	5.6	4.5
Sunshine duration (h)	131.9	132.9	171.3	192.2	249.6	281.8	336.3	312.5	234.7	183.2	129.3	101.2	2456.8
Number of days with precipitation	11	11	11	12	9	8	5	6	8	11	12	13	118

(continued)

Table C.1 (continued)

PODGORICA	Jan	Feb	Mar	Apr	May	June	July	Aug	Sep	Oct	Nov	Dec	Annual
Number of summer days	0	0	0	3	18	26	31	30	23	9	0	0	141
Number of tropical days	0	0	0	0	5	15	26	25	9	1	0	0	81
Number of tropical nights	0	0	0	0	1	8	22	21	3	0	0	0	55
Number of frost days	10	6	1	0	0	0	0	0	0	0	1	6	24
Number of icing days	0	0	0	0	0	0	0	0	0	0	0	0	0
KF m/a	27.7	23.0	14.8	9.6	4.2	2.7	1.0	2.1	6.8	10.6	23.2	34.0	103.1
Humidity	ph	ph	ph	h	sa	a	a	a	h	h	ph	ph	H
CETINJE	**Jan**	**Feb**	**Mar**	**Apr**	**May**	**June**	**July**	**Aug**	**Sep**	**Oct**	**Nov**	**Dec**	**Annual**
Hor.visibility (km)	14.5	14.6	15.7	16.0	17.2	18.1	18.8	18.4	17.4	15.9	14.6	13.6	16.2
Cloudiness	5.4	5.5	5.5	5.7	4.9	3.9	2.7	2.8	3.9	4.9	5.7	5.8	4.7
Number of days with precipitation	13	13	13	14	13	10	7	8	10	13	14	15	142
Number of summer days	0	0	0	0	6	17	26	25	11	2	0	0	86
Number of tropical days	0	0	0	0	0	4	12	13	2	0	0	0	31
Number of tropical nights	0	0	0	0	0	0	0	0	0	0	0	0	0
Number of frost days	24	21	16	4	0	0	0	0	0	4	14	21	103
Number of icing days	1	0	0	0	0	0	0	0	0	0	0	0	2
KF m/a	374.7	249.1	68.3	27.3	10.0	5.7	2.4	4.9	14.0	32.1	87.6	251.7	323.1
Humidity	ph	ph	ph	ph	h	sh	a	sa	ph	ph	ph	ph	PH
NIKŠIĆ	**Jan**	**Feb**	**Mar**	**Apr**	**May**	**June**	**July**	**Aug**	**Sep**	**Oct**	**Nov**	**Dec**	**Annual**
Hor.visibility (km)	17.0	16.7	17.9	18.8	19.8	20.6	21.3	20.8	20.0	18.9	17.4	16.7	18.8
Cloudiness	5.3	5.5	5.6	5.9	5.2	4.4	3.0	3.0	4.1	4.8	5.6	5.7	4.8
Sunshine duration (h)	123.0	128.6	157.4	170.7	224.4	257.9	311.4	291.3	213.1	170.9	123.2	108.5	2280.3
Number of days with precipitation	12	11	11	13	12	11	7	8	9	11	13	13	132
Number of summer days	0	0	0	0	4	15	24	24	10	1	0	0	79
Number of tropical days	0	0	0	0	0	3	10	12	1	0	0	0	27
Number of tropical nights	0.0	0.0	0.0	0.0	0.0	0.1	0.3	0.4	0.0	0.0	0.0	0.0	0.9
Number of frost days	20	18	9	2	0	0	0	0	0	1	7	16	74
Number of icing days	2	1	0	0	0	0	0	0	0	0	0	1	5
KF m/a	89.1	73.8	28.7	15.7	7.3	4.6	2.0	3.5	9.2	16.9	43.7	85.9	168.9
Humidity	ph	ph	ph	ph	h	sa	a	sa	h	ph	ph	ph	PH
KOLAŠIN	**Jan**	**Feb**	**Mar**	**Apr**	**May**	**June**	**July**	**Aug**	**Sep**	**Oct**	**Nov**	**Dec**	**Annual**
Hor.visibility (km)	11.5	11.7	12.8	13.8	14.8	15.5	16.4	15.9	13.5	12.8	11.3	9.9	13.3
Cloudiness	6.4	6.5	6.5	6.6	6.0	5.1	4.1	4.3	5.5	5.8	6.6	7.0	5.9
Sunshine duration (h)	67.0	87.4	125.3	141.5	190.5	200.9	238.1	227.2	167.2	132.9	77.1	47.6	1702.7
Number of days with precipitation	15	15	16	16	15	13	10	10	11	13	15	17	166
Number of summer days	0	0	0	0	2	8	15	17	5	0	0	0	48
Number of tropical days	0	0	0	0	0	1	3	4	0	0	0	0	9
Number of tropical nights	0	0	0	0	0	0	0	0	0	0	0	0	0

(continued)

Table C.1 (continued)

KOLAŠIN	Jan	Feb	Mar	Apr	May	June	July	Aug	Sep	Oct	Nov	Dec	Annual
Number of frost days	26	23	21	9	1	0	0	0	1	6	17	23	126
Number of icing days	7	5	2	0	0	0	0	0	0	0	2	7	23
KF m/a	–	–	73.5	26.9	10.7	7.1	3.8	4.8	13.9	25.2	88.8	–	279.6
Humidity	–	–	ph	ph	h	h	sa	sa	ph	ph	ph	–	PH

ŽABLJAK	Jan	Feb	Mar	Apr	May	June	July	Aug	Sep	Oct	Nov	Dec	Annual
Hor.visibility (km)	16.9	15.9	16.2	16.6	18.2	18.5	19.3	19.8	17.9	17.2	15.8	14.7	17.2
Cloudiness	6.3	6.6	6.7	6.8	6.5	5.9	5.0	4.9	5.5	5.7	6.3	6.7	6.1
Sunshine duration (h)	104.0	113.9	144.6	157.3	194.2	209.1	258.7	242.7	190.2	156.4	106.3	84.7	1962.1
Number of days with precipitation	16	16	17	17	17	15	11	11	12	13	15	18	180
Number of summer days	0	0	0	0	0	2	6	6	1	0	0	0	14
Number of tropical days	0	0	0	0	0	0	0	0	0	0	0	0	1
Number of tropical nights	0	0	0	0	0	0	0	0	0	0	0	0	0
Number of frost days	29	26	26	16	3	0	0	0	1	9	21	27	159
Number of icing days	9	10	6	1	0	0	0	0	0	0	3	8	37
KF m/a	–	–	–	32.0	10.6	6.8	4.4	4.7	11.5	25.1	152.2	–	268.0
Humidity	–	–	–	ph	h	h	sa	sa	h	ph	ph	–	PH

BERANE	Jan	Feb	Mar	Apr	May	June	July	Aug	Sep	Oct	Nov	Dec	Annual
Hor.visibility (km)	12.7	15.2	16.8	17.7	18.1	17.8	17.8	17.5	16.2	15.6	14.4	11.9	16.0
Cloudiness	6.8	6.4	6.2	6.2	5.8	5.1	4.3	4.2	5.4	5.8	6.8	7.5	5.9
Sunshine duration (h)	66.1	91.7	133.8	153.4	192.7	223.2	260.8	242.0	168.7	132.9	79.4	47.8	1792.7
Number of days with precipitation	10	10	10	11	11	10	8	7	9	9	10	12	117
Number of summer days	0	0	0	1	8	16	23	24	11	3	0	0	85
Number of tropical days	0	0	0	0	1	5	10	12	2	0	0	0	30
Number of tropical nights	0	0	0	0	0	0	0	0	0	0	0	0	0
Number of frost days	25	22	16	5	0	0	0	0	0	4	14	22	108
Number of icing days	6	3	1	0	0	0	0	0	0	0	1	6	16
KF m/a	–	88.9	13.8	8.5	4.8	3.6	2.8	2.7	5.8	7.5	24.9	227.1	92.9
Humidity	–	ph	ph	h	sa	sa	a	a	sh	h	ph	ph	H

BIJELO POLJE	Jan	Feb	Mar	Apr	May	June	July	Aug	Sep	Oct	Nov	Dec	Annual
Hor.visibility (km)	12.5	14.5	15.7	16.4	15.9	16.1	15.5	14.8	13.3	13.0	12.7	11.7	14.3
Cloudiness	7.3	6.6	6.5	6.5	6.4	5.9	5.2	5.4	6.2	6.5	7.3	8.0	6.5
Number of days with precipitation	12	12	11	13	13	11	10	9	11	11	12	14	138
Number of summer days	0	0	0	1	8	16	22	23	10	2	0	0	82
Number of tropical days	0	0	0	0	1	4	9	10	2	0	0	0	26
Number of tropical nights	0	0	0	0	0	0	0	0	0	0	0	0	0
Number of frost days	25	22	15	4	0	0	0	0	0	3	14	22	105
Number of icing days	7	3	1	0	0	0	0	0	0	0	1	6	17
KF m/a	–	122.6	12.9	8.3	5.0	4.1	3.2	3.0	6.2	8.0	27.3	548.3	98.6

(continued)

Table C.1 (continued)

BIJELO POLJE	Jan	Feb	Mar	Apr	May	June	July	Aug	Sep	Oct	Nov	Dec	Annual
Humidity	–	ph	h	h	sh	sa	a	a	sh	h	ph	ph	H

PLJEVLJA	Jan	Feb	Mar	Apr	May	June	July	Aug	Sep	Oct	Nov	Dec	Annual
Hor.visibility (km)	11.8	14.8	16.2	16.6	16.8	16.3	16.8	16.4	15.1	14.8	13.1	11.3	15.0
Cloudiness	7.0	6.4	6.1	6.3	5.9	5.5	4.5	4.7	5.7	6.0	6.8	7.6	6.0
Sunshine duration (h)	51.1	86.3	124.0	140.5	174.1	195.0	239.4	225.6	158.8	125.6	69.0	36.1	1625.4
Number of days with precipitation	11	12	13	13	13	13	10	10	11	10	11	13	140
Number of summer days	0	0	0	0	6	13	19	21	8	2	0	0	68
Number of tropical days	0	0	0	0	0	3	6	8	1	0	0	0	18
Number of tropical nights	0	0	0	0	0	0	0	0	0	0	0	0	0
Number of frost days	27	23	19	7	1	0	0	0	0	4	16	24	121
Number of icing days	9	4	1	0	0	0	0	0	0	0	1	8	24
KF m/a	–	–	12.2	7.2	4.8	5.0	3.6	3.5	5.5	6.8	21.7	–	89.6
Humidity	–	–	h	h	sa	sa	sa	sa	sh	h	ph	–	H

Data source RHMZ (KF m/a and humidity, calculation by M. Knežević)
Number of summer days—Number of days with a maximum daily temperature equal to or above 25 °C
Number of tropical days—Number of days with maximum daily temperature equal to or above 30 °C.
Number of tropical nights—Number of days with minimum daily temperature equal to or above 20 °C.
Number of frosty days—Number of days with minimum daily temperature equal to or lower than 0 °C.
Number of icing days—Number of days with maximum daily temperature equal to or lower than 0 °C.
Criteria for humidity assessment:
For monthly assessment—Monthly rain factor—KFm (Gračanin, 1950): a arid (1.6–3.3), sa semiarid (3.3–5.0), sh semihumid (5.0–6.6), h humid (6.6–13.3), ph perhumid (>13.3); and
For an annual assessment—Lang's rain factor Kfa, *A* arid (0–40), *SA* semiarid (40–60), *SH* (60–80), *H* humid (80–160), *PH* perhumid (>160).

Index

M. Knežević et al., *The Soils of Montenegro,* World Soils Book Series, https://doi.org/10.1007/978-3-031-82919-2

MIX
Papier aus verantwortungsvollen Quellen
Paper from responsible sources
FSC® C105338

If you have any concerns about our products,
you can contact us on
ProductSafety@springernature.com

In case Publisher is established outside the EU,
the EU authorized representative is:
Springer Nature Customer Service Center GmbH
Europaplatz 3, 69115 Heidelberg, Germany

Printed by Libri Plureos GmbH
in Hamburg, Germany